# Statistics for Biology and Health

*Series Editors:*
M. Gail
K. Krickeberg
J. Samet
A. Tsiatis
W. Wong

For other titles published in this series, go to
http://www.springer.com/series/2848

Thomas Hamelryck • Kanti Mardia
Jesper Ferkinghoff-Borg
Editors

# Bayesian Methods in Structural Bioinformatics

*Editors*

Thomas Hamelryck
University of Copenhagen
Department of Biology
Bioinformatics Centre
Copenhagen
Denmark

Jesper Ferkinghoff-Borg
Technical University of Denmark
Department of Electrical Engineering
Lyngby
Denmark

Kanti Mardia
University of Leeds
School of Mathematics
Department of Statistics
Leeds
United Kingdom

*Statistics for Biology and Health Series Editors*

M. Gail
National Cancer Institute
Bethesda, MD
USA

A. Tsiatis
Department of Statistics
North Carolina State University
Raleigh, NC
USA

Klaus Krickeberg
Le Châtelet
Manglieu
France

W. Wong
Department of Statistics
Stanford University
Stanford, CA
USA

Jonathan M. Samet
Department of Preventive Medicine
Keck School of Medicine
University of Southern California
Los Angeles, CA
USA

ISSN 1431-8776
ISBN 978-3-642-27224-0       e-ISBN 978-3-642-27225-7
DOI 10.1007/978-3-642-27225-7
Springer Heidelberg Dordrecht London New York

Library of Congress Control Number: 2012933773

© Springer-Verlag Berlin Heidelberg 2012
This work is subject to copyright. All rights are reserved, whether the whole or part of the material is concerned, specifically the rights of translation, reprinting, reuse of illustrations, recitation, broadcasting, reproduction on microfilm or in any other way, and storage in data banks. Duplication of this publication or parts thereof is permitted only under the provisions of the German Copyright Law of September 9, 1965, in its current version, and permission for use must always be obtained from Springer. Violations are liable to prosecution under the German Copyright Law.
The use of general descriptive names, registered names, trademarks, etc. in this publication does not imply, even in the absence of a specific statement, that such names are exempt from the relevant protective laws and regulations and therefore free for general use.

Printed on acid-free paper

Springer is part of Springer Science+Business Media (www.springer.com)

*Thomas Hamelryck dedicates his contributions to his mother Gib, and to the memory of his father Luc (1942–2011).*

*Kanti Mardia dedicates his contributions to his grandson Sashin Raghunath.*

# Foreword

The publication of this ground-breaking and thought-provoking book in a prestigious Springer series will be a source of particular pleasure and of stimulus for all scientists who have used Bayesian methods in their own specialized area of Bioinformatics, and of excitement for those who have wanted to understand them and learn how to use them but have never dared ask.

I met the lead author, Dr. Hamelryck, at the start of his career, when, as part of his PhD in Protein Crystallography, he determined and proceeded to analyze in great detail the 3D structures of several tight protein-carbohydrate complexes. His attention was drawn to Bayesian and related statistical methods by the profound impact they were having at the time on the two workhorses of macromolecular crystallography, namely experimental phasing and structure refinement. In both cases, recourse to the key Bayesian concept of marginalisation with respect to the phases (treated as nuisance parameters) freed those techniques from the limitations inherent in their old, least-squares based implementations, and gave access, through a shift to a maximum-likelihood approach, to much higher quality, much less biased electron-density maps and atomic models. Preferring the world of computational methods to the biochemist's bench, Dr. Hamelryck went for a post-doc in Copenhagen, in the laboratory where he subsequently developed a highly productive structural extension to the already prominent Bioinformatics Department headed by Prof. Anders Krogh. The contents of this book include a representative sample of the topics on which he and his collaborators have focussed their efforts since that time.

The use of advanced statistical methods in Bioinformatics is of course well established, with books such as *Biological Sequence Analysis – Probabilistic Models of Proteins and Nucleic Acids* by Durbin, Eddy, Krogh and Mitchison setting a very high standard of breadth, rigour and clarity in the exposition of the mathematical techniques involved and in the description of their implementations. The present book, however, is the first of its kind in directing the arsenal of Bayesian methods and their advanced computational tools towards the analysis, simulation and prediction of macromolecular (especially protein) structure in three dimensions.

vii

This is an extremely ambitious goal, as this area comprises no less than the Protein Folding Problem, arguably the most fundamental riddle in the whole of the Life Sciences.

The book begins with an up-to-date coverage, in Part I, of the concepts of Bayesian statistics and of the versatile computational tools that have been developed to make their use possible in practice. A particular highlight of the book is found in Part II, which presents a review of the time-honored use of knowledge-based potentials obtained by data mining, along with a critical reassessment of their exact relationship with potentials of mean force in statistical physics. Its conclusion is a highly original and welcome contribution to the solution of this long-standing problem. Another highlight is the combined use in Part V of the type of directional statistics that can be compiled by the methods of Part IV with the technology of Bayesian networks to produce very efficient methods for sampling the space of plausible conformations of protein and RNA molecules.

It is fitting that the book should return in its final Part VI to the interface between Bayesian methods of learning the structural regularities of macromolecules from known 3D structures on the one hand, and experimental techniques for determining new 3D structures on the other. As is well known, most of these techniques (with the exception perhaps of ultra-high resolution X-ray crystallography with plentiful sources of experimental phase information) need to be supplemented with some degree of low-level a priori knowledge about bond lengths and bond angles to enforce sensible stereochemistry in the resulting atomic models. The Bayesian picture turns this conventional approach on its head, making the process look instead like structure prediction assisted by X-ray (or NMR) data, with the measurements delivered by each experimental technique providing the likelihood factor to supplement the prior probability supplied by previous learning, in order to cast the final result of this inference into the form of a posterior distribution over an ensemble of possible structures. It is only by viewing structure determination in this overtly Bayesian framework that the classical problems of model bias and map overinterpretation can be avoided; and indeed crystallographers, for example, are still far from having made full use of the possibilities described here. Dr. Hamelryck is thus likely to have paved the way for future improvements in his original field of research, in spite of having since worked in a related but distant area, using tools first developed and applied to address problems in medical diagnosis. This instance of the interconnectedness of different branches of science through their methods, and this entire book, provide a splendid illustration of the power of mathematical approaches as the ultimate form of re-useable thought, and of Bayesian methods in particular as a repository of re-useable forms of scientific reasoning.

I am confident that the publication of this book will act as a rallying call for numerous investigators using specific subsets of these methods in various areas of Bioinformatics, who will feel encouraged to connect their own work to the viewpoints and computational tools presented here. It can therefore be expected that this will lead to a succession of enlarged new editions in the future. Last but not least, this book should act as a magnet in attracting young researchers to learn these advanced and broadly adaptable techniques, and to apply them across other fields of science as well as to furthering the subject matter of the book itself.

Global Phasing Ltd.,
Cambridge, UK

*Gerard Bricogne*

# Preface

The *protein folding problem* is the loose denominator for an amalgam of closely related problems that include protein structure prediction, protein design, the simulation of the protein folding process and the docking of small molecules and biomolecules. Despite some, in our view, overly optimistic claims,[1] the development of an insightful, well-justified computational model that routinely addresses these problems is one of the main open problems in biology, and in science in general, today [461]. Although there is no doubt that tremendous progress has been made in the conceptual and factual understanding of how proteins fold, it has been extraordinary difficult to translate this understanding into corresponding algorithms and predictions. Ironically, the introduction of CASP[2] [400], which essentially evaluates the current state of affairs every two years, has perhaps lead to a community that is more focussed on pragmatically fine-tuning existing methods than on conceptual innovation.

In the opinion of the editors, the field of structural bioinformatics would benefit enormously from the use of well-justified machine learning methods and probabilistic models that *treat protein structure in atomic detail.* In the last 5 years, many classic problems in structural bioinformatics have now come within the scope of such methods. For example, conformational sampling, which up to now typically involved approximating the conformational space using a finite set of main chain fragments and side chain rotamers, can now be performed in continuous space using graphical models and directional statistics; protein structures can now be compared and superimposed in a statistically valid way; from experimental data, Bayesian methods can now provide protein ensembles that reflect the statistical uncertainty. All of these recent innovations are touched upon in the book, together with some more cutting edge developments that yet have to prove the extent of their merits.

---

[1] See for example *Problem solved\* (\*sort of)* in the news section of the August 8th, 2008 issue of Science, and the critical reply it elicited.

[2] CASP stands for "Critical Assessment of protein Structure Prediction".

A comprehensive treatment of probabilistic methods in structural bioinformatics is, at first sight, something that would require several weighty book volumes. However, upon closer scrutiny, it becomes clear that the use of well-justified probabilistic methods in structural bioinformatics is currently in its infancy, and that their potential is enormous. Many knowledge based methods that claim to be firmly rooted in probability theory or statistical physics are at best heuristic methods that are often only partly understood. A classic example are the so called *potentials of mean force* that make use of pairwise distances in proteins. The validity and scope of these potentials have been topics of hot debate for over twenty years. Indeed, methods that both consider biomolecular structure in atomic detail and have a sound probabilistic justification are currently far and between.

In this book, we therefore focus on methods that have two important features in common. First, the focus lies on methods that are well justified from a probabilistic point of view, even if they are approximative. Quite a few chapters make use of point estimates, such as empirical Bayes, maximum likelihood or even moment estimates. However, in all cases, these are used as valid *approximations* of a true Bayesian treatment. Second, the methods deal with biomolecular structure in atomic detail. In that respect, classic applications of probabilistic reasoning in structural bioinformatics such as secondary structure prediction fall outside the scope of the book.

This book should be of use to both novices and experts, though we do assume knowledge of structural biology. Introductory chapters on Bayesian methods and Markov chain Monte Carlo methods, which play a key role in Bayesian inference, provide methodological background. These chapters will also be useful to experts in structural bioinformatics that are perhaps not so versed in probabilistic modelling. The remaining parts address various timely topics in structural bioinformatics; we give a short overview.

As mentioned before, the first two chapters provide the foundations. In the first chapter, Hamelryck gives a high level overview of the Bayesian interpretation of probability. The chapter also touches upon relevant topics in information theory and statistical mechanics, and briefly discusses graphical models. Despite the fact that Bayesian methods are now firmly established in statistics, engineering and science in general, most university courses on statistics in these disciplines still uniquely focus on frequentist statistics. The underlying reasons are clearly beyond the usual academic inertia, and can probably be identified with two main perceived problems [51]: Bayesian statistics is unjustly seen as inherently subjective and thus unsuitable for scientific research, and the difficulty of integrating the Bayesian view with the – for many applications still dominating – frequentist paradigm. In addition, many frequentist methods can be seen as perfectly valid approximations to Bayesian methods, thereby removing the apparent need for a paradigm shift. Therefore, we believe that the introductory chapter on Bayesian statistics is quite appropriate. Ferkinghoff-Borg provides an overview of Markov chain Monte Carlo methods. These sampling methods are vital for Bayesian inference, especially when it involves the exploration of the conformational space of proteins. Together, these two chapters should provide a decent start for anybody with some knowledge of structural bioinformatics, but a lacking background in Bayesian statistics. Both chapters focus on concepts rather than mathematical rigor and provide ample references for deeper study. However, these two chapters provide the bedrock on which the rest of the book is founded.

The second part addresses the estimation of so-called knowledge based potentials from data. Recently, dramatic progress has been made in the understanding of the statistics behind these potentials, which were previously justified using rather *ad hoc* physical arguments. The chapter by Borg et al. discusses knowledge based potentials from a physical viewpoint, and explains their statistical background. The chapter by Frellsen et al. overlaps slightly with the previous chapter, but has a more statistical slant. Knowledge based potentials can be formally understood in a Bayesian framework, which justifies, clarifies and extends them. The chapter by Frellsen et al. discusses the recently introduced *reference ratio method*, and highlights its theoretical and conceptual importance for one of the holy grails of structural bioinformatics: a rigorous and efficient probabilistic model of protein structure. Finally, the chapter by Podtelezhnikov and Wild discusses the application of another well-founded machine learning method to the construction of knowledge based potentials, namely *contrastive divergence learning*. The preliminary results reviewed in this chapter establish the method as promising for the future.

In Part III, we turn to directional statistics. Directional statistics concerns data on unusual manifolds such as the torus, the sphere or the real projective plane, which can be considered as a sphere with its antipodes identified. Examples of data from such manifolds include wind directions and dihedral angles in molecules. The need for directional statics can be understood by considering the fact that biomolecular structure is often expressed in terms of angles, and that the average of, for example, $1°$ and $359°$ is not 180 but zero. This is of course due to the fact that such data is naturally represented on the circle, rather than on the line. Hence, directional statistics is becoming vital for the formulation of probabilistic models of biomolecular structure. The two chapters by Mardia and Frellsen and by Kent are of a more technical nature and provide information on parameter estimation and sampling for two distributions from directional statistics that are of particular relevance for biomolecular structure. Kent's chapter discusses the Fisher-Bingham 5 (or Kent) distribution, which can be used to model data on the two-dimensional sphere, that is, data consisting of unit vectors. Frellsen and Mardia discuss the univariate, bivariate and multivariate von Mises distributions, for data on the circle and the torus, respectively. The latter case is of special relevance for modeling the $\phi$ and $\psi$ angles in proteins, which are well known from the celebrated Ramachandran plot.

Part IV explores the use of shape theory in comparing protein structures. Comparing and superimposing protein structures is one of the classic problems in structural bioinformatics. The chapter by Theobald discusses the superposition of proteins when their equivalent amino acids are known. Classically, this is done using a least squares criterion, but this leads to poor performance in many cases. Theobald describes a maximum likelihood alternative to the classic method, and also discusses a fully Bayesian extension. The problem of superimposing protein structures when the equivalent amino acids are *not* known is subsequently discussed in the chapter by Mardia and Nyirongo. They review a Bayesian model that is built upon a Poisson process and a proper treatment of the prior distributions of the nuisance variables.

Part V is concerned with the use of graphical models in structural bioinformatics. The chapter by Boomsma et al. introduces probabilistic models of RNA and protein structure that are based on the happy marriage of directionals statistics and dynamic Bayesian networks. These models can be used in conformational sampling, but are also vital elements for the formulation of a complete probabilistic description of protein structure. Yanover and Fromer discuss belief propagation in graphical models to solve the classic problem of side chain placement on a given protein main chain, which is a key problem in protein design.

In the sixth, final part, the inference of biomolecular structure from experimental data is discussed. This is of course one of the most fundamental applications of statistics in structural biology and structural bioinformatics. It is telling that currently most structure determination methods rely on a so-called *pseudo-energy*, which combines a physical force field with a heuristic force field that brings in the effect of the experimental data. Only recently methods have emerged that formulate this problem of inference in a rigorous, Bayesian framework. Habeck discusses Bayesian inference of protein structure from NMR data, while Hansen discusses the case of SAXS.

Many of the concepts and methods presented in the book are novel, and have neither been tested, honed or proven in large scale applications. However, the editors have little doubt that many concepts presented in this book will have a profound effect on the incremental solution of one of the great challenges in science today.

Copenhagen, Leeds

*Thomas Hamelryck*
*Kanti V. Mardia*
*Jesper Ferkinghoff-Borg*

# Acknowledgements

We thank Justinas V. Daugmaudis for his invaluable help with typesetting the book in LaTeX. In addition, we thank Christian Andreetta, Joe Herman, Kresten Lindorff-Larsen, Simon Olsson and Jan Valentin for comments and discussion.

This book was supported by the Danish Research Council for Technology and Production Sciences (FTP), under the project "Protein structure ensembles from mathematical models – with application to Parkinson's $\alpha$-synuclein", and the Danish Research Council for Strategic Research (NaBiIT), under the project "Simulating proteins on a millisecond time-scale".

# Contents

**Part I  Foundations**

**1   An Overview of Bayesian Inference and Graphical Models** .......... 3
Thomas Hamelryck

**2   Monte Carlo Methods for Inference in High-Dimensional Systems** ................................................................. 49
Jesper Ferkinghoff-Borg

**Part II   Energy Functions for Protein Structure Prediction**

**3   On the Physical Relevance and Statistical Interpretation of Knowledge-Based Potentials** ........................................ 97
Mikael Borg, Thomas Hamelryck, and Jesper Ferkinghoff-Borg

**4   Towards a General Probabilistic Model of Protein Structure: The Reference Ratio Method** ............................... 125
Jes Frellsen, Kanti V. Mardia, Mikael Borg, Jesper Ferkinghoff-Borg, and Thomas Hamelryck

**5   Inferring Knowledge Based Potentials Using Contrastive Divergence** ................................................................ 135
Alexei A. Podtelezhnikov and David L. Wild

**Part III   Directional Statistics for Biomolecular Structure**

**6   Statistics of Bivariate von Mises Distributions** ......................... 159
Kanti V. Mardia and Jes Frellsen

**7   Statistical Modelling and Simulation Using the Fisher-Bingham Distribution** ....................................... 179
John T. Kent

xvii

xviii

**Part IV  Shape Theory for Protein Structure Superposition**

**8  Likelihood and Empirical Bayes Superposition of Multiple Macromolecular Structures** ............................................... 191
Douglas L. Theobald

**9  Bayesian Hierarchical Alignment Methods** ............................ 209
Kanti V. Mardia and Vysaul B. Nyirongo

**Part V  Graphical Models for Structure Prediction**

**10  Probabilistic Models of Local Biomolecular Structure and Their Applications** ....................................................... 233
Wouter Boomsma, Jes Frellsen, and Thomas Hamelryck

**11  Prediction of Low Energy Protein Side Chain Configurations Using Markov Random Fields** ......................... 255
Chen Yanover and Menachem Fromer

**Part VI  Inferring Structure from Experimental Data**

**12  Inferential Structure Determination from NMR Data** ................ 287
Michael Habeck

**13  Bayesian Methods in SAXS and SANS Structure Determination** ..... 313
Steen Hansen

**References** ............................................................. 343

**Index** ................................................................. 377

# Contributors

**Wouter Boomsma** Department of Astronomy and Theoretical Physics, Lund University, Lund, Sweden, wouter@thep.lu.se

Department of Biomedical Engineering, DTU Elektro, Technical University of Denmark, Lyngby, Denmark, wb@elektro.dtu.dk

**Mikael Borg** The Bioinformatics Centre, Department of Biology, University of Copenhagen, Ole Maaløesvej 5, 2200 Copenhagen, Denmark, borg@binf.ku.dk

**Jesper Ferkinghoff-Borg** Department of Biomedical Engineering, DTU Elektro, Technical University of Denmark, Lyngby, Denmark, jfb@elektro.dtu.dk

**Jes Frellsen** The Bioinformatics Centre, Department of Biology, University of Copenhagen, Ole Maaløesvej 5, 2200 Copenhagen, Denmark, frellsen@binf.ku.dk

**Menachem Fromer** The Hebrew University of Jerusalem, Jerusalem, Israel, fromer@cs.huji.ac.il

**Michael Habeck** Department of Protein Evolution, Max-Planck-Institute for Developmental Biology, Spemannstr. 35, Tübingen, Germany, michael.habeck@tuebingen.mpg.de

Department of Empirical Inference, Max-Planck-Institute for Intelligent Systems, Spemannstr. 38, Tuebingen, Germany

**Thomas Hamelryck** The Bioinformatics Centre, Department of Biology, University of Copenhagen, Ole Maaløesvej 5, 2200 Copenhagen, Denmark, thamelry@binf.ku.dk

**Steen Hansen** Department of Basic Sciences and Environment, University of Copenhagen, Faculty of Life Sciences, Thorvaldsensvej 40, DK-1871 FRB C, Frederiksberg, Denmark, slh@life.ku.dk

**John T. Kent** Department of Statistics, University of Leeds, Leeds LS2 9JT, UK

**Kanti V. Mardia** Department of Statistics, University of Leeds, Leeds LS2 9JT, UK, k.v.mardia@leeds.ac.uk

**Vysaul B. Nyirongo** Statistics Division, United Nations, New York, NY 10017, USA, nyirongov@un.org

**Alexei A. Podtelezhnikov** Michigan Technological University, Houghton, MI, USA, apodtele@gmail.com

**Douglas L. Theobald** Brandeis University, 415 South St, Waltham MA, USA, dtheobald@brandeis.edu

**David L. Wild** University of Warwick, Coventry, CV4 7AL, UK, d.l.wild@warwick.ac.uk

**Chen Yanover** Fred Hutchinson Cancer Research Center, Seattle, WA, USA, cyanover@fhcrc.org

# Acronyms

| | |
|---|---|
| AIC | Akaike information criterion |
| BIC | Bayesian information criterion |
| BMMF | Best max-marginal first |
| BN | Bayesian network |
| BP | Belief propagation |
| CASP | Critical Assessment of Techniques for Protein Structure Prediction |
| CD | Contrastive divergence |
| CPT | Conditional probability table |
| DBN | Dynamic Bayesian network |
| DEE | Dead end elimination |
| DIC | Deviance information criterion |
| EM | Expectation maximization |
| GA | Genetic algorithm |
| GBP | Generalized belief propagation |
| GE | Generalized ensemble |
| GMEC | Global minimum energy configuration |
| GMH | Generalized multihistogram equations |
| GIFT | Generalized indirect Fourier transformation |
| HMM | Hidden Markov model |
| IFT | Indirect Fourier transformation |
| ILP | Integer linear programming |
| ISD | Inferential structure determination |
| JT | Junction tree |
| KBP | Knowledge based potential |
| LP | Linear programming |
| MAP | Maximum a posteriori |
| MCMC | Markov chain Monte Carlo |
| MCSA | Monte Carlo simulated annealing |
| ML | Maximum likelihood |
| MM | Max-marginal |
| MPLP | Max-product linear programming |

| | |
|---|---|
| MRF | Markov random field |
| MUCA | Multicanonical ensemble |
| NMR | Nuclear magnetic resonance spectroscopy |
| OLS | Ordinary least squares |
| PDB | Protein data bank |
| PDF | Probability density function |
| PMF | Potentials of mean force |
| REMD | Replica-exchange molecular dynamics |
| RMSD | Root mean square deviation |
| SANS | Small angle neutron scattering |
| SAS | Small angle scattering |
| SAXS | Small angle X-ray scattering |
| SCMF | Self-consistent mean field |
| SCWRL | Side-chains with a rotamer library |
| SLS | Static light scattering |
| SPRINT | Side-chain prediction inference toolbox |
| STRIPES | Spanning tree inequalities and partitioning for enumerating solutions |
| TRMP | Tree-reweighted max-product |
| tBMMF | Type-specific BMMF |
| WHAM | Weighted histogram analysis method |
| WL algorithm | Wang-Landau algorithm |

# Part I
# Foundations

# Chapter 1
# An Overview of Bayesian Inference and Graphical Models

**Thomas Hamelryck**

## 1.1 Introduction

The Bayesian view of statistics interprets probability as *a measure of a state of knowledge* or *a degree of belief*, and can be seen as an extension of the rules of logic to reasoning in the face of uncertainty [342]. The Bayesian view has many advantages [48, 342, 428, 606]: it has a firm axiomatic basis, coincides with the intuitive idea of probability, has a wide scope of applications and leads to efficient and tractable computational methods. The main aim of this book is to show that a Bayesian, probabilistic view on the problems that arise in the simulation, design and prediction of biomolecular structure and dynamics is extremely fruitful.

This book is written for a mixed audience of computer scientists, bioinformaticians, and physicists with some background knowledge of protein structure. Throughout the book, the different authors will use a Bayesian viewpoint to address various questions related to biomolecular structure. Unfortunately, Bayesian statistics is still not a standard part of the university curriculum; most scientists are more familiar with the frequentist view on probability. Therefore, this chapter provides a quick, high level introduction to the subject, with an emphasis on introducing ideas rather than mathematical rigor.

In order to explain the rather strange situation of two mainstream paradigms of statistics and two interpretations of the concept of probability existing next to each other, we start with explaining the historical background behind this schism, before sketching the main aspects of the Bayesian methodology. In the second part of this chapter, we will give an introduction to graphical models, which play a central role in many of the topics that are discussed in this book. We also discuss some useful concepts from information theory and statistical mechanics, because of their close ties to Bayesian statistics.

---

T. Hamelryck (✉)
The Bioinformatics Centre, University of Copenhagen, Copenhagen, Denmark
e-mail: thamelry@binf.ku.dk

T. Hamelryck et al. (eds.), *Bayesian Methods in Structural Bioinformatics*,
Statistics for Biology and Health, DOI 10.1007/978-3-642-27225-7_1,
© Springer-Verlag Berlin Heidelberg 2012

## 1.2 Historical Background

The term *Bayesian* refers to Thomas Bayes (1701?–1761; Fig. 1.1), a Presbyterian minister who proved a special case of what is now called Bayes' theorem [27, 31, 691, 692]. However, it was the French mathematician and astronomer Pierre-Simon Laplace (1749–1827; Fig. 1.1) who introduced the general form of the theorem and used it to solve problems in many areas of science [415, 692]. Laplace's memoir from 1774 had an enormous impact, while Bayes' paper posthumously published paper from 1763 did not address the same general problem, and was only brought to the general attention much later [183, 692]. Hence, the name Bayesian statics follows the well known Stigler's law of eponymy, which states that "no scientific discovery is named after its original discoverer." Ironically, the iconic portrait of Thomas Bayes (Fig. 1.1), that is often used in textbooks on Bayesian statistics, is of very doubtful authenticity [31].

For more than 100 years, the Bayesian view of statistics reigned supreme, but this changed drastically in the first half of the twentieth century with the emergence of the so-called frequentist view of statistics [183]. The frequentist view of probability gradually overshadowed the Bayesian view due to the work of prominent figures such as Ronald Fisher, Jerzy Neyman and Egon Pearson. They viewed probability not as a measure of a state of knowledge or a degree of belief, but as a frequency: an event's probability is the frequency of observing that event in a large number of trials.

**Fig. 1.1** *(Left)*: An alleged picture of Thomas Bayes (1701?–1761). The photograph is reproduced from the Springer Statistics Calendar, December issue, 1981, by Stephen M. Stigler (Springer-Verlag, New York, 1980). The legend reads: "This is the only known portrait of him; it is taken from the 1936 History of Life Insurance (by Terence O'Donnell, American Conservation Co., Chicago). As no source is given, the authenticity of even this portrait is open to question". The book does not mention a source for this picture, which is the only known picture of Thomas Bayes. The photo appears on page 335 with the caption "Rev. T. Bayes: Improver of the Columnar Method developed by Barrett." *(Right)*: Pierre-Simon Laplace (1749–1827). Engraved by J. Pofselwhite

The difference between the two views is illustrated by the famous sunrise problem, originally due to Laplace [342]: what is the probability that the sun will rise tomorrow? The Bayesian approach is to construct a probabilistic model of the process, estimate its parameters using the available data following the Bayesian probability calculus, and obtain the requested probability from the model. For a frequentist, the question is meaningless, as there is no meaningful way to calculate its probability as a frequency in a large number of trials.

During the second half of the twentieth century the heterogeneous amalgam of methods known as frequentist statistics became increasingly under pressure. The Bayesian view of probability was kept alive – in various forms and disguises – in the first part of the twentieth century by figures such as John Maynard Keynes, Frank Ramsey, Bruno de Finetti, Dorothy Wrinch and Harold Jeffreys. Jeffreys' seminal book "Theory of probability" first appeared in 1939, and is a landmark in the history of the Bayesian view of statistics [343]. The label "Bayesian" itself appeared in the 1950s [183], when Bayesian methods underwent a strong revival. By the 1960s it became the term preferred by people who sought to escape the limitations and inconsistencies of the frequentist approach to probability theory [342,444,509]. Before that time, Bayesian methods were known under the name of *inverse probability*, because they were often used to infer what was seen as "causes" from "effects" [183]; the causes are tied to the parameters of a model, while the effects are evident in the data.[1]

The emergence of powerful computers, the development of flexible Markov chain Monte Carlo (MCMC) methods (see Chap. 2), and the unifying framework of graphical models, has brought on many practical applications of Bayesian statistics in numerous areas of science and computing. One of the current popular textbooks on machine learning for example, is entirely based on Bayesian principles [62]. In physics, the seminal work of Edwin T. Jaynes showed that statistical mechanics is nothing else than Bayesian statistics applied to physical systems [335, 336]. Physicists routinely use Bayesian methods for the evaluation of experimental data [154]. It is also becoming increasingly accepted that Bayesian principles underlie human reasoning and the scientific method itself [342], as well as the functioning of the human brain [334,391]. After this short historical introduction, we now turn to the theory and practice of the Bayesian view on probability.

## 1.3 The Bayesian Probability Calculus

### 1.3.1 Overview

Statistical inference is the process of taking decisions or answering questions based on data that are affected by random variations. An examples of statistical inference

---

[1]It should be noted that this is a very naive view. Recently, great progress has been made regarding causal models and causal reasoning [569].

is the prediction of protein structure based on sequence or experimental data. As discussed above, there are several "schools" in statistics that adopt different methods and interpretations with respect to statistical inference. In this book, we adopt the Bayesian view on statistical inference. One of the characteristics of the Bayesian form of statistical inference is that its outcome is typically a probability distribution – called the *posterior probability distribution* or in short the *posterior distribution* – as opposed to a single value. For example, in protein structure prediction or determination from experimental data, we would obtain a posterior distribution over the space of relevant protein structures; this can be accomplished by sampling an ensemble of protein structures. Chapters 12 and 13 discuss the application of Bayesian inference to protein structure determination from nuclear magnetic resonance (NMR) and small angle X-ray scattering data (SAXS), respectively.

The posterior distribution results from the application of the full Bayesian probability calculus, as explained below. From a more practical point of view, the Bayesian approach is also characterized by the use of *hierarchical models*, and the treatment of so-called *nuisance parameters*. These concepts will be explained in more detail below. Hierarchical models and nuisance variables often lead to posterior distributions that cannot be expressed explicitly in a handy formula; in most cases of practical relevance, the posterior distribution is represented by a set of samples. Markov chain Monte Carlo methods to obtain such samples are discussed in Chap. 2. In some cases, working with posterior distributions becomes inconvenient, unnecessary or problematic. On the one hand, one might be interested in the best prediction, such as for example the protein structure that fits the data best. Such a "best" prediction is implied by the posterior distribution, but can often be obtained without actually needing to calculate the posterior distribution itself. Picking the best prediction is the goal of decision theory, which makes use of the rules of the Bayesian probability calculus. On the other hand, constructing the distribution itself might be problematic, because it is computationally intractable or simply because the process is too time consuming. In these cases, one might settle for a point estimate – that is, a single value – instead of a posterior distribution. Often, the use of such a point estimate is a good approximation to a full Bayesian approach with respect to the obtained results, while also being computationally much more efficient than a full Bayesian approach. For example, a probabilistic model of protein structure that is to be used in conformational searching needs to be computationally efficient, as typically many millions of structures will need to be evaluated (see Chap. 10). Popular methods to obtain point estimates include:

- Maximum likelihood (ML) estimation
- Maximum a posteriori (MAP) estimation
- Pseudolikelihood estimation
- Moment estimation
- Shrinkage estimation
- Empirical Bayes methods

All these methods will be briefly discussed in this section. As with all approximations, they work well in many cases, and fail spectacularly in others. We now

# 1 An Overview of Bayesian Inference and Graphical Models

introduce the elements of the full Bayesian approach to statistical inference, before we introduce the various approximations.

## 1.3.2 Full Bayesian Approach

The goal of the Bayesian probability calculus is to obtain the probability distribution over a set of hypotheses, or equivalently, of the parameters of a probabilistic model, given a certain set of observations. This quantity of interest is called the *posterior distribution*. One of the hallmarks of the Bayesian approach is that the posterior distribution is a probability distribution over all possible values of the parameters of a probabilistic model. This is in contrast to methods such as maximum likelihood estimation, which deliver one 'optimal' set of values for the parameters, called a *point estimate*.

The posterior distribution is proportional to the product of the *likelihood*, which brings in the information in the data, and the *prior distribution* or, in short, *prior*, which brings in the knowledge one had before the data was observed. The Bayesian probability calculus makes use of Bayes' theorem:

$$p(\mathbf{h} \mid \mathbf{d}) = \frac{p(\mathbf{d} \mid \mathbf{h}) p(\mathbf{h})}{p(\mathbf{d})} \tag{1.1}$$

where:

- $\mathbf{h}$ is a hypothesis or the parameters of a model
- $\mathbf{d}$ is the data
- $p(\mathbf{h} \mid \mathbf{d})$ is the *posterior distribution,* or in short the *posterior.*
- $p(\mathbf{d} \mid \mathbf{h})$ is the *likelihood.*
- $p(\mathbf{h})$ is the *prior distribution,* or in short the *prior.*
- $p(\mathbf{d})$ is the *marginal probability of the data* or the *evidence*, with $p(\mathbf{d}) = \int p(\mathbf{d}, \mathbf{h}) d\mathbf{h}$.

$p(\mathbf{d})$ is a normalizing constant that only depends on the data, and which in most cases does not need to be computed explicitly. As a result, Bayes' theorem is often applied in practice under the following form:

$$p(\mathbf{h} \mid \mathbf{d}) \propto p(\mathbf{d} \mid \mathbf{h}) p(\mathbf{h})$$

$$\text{posterior} \propto \text{likelihood} \times \text{prior}$$

Conceptually, the Bayesian probability calculus updates the prior belief associated with a hypothesis in the light of the information gained from the data. This prior information is of course embodied in the prior distribution, while the influence of the data is brought in by the likelihood. Bayes' theorem (Eq. 1.1) is a perfectly acceptable and rigorous theorem in both Bayesian and frequentist statistics, but this conceptual interpretation is specific to the Bayesian view of probability.

Another aspect of the Bayesian approach is that this process can be applied *sequentially*, or *incrementally*. Suppose that one has obtained a posterior distribution from one data set, and that a second, additional data set that is independent of the first data set now needs to be considered. In that case, the posterior obtained from the first data set serves as the prior for the calculation of the posterior from the second data set. This can be easily shown. Suppose that in addition to the data $\mathbf{d}$, we also have data $\mathbf{d}'$. If $\mathbf{d}$ and $\mathbf{d}'$ are indeed independent conditional on $\mathbf{h}$, applying the the usual rules of the Bayesian probability calculus leads to:

$$
\begin{aligned}
p(\mathbf{h} \mid \mathbf{d}, \mathbf{d}') &\propto p(\mathbf{d}, \mathbf{d}' \mid \mathbf{h}) p(\mathbf{h}) \\
&= p(\mathbf{d}' \mid \mathbf{d}, \mathbf{h}) p(\mathbf{d} \mid \mathbf{h}) p(\mathbf{h}) \\
&= p(\mathbf{d}' \mid \mathbf{h}) p(\mathbf{d} \mid \mathbf{h}) p(\mathbf{h}) \\
&= \text{likelihood from } \mathbf{d}' \times \text{posterior from } \mathbf{d}
\end{aligned}
$$

Obviously, the last two factors correspond to the posterior distribution obtained from $\mathbf{d}$ alone, and hence the 'posterior becomes the prior'. The second step is due to the so-called product rule of probability theory, which states for any probability distribution over $M$ variables $x_1, x_2, \ldots, x_M$:

$$
p(x_1, x_2, \ldots, x_M) = p(x_1 \mid x_2, \ldots, x_M) p(x_2, \ldots, x_M)
$$

The third step is due to the assumed independence between $\mathbf{d}$ and $\mathbf{d}'$ given $\mathbf{h}$; in general, if random variables $a$ and $b$ are independent given $c$, by definition:

$$
\begin{aligned}
p(a, b \mid c) &= p(a \mid c) p(b \mid c) \\
p(a \mid b, c) &= p(a \mid c)
\end{aligned}
$$

Note that it does not matter whether $\mathbf{d}$ or $\mathbf{d}'$ was observed first; all that matters is that the data are conditionally independent given $\mathbf{h}$.

In the next section, a simple example illustrates these concepts. At this point, we also need to recall the two fundamental rules of probability theory:

$$
\text{Sum rule: } p(a) = \sum_b p(a, b)
$$

$$
\text{Product rule: } p(a, b) = p(a \mid b) p(b) = p(b \mid a) p(a)
$$

These two rules lead directly to Bayes' theorem. Their justification will be discussed briefly in Sect. 1.4.

# 1 An Overview of Bayesian Inference and Graphical Models

### 1.3.3 Example: The Binomial Distribution

Let us consider the Bayesian estimation of the parameter $\theta$ of a binomial distribution – a problem that was originally first addressed by Bayes and Laplace. The binomial distribution is a natural model for data that consists of trials with two possible outcomes, often labeled "success" and "failure". The distribution has a single parameter $\theta \in [0, 1]$. According to the binomial distribution, the probability of observing $k$ successes in $n$ trials is given by:

$$p(k \mid \theta, n) = \binom{n}{k} \theta^k (1 - \theta)^{n-k} \qquad (1.2)$$

In practice, $n$ is typically given as part of the experimental design, and left out of the expression. Now, suppose we observe $k$ successful trials out of $n$ draws, and we want to infer the value of $\theta$. Following the Bayesian probability calculus, we need to obtain the posterior distribution, which is equal to:

$$p(\theta \mid k) \propto p(k \mid \theta) p(\theta)$$

Obviously, the first factor, which is the likelihood, is the binomial distribution given in Eq. 1.2. Now, we also need to specify the second factor, which is the prior distribution, that reflects our knowledge about $\theta$ before $k$ was observed. Following Laplace and Bayes, we could adopt a uniform distribution on the interval $[0, 1]$. In that case, we obtain the following result for the posterior:

$$p(\theta \mid k) \propto \theta^k (1 - \theta)^{n-k}$$

The shape of the posterior distribution corresponds to a Beta distribution $Be(\theta \mid \alpha_1, \alpha_2)$, with parameters $\alpha_1 = k + 1$ and $\alpha_2 = n - k + 1$. The Beta distribution is, as expected, a distribution on the interval $[0, 1]$. The result for $k = n - k = 5$ is shown in Fig. 1.2; as expected, the posterior density of $\theta$ reaches a maximum at 0.5.

One might think that a uniform prior on the unit interval reflects a state of "complete ignorance", but this is, in fact, not the case. As we will see in Sect. 1.6.1.3, the prior that reflects completely ignorance in this case is a so-called Jeffreys' prior.

### 1.3.4 Hierarchical Models and Nuisance Variables

A characteristic of Bayesian statistics is the use of a hierarchical approach: observations depend on certain parameters which in turn depend on further parameters, called *hyperparameters*. Naturally, the chain ends at some point: all parameters are ultimately dependent on prior distributions with given, fixed parameters. Such

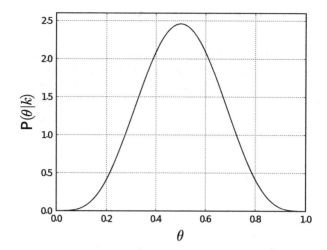

**Fig. 1.2** The Beta distribution $\text{Be}(\theta \mid \alpha_1, \alpha_2)$ for $\alpha_1 = \alpha_2 = 6$. This corresponds to the posterior distribution of the binomial parameter $\theta$ resulting from five successes in ten trials ($k = n - k = 5$) combined with a uniform prior on $\theta$. As expected, the posterior density of $\theta$ is symmetric and reaches a maximum at 0.5

hierarchical models typically also contain *nuisance parameters*; parameters whose values are not of direct interest. These nuisance parameters are typically unobserved; unobserved variables are also called *latent* variables. In a typical Bayesian approach, such latent nuisance parameters are dealt with by simply integrating them out.

A classic example of integrating away nuisance parameters in a hierarchical model is found in the *mixture model*, where a single discrete, latent variable can adopt a finite set of values, and each value specifies the parameters of a continuous distribution. Mixture models go beyond fitting a single distribution to the data by using a finite number of weighted distributions, called *mixture components*. In that way, data that is multimodal can still be modelled using standard unimodal distributions such as the Gaussian distribution. If we neglect the use of priors for the moment, such a model looks like Fig. 1.3a, where $h$ is the discrete variable and $x$ is the continuous variable. Often, one models multimodal distributions as mixtures of Gaussian distributions, and the parameters specified by the discrete variable value $h$ are the mean $\mu_h$ and standard deviation $\sigma_h$:

$$p(x \mid h) = \mathcal{N}(x \mid \mu_h, \sigma_h)$$

In such a model the index of the mixture component $h$ is often a nuisance parameter; it serves to construct a tractable probabilistic model, but its value is not of direct interest. However, there are cases where the mixture components are of interest, for example in clustering. In order to calculate the probability of an observation according to the mixture model, the mixture component is therefore

# 1  An Overview of Bayesian Inference and Graphical Models

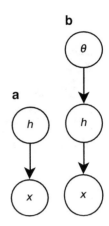

**Fig. 1.3** (**a**) A mixture model. $p(h)$ is a multinomial distribution with given parameter $\boldsymbol{\theta}$. $p(x \mid h)$ is a Gaussian distribution whose mean and variance are determined by $h$. (**b**) A hierarchical model with three levels for Bayesian inference of $\boldsymbol{\theta}$. $p(\boldsymbol{\theta})$ is the prior distribution for the parameter $\boldsymbol{\theta}$; it is a Dirichlet distribution with given parameter vector $\boldsymbol{\alpha}$. $p(h \mid \boldsymbol{\theta})$ is a multinomial distribution

simply integrated away, by summing over all possible components. For a Gaussian mixture model with $H$ components, this corresponds to:

$$p(x) = \sum_{h=1}^{H} \mathcal{N}(x \mid \mu_h, \sigma_h) p(h)$$

For $p(h)$, one typically uses a multinomial distribution. In practice, this distribution is parameterized by a vector $\boldsymbol{\theta}$ that assigns a probability to each value of $h$. This simple variant of the multinomial distribution – where $h$ is a single outcome and not a vector of counts – is often called the *categorical* or the *discrete* distribution.

Let us now consider Bayesian parameter estimation for such a model, and focus on the parameters of the multinomial distribution, $\boldsymbol{\theta}$ (see Fig. 1.3b). The goal of the Bayesian inference is to obtain a probability distribution for $\boldsymbol{\theta}$ [697]. Following the rules of the Bayesian probability calculus, we need to specify a prior distribution over $\boldsymbol{\theta}$, with given, fixed parameters. As we will discuss in Sect. 1.6.2, the Dirichlet distribution [164] is typically used for this purpose:

$$\text{Di}(\boldsymbol{\theta} \mid \boldsymbol{\alpha}) = \frac{1}{Z(\boldsymbol{\alpha})} \prod_{h=1}^{H} \theta_h^{\alpha_h - 1}$$

where $Z(\boldsymbol{\alpha})$ is a normalization factor. Recall that the parameter $\boldsymbol{\theta}$ is a *probability vector*: a vector of positive real numbers that sum to one. The parameter vector $\boldsymbol{\alpha} = \{\alpha_1, \ldots, \alpha_H\}$ of the Dirichlet prior, which reflects our prior beliefs on $\boldsymbol{\theta}$, can be interpreted as a set of prior observations of the values of $h$. The resulting model is a hierarchical model with three levels, and is shown in Fig. 1.3b.

A Dirichlet distribution is a probability distribution over all such probability vectors with a fixed length. Its *support* – that is, the space where probability mass can be present – is the *simplex*, which is a generalization of the triangle in higher dimensions (see Fig. 1.4). The Dirichlet distribution is an example of a distribution

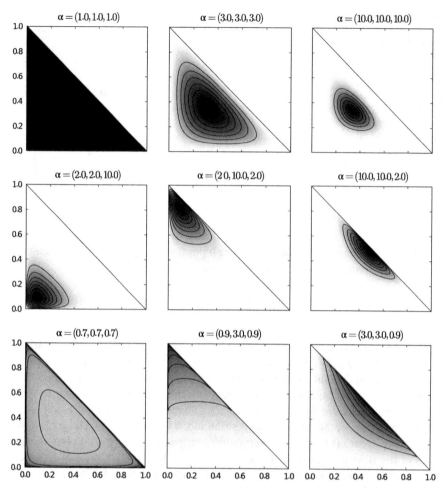

**Fig. 1.4** The density of the Dirichlet distribution for $H = 3$, for different values of $\boldsymbol{\alpha} = (\alpha_1, \alpha_2, \alpha_3)$. For $H = 3$, the Dirichlet distribution is a probability distribution over three-dimensional probability vectors $\boldsymbol{\theta} = (\theta_1, \theta_2, \theta_3)$; the support of the distribution is thus a triangle with vertices at $(1, 0, 0)$, $(0, 1, 0)$ and $(0, 0, 1)$. The figure shows $\theta_1$ on the $X$-axis and $\theta_2$ on the $Y$-axis; $\theta_3$ is implied by the constraint $\theta_1 + \theta_2 + \theta_3 = 1$. Increasing probability density is shown as *darker shades of gray*, and the *black lines* indicate equiprobability contours. Note that the distribution is uniform on the triangle for $\boldsymbol{\alpha} = (1, 1, 1)$

whose support is *not* the familiar infinite $N$-dimensional Euclidean space, for which the prominent Gaussian distribution is commonly used. In Chaps. 6, 7 and 10 we will see that probability distributions on unusual manifolds such as the sphere or the torus are important for the formulation of probabilistic models of biomolecular structure. Such distributions also arise in probabilistic methods to superimpose protein structures, as discussed in Chaps. 8 and 9.

We only discussed the inference of $\theta$; of course, one would also use priors for the means and the standard deviations of the Gaussian distributions. Hierarchical models are simple examples of graphical models, which will be discussed in more detail in Sect. 1.9.

## 1.3.5 Point Estimates

### 1.3.5.1 MAP and ML Estimation

The posterior distribution contains all the information of interest on the parameters or the hypothesis under consideration. However, sometimes one may prefer to use a *point estimate*: a specific estimate of the parameter(s) that is in some way optimal. There are two possible reasons for this way of proceeding.

First, the calculation of the posterior might be computationally intractable or impractical, and therefore one settles for a simple point estimate. Whether this produces meaningful results depends on the shape of the posterior, and the intended use of the point estimate. Roughly speaking, point estimates will make most sense when the posterior is sharply peaked and unimodal.

Second, one can be facing a *decision problem*: an optimal decision needs to be made. In the latter case, consider buying a bottle of wine: having an idea of which wines in a shop are good and which aren't is a good start, but in the end one wants to bring home a specific bottle that goes well with the dinner in the evening. For such decision problems, one needs to define a *loss function* $L(\hat{\theta}, \theta)$, that measures the price of acting as if the point estimate $\hat{\theta}$ is true, when the real value is actually $\theta$. The point estimate that minimizes the *expected loss* is called a *Bayes estimator* for that loss function. The expected loss $\bar{L}(\hat{\theta})$ is for given data $\mathbf{d}$ is:

$$\bar{L}(\hat{\theta}) = \int L(\hat{\theta}, \theta) p(\theta \mid d) d\theta$$

Two point estimates are especially common. First, one can simply use the maximum of the posterior distribution, in which case one obtains the *maximum a posteriori* (MAP) estimate:

$$\theta_{MAP} = \arg \max_{\theta} p(\mathbf{d} \mid \theta) p(\theta)$$

The MAP estimate essentially follows from a specific loss function, called the *zero-one loss function*:

$$L(\hat{\theta}, \theta) = 0, \text{ if } \left\| \hat{\theta} - \theta \right\| \leq \varepsilon$$

$$L(\hat{\theta}, \theta) = 1, \text{ if } \left\| \hat{\theta} - \theta \right\| > \varepsilon$$

This loss function is zero if the point estimate $\hat{\theta}$ is in a ball with radius $\varepsilon$ close to 0 around $\theta$, and one otherwise.

Second, if one assumes that the prior is uniform, one obtains the *maximum likelihood* (ML) estimate:

$$\theta_{ML} = \arg\max_{\theta} p(\mathbf{d} \mid \theta)$$

The ML estimate is very commonly used, and is also an established part of frequentist statistics. However, the fact that one assumes a uniform prior in this case can lead to problems, notably when the data are sparse or when a uniform prior in fact imposes an overly strong and unsuited prior belief.

When will the use of MAP and ML point estimates instead of true posterior distributions resulting from Bayesian inference be justified? If the posterior distribution is unimodal and highly peaked, a MAP estimate will perform well. Notably, the use of ML and MAP point estimates is quite common for graphical models such as Bayesian networks that are trained from large amounts of data (see Sect. 1.9), for reasons of computational performance. This is for example the case for the models presented in Chap. 10.

### 1.3.5.2 Asymptotic Properties

ML and MAP estimation provide a point estimate, and not a full posterior distribution over the parameters. However, under some general assumptions, it is still possible to obtain an approximation of the posterior distribution from these point estimates. For this, we need the *Fisher information* [211].

For large datasets and under some general assumptions, the posterior distribution will converge to a Gaussian distribution, whose mean is the ML estimate. The variance $\sigma^2$ of this Gaussian distribution is proportional to the inverse of the Fisher information, $\mathcal{I}(\theta)$:

$$\sigma^2 = \frac{1}{M \mathcal{I}(\theta)}$$

where $M$ is the number of observations, for $M \to \infty$. The Fisher information $\mathcal{I}(\theta)$ is minus the expectation of the second derivative with respect to $\theta$ of the log-likelihood

$$\mathcal{I}(\theta) = -\mathbb{E}_{p(d|\theta)} \left\{ \frac{d^2 \log p(\mathbf{d} \mid \theta)}{d\theta^2} \right\}$$

The expectation is with respect to $\mathbf{d}$ under the likelihood $p(\mathbf{d} \mid \theta)$. The Fisher information has an interesting interpretation: it measures how much information is contained in the data about the parameter $\theta$ by considering the curvature of the log-likelihood landscape around the ML estimate. A low value indicates that the peak of the log-likelihood function around the maximum likelihood estimate is blunt,

# 1 An Overview of Bayesian Inference and Graphical Models

corresponding to a low information content in the data. A high value indicates a sharp peak around the estimate, corresponding to high information content.

In the generalization to the multidimensional case, the *Fisher information matrix* comes into play. For an $N$-dimensional parameter $\boldsymbol{\theta}$, $\mathcal{I}(\boldsymbol{\theta})$ is an $N \times N$ matrix with elements $i, j$:

$$\mathcal{I}(\boldsymbol{\theta})_{ij} = -\mathbb{E}_{p(\mathbf{d}|\boldsymbol{\theta})} \left\{ \frac{\partial^2 \log p(\mathbf{d} \mid \boldsymbol{\theta})}{\partial \theta_i \, \partial \theta_j} \right\}$$

We will encounter the Fisher information again in Sect. 1.6.1.3, where it shows up in the calculation of the Jeffreys' prior, and in Chap. 2, where it corresponds to an important physical quantity called the *heat capacity*.

### 1.3.5.3 Empirical Bayes and Shrinkage Estimators

Empirical Bayes methods are nothing else than ML estimation applied to certain hierarchical models, so the name is rather misleading. Essentially, the probabilistic model consists of what looks like a likelihood function and a corresponding prior distribution. In a true Bayesian approach, the parameters of the prior would be specified *beforehand*, reflecting the prior state of knowledge and without being influenced by the data. In the empirical Bayes approach however, a ML point estimate obtained from the data is used for the parameters of the prior. In Chap. 8, an empirical Bayes approach is used to develop a probabilistic method to superimpose protein structures.

A classic example of empirical Bayes is the *Beta-binomial distribution*, which represents the probability of observing a pair of counts $a, b$ for two events $A$ and $B$. The nature of this model (see Fig. 1.3b) is probably best understood by considering the process of obtaining samples from it. First, the binomial parameter $\theta$ is drawn from the Beta distribution with parameters $\alpha_1, \alpha_2$. Note that the Beta distribution is nothing else then the two-dimensional Dirichlet distribution. Then, a set of $n$ outcomes – falling into the two events $A$ and $B$ – are drawn from the binomial distribution with parameter $\theta$. More explicitly, this can be written as:

$$\theta \sim p(\theta \mid \alpha_1, \alpha_2) = \frac{1}{Z(\alpha_1, \alpha_2)} \theta^{\alpha_1 - 1} (1 - \theta)^{\alpha_2 - 1}$$

$$a, b \sim p(a, b \mid \theta) = \binom{a + b}{a} \theta^a (1 - \theta)^b$$

where $Z(\alpha_1, \alpha_2)$ is a normalization constant, and $n = a + b$.

The probability $p(a, b)$, which according to the beta binomial distribution is $p(a, b \mid \alpha_1, \alpha_2)$, is then obtained by integrating out the nuisance parameter $\theta$:

$$p(a, b \mid \alpha_1, \alpha_2) = \int p(a, b \mid \theta) p(\theta \mid \alpha_1, \alpha_2) d\theta$$

In the empirical Bayes method applied to this hierarchical model, the parameters $\alpha_1, \alpha_2$ are estimated from the data by maximum likelihood [510]. In a true Bayesian treatment, the values of $\alpha_1, \alpha_2$ would be chosen based on the prior belief, independent of the data.

Empirical Bayes methods also lie at the theoretical heart of another type of point estimation methods: the *shrinkage estimators*, of which the James-Stein estimator is the most well known [169, 170, 688, 689]. The statistical community was shocked when Charles Stein showed in 1955 that the conventional ML estimation methods for Gaussian models are suboptimal – in dimensions higher than two – in term of the expected squared error.

Suppose we have a set of observations $\mathbf{y}_1, \mathbf{y}_2, \ldots, \mathbf{y}_m$ with mean $\bar{\mathbf{y}}$ drawn from an $N$-dimensional Gaussian distribution ($N > 2$) with unknown mean $\boldsymbol{\mu}$ and given covariance matrix $\sigma^2 \mathbf{I}_N$. The maximum likelihood estimate $\hat{\boldsymbol{\mu}}_{ML}$ of $\boldsymbol{\mu}$ is simply the mean $\bar{\mathbf{y}}$, while the James-Stein estimate $\hat{\boldsymbol{\mu}}_{JS}$ is:

$$\hat{\boldsymbol{\mu}}_{JS} = \left(1 - \frac{\sigma^2(N-2)}{\|\bar{\mathbf{y}}\|^2}\right)\bar{\mathbf{y}}$$

Stein and James proved in 1961 that the latter estimate is superior to the ML estimate in terms of the expected total squared error loss [689].

James-Stein estimation is justified by an empirical Bayes approach in the following way [169, 170]. First, one assumes that the prior distribution of the mean $\boldsymbol{\mu}$ is a Gaussian distribution with mean $\mathbf{0}$ and unknown covariance matrix $\tau^2 \mathbf{I}_N$. Second, following the empirical Bayes approach, one estimates the unknown parameter $\tau^2$ of the prior from the data. Finally, the James-Stein estimate is obtained as the Bayes estimator under a quadratic loss function for the resulting empirical Bayes "posterior". The poor performance of the ML estimate can be understood as the result of the *de facto* use of a uniform prior over the mean, which in this case is actually overly informative.

Because the prior distribution has mean $\mathbf{0}$, the James-Stein estimate "shrinks" the ML estimate towards $\mathbf{0}$, hence the name *shrinkage estimators*. Shrinkage estimators can be extended beyond the Gaussian case discussed above, and have proven to be very useful for obtaining point estimates in high dimensional problems [90, 628].

### 1.3.5.4 Pseudolikelihood and Moment Estimation

For point estimation, ML and MAP estimation are by far the most commonly used methods. However, even these approximations to a full Bayesian analysis can be intractable. In such cases, estimation based on the pseudolikelihood or the method of moments can be an alternative. These methods are for example used in Chaps. 6 and 7, for parameter estimation of probability distributions over angular variables.

In the pseudolikelihood approach [55, 480, 696], the likelihood is approximated by the product of the marginal, conditional probabilities. For example, suppose the

# 1 An Overview of Bayesian Inference and Graphical Models

data $\mathbf{d}$ consists of observations $d_1, d_2, \ldots, d_M$ that are generated by a probabilistic model with parameter $\boldsymbol{\theta}$. In that case, the likelihood is approximated by the pseudolikelihood in the following way:

$$
\begin{aligned}
p(\mathbf{d} \mid \boldsymbol{\theta}) &= p(d_1, d_2, \ldots, d_M \mid \boldsymbol{\theta}) \\
&\approx p(d_1 \mid d_2, \ldots, d_M, \boldsymbol{\theta}) p(d_2 \mid d_1, d_3, \ldots, d_M, \boldsymbol{\theta}) \\
&\quad \ldots p(d_M \mid d_1, \ldots, d_{M-1}, \boldsymbol{\theta}) \\
&= \prod_{m=1}^{M} p(d_m \mid \{d_1, \ldots, d_M\} \setminus d_m, \boldsymbol{\theta})
\end{aligned}
$$

Estimation by maximum pseudolikelihood is done by finding the values for the parameters that maximize the pseudolikelihood, similar to ML estimation. For some distributions, such as the multivariate Gaussian distribution or the von Mises distribution on the circle, this type of point estimation performs essentially as well as ML estimation [480]. The pseudolikelihood approach is also useful for fast, heuristic parameter estimation in intractable graphical models [696].

Parameter estimation using the method of moments is done by making use of functions that relate the model's parameters to the moments – typically the mean and the variance. The moments are simply calculated from the data. For example [733], the Gamma distribution $\Gamma(x)$ with parameters $\alpha, \beta$ has the following density function:

$$
\Gamma(x) \propto x^{\alpha-1} \exp(-\frac{x}{\beta})
$$

The mean $\mu$ and variance $\sigma^2$ of this distribution can be written in function of $\alpha, \beta$:

$$
\mu = \alpha\beta
$$
$$
\sigma^2 = \alpha\beta^2
$$

Hence, we can estimate $\alpha, \beta$ by making use of the mean and variance calculated from the data:

$$
\alpha = \frac{\mu^2}{\sigma^2}
$$
$$
\beta = \frac{\sigma^2}{\mu}
$$

Often, ML estimation needs to be done by numerical optimization methods such as the Newton-Raphson method, that require a reasonable set of starting values. Finding these values is often done using moment estimation.

## 1.4 Foundational Arguments

The Bayesian view of probability can be justified in several ways. Perhaps the most convincing justification is its firm axiomatic basis: if one adopts a small set of requirements regarding beliefs including respecting the rules of logic, the rules of probability theory[2] necessarily follow [127]. Secondly, de Finetti's theorem states that if a dataset follows certain common conditions, an appropriate probabilistic model for the data necessarily consists of a likelihood and a prior [49]. Finally, another often used justification, also due to de Finetti, is based on gambling, where the use of beliefs that respect the probability calculus avoids situations of certain loss for a bookmaker. We will take a look at these three justifications in a bit more detail.

### 1.4.1 The Cox Axioms

The Cox axioms [127,342], first formulated by Richard T. Cox in 1946, emerge from a small set of requirements; properties that clearly need to be part of any consistent calculus involving degrees of belief. From these axioms, the Bayesian probability calculus follows. Informally, the Cox axioms and their underlying justifications correspond to:

1. Degrees of belief are expressed as real numbers. Let's say the belief in event $a$ is written as $\mathcal{B}(a)$.
2. Degrees of belief are ordered: if $\mathcal{B}(a) > \mathcal{B}(b)$ and $\mathcal{B}(b) > \mathcal{B}(c)$ then $\mathcal{B}(a) > \mathcal{B}(c)$.
3. There is a function $\mathcal{F}$ that connects the beliefs in a proposition $a$ and its negation $\sim a$:

$$\mathcal{B}(a) = \mathcal{F}[\mathcal{B}(\sim a)]$$

4. If we want to calculate the belief that two propositions $a$ and $b$ are true, we can first calculate the belief that $b$ is true, and then the belief that $b$ is true given that $a$ is true. Since the labelling is arbitrary, we can switch $a$ and $b$ around in this statement, which leads to the existence of a function $\mathcal{G}$ that has the following property:

$$\mathcal{B}(a,b) = \mathcal{G}[\mathcal{B}(a \mid b), \mathcal{B}(b)] = \mathcal{G}[\mathcal{B}(b \mid a), \mathcal{B}(a)]$$

These axioms imply a set of important requirements, including:

---

[2]In 1933, Kolomogorov formulated a set of axioms that form the basis of the mathematical theory of probability. Most interpretations of probability, including the frequentist and Bayesian interpretations, follow these axioms. However, the Kolomogorov axioms are compatible with many *interpretations* of probability.

# 1 An Overview of Bayesian Inference and Graphical Models

- Consistency with logic, when beliefs are absolute (that is, true or false).
- Different ways of reasoning for the same problem within the rules of the calculus lead to the same result.
- Identical states of knowledge, differing by labelling only, lead to the assignment of identical degrees of belief.

Surprisingly, this simple set of axioms is sufficient to pinpoint the rules of probabilistic inference completely. The functions $\mathcal{F}$ and $\mathcal{G}$ turn out to be $\mathcal{F}(x) = 1 - x$ and $\mathcal{G}(x, y) = xy$, as expected. In particular, the axioms lead to the two central rules of probability theory. To recall, these rules are the *product rule*:

$$p(a, b) = p(a \mid b)p(b) = p(b \mid a)p(a)$$

which directly leads to Bayes' theorem, and the *sum rule*:

$$p(a) = \sum_b p(a, b)$$

## 1.4.2 Other Arguments

Another argument, due to de Finetti and Ramsey [705], is based on linking beliefs with a willingness to bet. If a bookmaker isn't careful, he might propose a set of bets and odds that make it possible for a gambler to make a so-called *Dutch book*. A Dutch book guarantees a profit for the gambler, and a corresponding loss for the bookmaker, regardless of the outcome of the bets. It can be shown that if the bookmaker respects the rules of the probability calculus in the construction of the odds, the making of a Dutch book is impossible.

A second argument often invoked to justify the Bayesian view of probability is also due to de Finetti, and is called *de Finetti's representation theorem* [49]. The theorem deals with data that are *exchangeable*: that is, any permutation of the data does not alter the joint probability distribution. In simple words, the ordering of the data does not matter.

Let us consider the case of an exchangeable series of $N$ Bernoulli random variables, consisting of zeros and ones. For those data, de Finetti's theorem essentially guarantees that the joint probability distribution of the data can be written as:

$$p(x_1, \ldots, x_N) \propto \int_0^1 \left\{ \prod_{n=1}^N \theta^{x_n}(1 - \theta)^{1-x_n} \right\} p(\theta) d\theta$$

The two factors in the integral can be interpreted as a likelihood and a prior. The interpretation is that exchangeability leads to the existence of a likelihood and a prior, and can thus be interpreted as an argument in favor of the Bayesian viewpoint.

Although the theorem is stated here as it applies to binomial data, the theorem extends to many other cases.

## 1.5 Information Theory

Information theory was developed by Claude Shannon at the end of the 1940s [342, 645]. It has its origins in the study of the transmission of messages through noisy communication channels, but quickly found applications in nearly every branch of science and engineering. As it has many fundamental applications in probability theory – such as for example in the construction of suitable prior distributions [52] – we give a quick overview.

Informally speaking, information quantifies the "surprise" that is associated with gaining knowledge about the value of a certain variable. If the value was "expected" because its probability was high, the information gain is minimal. However, if the value $x$ was "unexpected" because its probability was low, the information gain is high. Several considerations lead to the following expression for the information $I$ associated with learning the value of a discrete variable $x$:

$$I(x) = -\log p(x)$$

### 1.5.1 Information Entropy

The *information entropy* $S_x$ is the expectation of the information of a discrete variable $x$:

$$S_x = -\sum_x p(x) \log p(x) = -\mathbb{E}_{p(x)} \{\log p(x)\}$$

The entropy is at its maximum when $p(x)$ is the uniform distribution. The information entropy becomes zero when $x$ adopts one specific value with probability one. Shannon's information entropy applies to discrete random variables: it cannot be extended to the continuous case by simply replacing the sum with an integral [337]. For the continuous case, a useful measure is the Kullback-Leibler divergence, which is discussed in the next section.

### 1.5.2 Kullback-Leibler Divergence

The Kullback-Leibler (KL) divergence is a natural measure of a distance between two probability distributions [408]. Strictly speaking, it is not a distance as it is not

# 1 An Overview of Bayesian Inference and Graphical Models

symmetric and does not respect the triangle inequality, which is why it is called a 'divergence'. In the discrete case, the KL divergence is defined as:

$$KL[p \parallel q] = \sum_x p(x) \log \frac{p(x)}{q(x)}$$

In the continuous case, the sum is replaced by an integral:

$$KL[p \parallel q] = \int p(x) \log \frac{p(x)}{q(x)} dx$$

A distance measure for probability distributions is useful in many different ways. For example, consider a "correct" but computationally intractable probability distribution, and a set of possible approximations. The KL divergence could then be used to pick the best approximation. We will encounter the KL divergence again in Sect. 1.8.3, where it will reveal a deep link between probability theory and statistical physics.

## 1.5.3 Mutual Information

The mutual information measures the mutual dependence of the two random variables. Intuitively, it tells you how much information you gain about the first random variable if you are given the value of the other one. The mutual information $I_{x,y}$ of two random variables $x$ and $y$ is:

$$I_{x,y} = \sum_x \sum_y p(x, y) \log \frac{p(x, y)}{p(x)p(y)}$$

In the corresponding version of the mutual information for continuous random variables, the sums are again replaced by integrals. The mutual information is nonnegative and symmetric. If $x$ and $y$ are independent, their mutual information is equal to zero.

## 1.6 Prior Distributions

### 1.6.1 Principles for the Construction of Priors

One of the most disputed and discussed aspect of the Bayesian view is the construction of the prior distribution. Typically, one wants to use a so-called non-informative prior. That is, a prior that correctly represents the "ignorance" in a particular situation and does not obscure the information present in the data. In many cases the choice of a non-informative prior is clear. For the finite, discrete case, the maximum entropy principle and the related principle of indifference apply. For the univariate, continuous case, Jeffreys' priors are appropriate. In many other situations, the choice of a suitable prior is less clear. The construction of suitable non-informative priors is still the object of much discussion and research. We will now briefly look at some of these methods to decide which prior distributions to use, including some of a more pragmatic nature.

### 1.6.1.1 Principle of Indifference

The earliest principle for the construction of a prior is due to Laplace and Bayes, and is called the *principle of insufficient reason* or the *principle of indifference*. If a variable of interest can adopt a finite number of values, and all of these values are indistinguishable except for their label, then the principle of indifference suggests to use the discrete uniform distribution as a prior. Hence, if the variable can adopt $K$ values, each value is assigned a probability equal to $\frac{1}{K}$. For continuous variables the principle often produces unappealing results, and alternative approaches are necessary (see Sect. 1.6.1.3). The principle of indifference can be seen as a special case of the maximum entropy principle, which we consider next.

### 1.6.1.2 Maximum Entropy

The principle of indifference can be seen as the result of a more general principle: the principle of maximum entropy [337, 342], often called *MaxEnt*. Let's again consider a variable $A$ that can adopt a finite number of discrete values $a_1, \ldots, a_K$. Suppose now that some information is available about $A$; for example its mean value $\bar{a}$:

$$\bar{a} = \sum_{k=1}^{K} p(a_k) a_k$$

The classic illustration of this problem is known as the *Brandeis dice problem* [339]. Suppose we are given the following information, and nothing else, about a certain dice: the average outcome $\bar{a}$ of throwing the dice is 4.5, instead of the average of an "honest" dice, which is 3.5. The question is now, what are the probabilities we assign to throwing any of the $K = 6$ values? Clearly, these probabilities will differ from the honest case, where the probability of throwing any of the six values is 1/6.

The problem is to come up with a plausible distribution $p(A)$ that is compatible with the given information. The solution is to find the probability distribution with maximum entropy that is compatible with the given information – in this case the mean $\bar{a}$. The problem can be easily solved using the method of *Lagrange multipliers*, which can be used to maximize a function under a given set of constraints. In this case we want to maximize the information entropy:

$$S_A = -\sum_{k=1}^{K} p(a_k) \log p(a_k)$$

subject to the constraints:

$$\sum_{k=1}^{K} p(a_k) a_k = \overline{a} \tag{1.3}$$

$$\sum_{k=1}^{K} p(a_k) = 1 \tag{1.4}$$

The resulting *Lagrangian function* for this problem is then, using $p_k \equiv p(a_k)$ and $\mathbf{p} \equiv (p_1, \ldots, p_K)$ for simplicity:

$$L(\mathbf{p}, \alpha, \beta) = -\sum_{k=1}^{K} p_k \log p_k - \alpha \left( \sum_{k=1}^{K} p_k - 1 \right) - \beta \left( \sum_{k=1}^{K} p_k a_k - \overline{a} \right) \tag{1.5}$$

where $\alpha$ and $\beta$ are the Lagrange multipliers. The second and third term impose the constraints of Eqs. 1.3 and 1.4, respectively. The solution is found by setting the partial derivatives with respect to $p_1, \ldots, p_K$ to zero:

$$\frac{\partial L(\mathbf{p}, \alpha, \beta)}{\partial p_k} = 0 = -\log(p_k) - 1 - \alpha - \beta a_k$$

which results in:

$$p_k = \exp(-1 - \alpha - \beta a_k)$$

By taking the normalization constraint (Eq. 1.4) into account, $\alpha$ can be eliminated. The final result is an exponential distribution law that depends on $\beta$, but not on $\alpha$:

$$p_k = \frac{1}{Z} \exp(-\beta a_k)$$

$$Z = \sum_{k=1}^{K} \exp(-\beta a_k)$$

where $Z$ is a normalization factor. The value of $\beta$ can be obtained numerically by combining the result with the constraints given by Eqs. 1.3 and 1.4. Surprisingly, we will encounter this probability distribution and its underlying derivation again, in the form of the famous Boltzmann equation in Sect. 1.8.2.

For the honest dice with an average equal to 3.5, the maximum entropy method delivers the expected result, namely a probability equal to $1/6 \approx 0.17$ for all values. If nothing is known about the average of the dice we obtain the same solution, because it maximizes the entropy in the absence of any constraints. This situation corresponds to the principle of indifference as discussed above. For the dishonest dice with an average equal to 4.5, we obtain the following (approximate) probabilities [339]:

$$\{p_1, \ldots, p_6\} = \{0.05, 0.08, 0.11, 0.16, 0.24, 0.35\}$$

For discrete, finite cases, the maximum entropy framework provides a convenient method to construct prior distributions that correctly reflect any prior knowledge. For the continuous case, the situation is not so clear [337]. The definition of information entropy does not simply generalize to the continuous case, and other ways to construct priors are typically needed.

### 1.6.1.3 Jeffreys' Prior

The principle of indifference and the MaxEnt method are intuitively reasonable, but applying the principle to cases that are not finite and discrete quickly leads to problems. For example, applying the principle to a variable $x$ on the positive real line $\mathbb{R}^+$ leads to a uniform distribution over $\mathbb{R}^+$. However, consider a non-linear monotone $y(x)$ transformation of $x$. Surely, ignorance on $x$ should imply equal ignorance on $y$, which leads to the demand that $p(y)dy$ should be equal to $p(x)dx$. Clearly, the uniform distribution on $\mathbb{R}^+$ does not fulfill the demands associated with these invariance considerations. In 1946, Jeffreys proposed a method to obtain priors that take these invariance considerations into account. In the univariate case, Jeffreys' prior is equal to:

$$p(x) \propto \sqrt{\mathcal{I}(x)}$$

where $\mathcal{I}(x)$ is the *Fisher information of the likelihood*. We also encountered the Fisher information in Sect. 1.3.5.2. The Jeffreys' priors are invariant under one-to-one reparameterization.

If the likelihood is a univariate Gaussian with known mean $\mu$ and unknown standard deviation $\sigma$, Jeffreys' rule leads to a prior over $\sigma$ equal to $p(\sigma) \propto \frac{1}{\sigma}$ [428]. If the likelihood is the binomial distribution with parameter $\theta$, the rule leads to a prior over $\theta$ equal to $p(\theta) \propto \frac{1}{\sqrt{\theta(1-\theta)}}$ [428]. Jeffreys' rule typically leads to good results for the continuous univariate case, but the extension to the multivariate case is problematic. We will encounter Jeffreys' priors in Chap. 12, where they serve as priors over nuisance variables in protein structure determination from NMR data.

### 1.6.1.4 Reference Priors

Maximum entropy and Jeffreys' priors can be justified in a more general framework: the *reference prior* [52]. Reference priors were proposed by José Bernardo [43, 50], and provide an elegant and widely applicable procedure to construct non-informative priors.

# 1 An Overview of Bayesian Inference and Graphical Models

The central idea behind reference priors is that one aims to find a prior that has a minimal effect on the posterior distribution, resulting in a *reference posterior* where the influence of the prior is minimal. Again, information theory comes to the rescue. Recall that the Kullback-Leibler divergence is a natural measure of the distance between two probability distributions. The reference prior is defined [43] as the prior distribution that *maximizes* the *expected* KL divergence between the prior $p(\theta)$ and the posterior distribution $p(\theta \mid D)$. The expectation is with respect to marginal probability of the data $p(d)$:

$$\mathbb{E}_{p(d)}\left[\int p(\theta) \log \frac{p(\theta \mid d)}{p(\theta)} d\theta\right]$$

An appealing property of reference prior approach is that the resulting prior is invariant under one-to-one transformations of $\theta$. Reference priors are often not properly normalized and thus not true probability distributions: they should be considered as mathematical devices to construct *proper* posterior distributions. We will encounter the concept of an expected KL divergence again in Sect. 1.7, where it justifies model selection using the Akaike information criterion.

## 1.6.2 Conjugate Priors

Conjugate priors are priors that can arise from any of the above considerations, but that have the computationally convenient property that the resulting posterior distribution has the same form as the prior. A well known and widely used example of a conjugate prior is the Dirichlet prior for the multinomial distribution, which we already encountered in Sect. 1.3.5.3 as the Beta-binomial distribution for two dimensions. Conjugate priors are typically used because they make the calculation of the posterior *easier*, not because this is the prior that necessarily best describes the prior state of knowledge.

The binomial distribution with parameters $n, \theta$ (with $n$ a positive integer and $0 < \theta < 1$) gives the probability of a discrete variable $x$ (with $0 \leq x \leq n$) according to:

$$\mathrm{Bi}(x \mid \theta, n) = \binom{n}{x} \theta^x (1 - \theta)^{n-x}$$

The Binomial distribution is typically interpreted as giving the probability of $x$ successes in $n$ trials, where each trial can either result in success or failure.

Now, we want to infer $\theta$ given $x$ and $n$, where $n$ is the length of the sequence, and thus typically known. Following the standard approach, and using the Binomial distribution as the likelihood, we obtain:

$$p(\theta \mid x, n) \propto \mathrm{Bi}(x \mid \theta, n) p(\theta)$$

For the prior, we now use a Beta distribution. The Beta distribution use parameters $\alpha_1 > 0, \alpha_2 > 0$ is the two-dimensional version of the Dirichlet distribution. The latter distribution is a probability distribution on the $N$-dimensional simplex, or alternatively, a probability distribution over the space of probability vectors – $p$ strictly positive real numbers that sum to one. In the case of the Beta distribution used as a prior for $\theta$, this becomes:

$$\mathrm{Be}(\theta \mid \alpha_1, \alpha_2) \propto \theta^{\alpha_1-1}(1-\theta)^{\alpha_2-1}$$

Hence, the posterior becomes:

$$p(\theta \mid x, n) \propto \mathrm{Bi}(x \mid n, \theta)\mathrm{Be}(\theta \mid \alpha_2, \alpha_2)$$
$$\propto \theta^x(1-\theta)^{n-x}\theta^{\alpha_1-1}(1-\theta)^{\alpha_2-1}$$
$$= \theta^{x+\alpha_1-1}(1-\theta)^{n-x+\alpha_2-1}$$

Evidently, the obtained posterior has the same functional form as the prior.

In addition, this prior has an interesting interpretation as a so-called *pseudocount*. Given $x$ and $n$, the ML estimate for $\theta$ is simply $\frac{x}{n}$. Using a Beta prior with parameters $\alpha_1, \alpha_2$ the MAP estimate for $\theta$ becomes $\frac{x+\alpha_1}{n+\alpha_1+\alpha_2}$. Hence, the use of a Beta prior can be interpreted as adding extra counts – often called *pseudocounts* – to the number of observed failures and successes. The whole approach can be easily extended to the higher dimensional case, by combining a multinomial likelihood with a Dirichlet prior.

### 1.6.3 Improper Priors

Improper priors are priors that are not probability distributions, because they are not properly normalized. Consider for example a parameter that can adopt any value on the real line (from $-\infty$ to $+\infty$). A uniform prior on such a parameter is an improper prior, since it cannot be properly normalized and hence is not a true probability density function. Nonetheless, such priors can be useful, provided the resulting posterior distribution is still a well defined and properly normalized probability distribution. Some of the above methods to construct priors, such as the reference prior method or Jeffreys' rule, often result in improper priors.

## 1.7 Model Selection

So far we have assumed that the problem of inference is limited to the parameters of a given model. However, typically it is not known which model is best suited for the data. A classic example is the mixture model, as discussed in Sect. 1.3.4.

# 1 An Overview of Bayesian Inference and Graphical Models

How many mixture components should be chosen? Underestimating the number of mixture components results in a poor model; overestimating results in overfitting and a model that generalizes poorly to new data sets. The problem of inferring the model in addition to the model parameters is called the *model selection problem*. As usual, this problem can be treated in a rigorous Bayesian framework, using the *Bayes factor*. However, in practice various approximations are often used, including the *Bayesian information criterion* and the *Akaike information criterion*. These criteria balance the requirement of fitting the data against the demand for a model that is not overly complex. Complexity in this case refers to the number of free parameters. Finally, we will briefly touch upon the *reversible jump MCMC* method, which makes it possible to infer both models and their respective parameters directly in a fully Bayesian way.

## 1.7.1 Bayes Factor

The Bayes factor $B$ is the ratio of the respective probabilities of the data $\mathbf{d}$ given the two different models $M_1$ and $M_2$:

$$B = \frac{p(\mathbf{d} \mid M_1)}{p(\mathbf{d} \mid M_2)}$$

Note that this expression can also be written as:

$$B = \frac{p(M_1 \mid \mathbf{d})}{p(M_2 \mid \mathbf{d})} \Big/ \frac{p(M_1)}{p(M_2)}$$

Hence, the Bayes factor gives an idea how the data affects the belief in $M_1$ and $M_2$ relative to the prior information on the models.

This Bayes factor is similar to the classic likelihood ratio, but involves integrating out the model parameters, instead of using the maximum likelihood values. Hence:

$$B = \frac{p(\mathbf{d} \mid M_1)}{p(\mathbf{d} \mid M_2)} = \frac{\int p(\mathbf{d} \mid \boldsymbol{\theta}_1, M_1)\pi_1(\boldsymbol{\theta}_1 \mid M_1)\mathrm{d}\boldsymbol{\theta}_1}{\int p(\mathbf{d} \mid \boldsymbol{\theta}_2, M_2)\pi_2(\boldsymbol{\theta}_2 \mid M_2)\mathrm{d}\boldsymbol{\theta}_2} \tag{1.6}$$

where $\boldsymbol{\theta}_1$ and $\boldsymbol{\theta}_2$ are the respective model parameters and and $\pi_1(\cdot)$ and $\pi_2(\cdot)$ are the corresponding priors over the model parameters.

The Bayes factor is typically interpreted using a scale introduced by Jeffreys [343]:

- If the logarithm of $B$ is below 0, the evidence is against $M_1$ and in favor of $M_2$.
- If it is between 0 and 0.5, the evidence in favor of $M_1$ and against $M_2$ is *weak*.
- If it is between 0.5 and 1, it is *substantial*.
- If it is between 1 and 1.5, it is *strong*.

- If it is between 1.5 and 2, it is *very strong*.
- If it is above 2, it is *decisive*.

### 1.7.2 Bayesian Information Criterion

Often, the calculation of the Bayes factor is intractable. In that case, one resorts to an approximation. In order to calculate the Bayes factors, we need to calculate $p(\mathbf{d} \mid M)$ for the models involved. This can be done using the following rough approximation:

$$\log p(\mathbf{d} \mid M) \approx \log p(\mathbf{d} \mid \boldsymbol{\theta}_{\text{ML}}, M) - \frac{1}{2} Q \log R \qquad (1.7)$$

where $\boldsymbol{\theta}_{\text{ML}}$ is the ML point estimate, $R$ is the number of data points and $Q$ is the number of free parameters in $\boldsymbol{\theta}_{\text{ML}}$. This approximation is based on the Laplace approximation [62], which involves representing a distribution as a Gaussian distribution centered at its mode.

Equation 1.7 is called the *Bayesian Information Criterion (BIC)* or the *Schwartz criterion* [638]. Model selection is done by calculating the BIC value for all models, and selecting the model with the largest value. Conceptually, the BIC corresponds to the likelihood penalized by the complexity of the model, as measured by the number of parameters. The use of the ML estimate $\boldsymbol{\theta}_{\text{ML}}$ in the calculation of the BIC is of course its weak point: the prior is assumed to be uniform, and any prior information is not taken into account.

### 1.7.3 Akaike Information Criterion

The Akaike information criterion (AIC), introduced by Hirotsugu Akaike in 1974 [2, 91], is used in the same way as the BIC and also has a similar mathematical expression:

$$\text{AIC} = 2 \log p(\mathbf{d} \mid \boldsymbol{\theta}_{\text{ML}}, M) - 2Q$$

Like the BIC, the AIC can be considered as a penalized likelihood, but the latter does not penalize model complexity as severely as the BIC.

Despite their similar mathematical expressions, the AIC and BIC are justified in different ways. The AIC is based on information theory: the idea is to use the KL divergence as a basis for model selection. The probability distribution embodied in the best model minimizes the KL divergence with respect to a "true", but unknown probability distribution $q(\cdot)$. Performing this selection based on the KL divergence requires an estimate of the following quantity, of which the AIC is an approximation:

$$\mathbb{E}_{q(\mathbf{y})} \left[ \mathbb{E}_{q(\mathbf{x})} \left[ \log p(\mathbf{x} \mid \boldsymbol{\theta}_{\text{ML}}(\mathbf{y})) \right] \right] \approx \text{AIC}$$

where $\boldsymbol{\theta}_{\text{ML}}(\mathbf{y})$ is the ML estimate for $\boldsymbol{\theta}$ obtained from $\mathbf{y}$.

# 1.7.4 Deviance Information Criterion

Often, the posterior distribution is only available in the form of a set of samples, typically obtained using an MCMC algorithm. For the calculation of the AIC and the BIC, the ML or MAP estimate of $\theta$ is needed. These estimates are not readily obtained from MCMC sampling. In that case, the *deviance information criterion (DIC)* [683] can be used for model selection. All quantities needed for the calculation of the DIC can be readily obtained from a set of samples from the posterior distribution. Another advantage of the DIC is that it takes the prior into account.

Like the BIC and AIC, the DIC balances the complexity of the model against the fit of the data. In the case of the DIC, these qualities are numerically evaluated as the *effective number of free parameters* $p_D$ and the *posterior mean deviance* $\overline{D}$. The DIC is subsequently given by:

$$\mathrm{DIC} = p_D + \overline{D}$$

These quantities a calculated as follows. First, the *deviance D* is defined as:

$$D(\theta) = -2 \log p(\mathbf{d} \mid \theta)$$

The *posterior mean deviance* $\overline{D}$ is given by the expectation of the deviance under the posterior:

$$\overline{D(\theta)} = \mathbb{E}_{p(\theta|\mathbf{d})} [D(\theta)]$$

The *effective number of parameters* is defined as:

$$p_D = \overline{D(\theta)} - D(\overline{\theta})$$

where $\overline{\theta}$ is the expectation of $\theta$ under the posterior. The effective number of parameters improves on the naive counting of free parameters as used in the BIC and AIC. The reason is twofold. First, it takes into account the effect of the prior information, which can reduce the effective dimensionality. Second, for hierarchical models, it is not always clear how many free parameters are actually involved from simply considering the model.

# 1.7.5 Reversible Jump MCMC

The Bayes factor, BIC, AIC or DIC can be used to select the best model among a set models whose parameters were inferred previously. However, it is also possible to sample different models, together with their parameters, following the Bayesian probability calculus using an MCMC algorithm. This can be done with a method introduced by Peter Green in 1995 called *reversible jump MCMC* [253].

A classic application is inference of mixture models without fixing the number of components. The main problem is that a transition from one model to another typically changes the dimensionality of the parameter space. The trick is thus to supplement the respective parameter spaces such that a one-to-one mapping is obtained. For more information we refer to [253].

## 1.8 Statistical Mechanics

### 1.8.1 Overview

Statistical mechanics can be viewed as probability theory applied to certain physical systems [150, 801]. It was developed largely independent of statistics, but lately the two fields are becoming more and more intertwined [537, 782–785]. Concepts from statistical physics, such as Bethe and Kikuchi approximations [60, 372], are now widely used for inference in graphical models [784, 785]. Conversely, probabilistic machine learning methods are increasingly used to handle problems in statistical physics [782]. Essentially, statistical physics makes use of *free energies* for inference, instead of probabilities [150, 801]. The fascinating connections between probability theory and various energy functions originating from statistical mechanics are further discussed in Chaps. 3, 5 and 4.

### 1.8.2 Boltzmann Distribution

Suppose a physical system can exist in a number of states, $s_1, s_2, \ldots, s_K$, called *microstates*. Each microstate $s_k$ occurs with a probability $p(s_k)$. Suppose there is an energy function $E$ that assigns an energy $E(s_k)$ to each state $s_k$, and that we are given the average of this energy $\bar{E}$, called the *internal energy* $U$, over all the states:

$$U = \bar{E} = \sum_{k=1}^{K} E(s_k) p(s_k)$$

Given *only* the average energy over all the states, and with the probabilities of the states unknown, the question is now how do we assign probabilities to all the states? A simple solution to this question makes use of the maximum entropy principle [150, 336]. The solution is a probability distribution over the states which results in the correct average value for the energy and which maximizes the entropy. Computationally, the solution can be found by making use of the following Lagrangian function, using $p_k \equiv p(s_k)$, $\mathbf{p} \equiv (p_1, \ldots, p_K)$ and $E(s_k) \equiv E_k$ for simplicity:

# 1 An Overview of Bayesian Inference and Graphical Models

$$L(\mathbf{p}, \alpha, \beta) = -\sum_{k=1}^{K} p_k \log p_k - \alpha \left\{ \sum_{k=1}^{K} p_k - 1 \right\} - \beta \left\{ \sum_{k=1}^{K} p_k E_k - \bar{E} \right\}$$

The first term concerns the maximum entropy demand, the second term ensures the probabilities sum to one, and the last term ensures that the correct average $\bar{E}$ is obtained. Of course, this Lagrangian function and its solution is formally identical to the one obtained for the Brandeis dice problem in Sect. 1.8.2 and Eq. 1.5. As mentioned before, the solution of the Lagrangian turns out to be an exponential distribution, called the *Boltzmann distribution* that is independent of $\alpha$:

$$p_k = \frac{1}{Z} \exp(-\beta E_k)$$

where $Z$ is a normalization factor:

$$Z = \sum_{k=1}^{K} \exp(-\beta E_k)$$

In statistical mechanics, $Z$ is called the *partition function*; the letter stems from the German *Zustandssumme*. In physical systems, $\beta$ has a clear interpretation: it is the inverse of the temperature.[3] In addition, $\beta$ is determined by its relation to $\bar{E}$:

$$\bar{E} = \sum_{k=1}^{K} p_k E_k = \frac{1}{Z} \sum_{k=1}^{K} \exp(-\beta E_k) E_k$$

In simple words, the final result can be summarized as:

$$\text{probability} = \frac{1}{\text{partition function}} \exp\left(\frac{-\text{energy}}{\text{temperature}}\right)$$

Any probability distribution can be cast in the form of a Boltzmann distribution, by choosing $\beta E_k$ equal to $-\log(p_k)$.

## 1.8.3 Free Energy

Working directly with the Boltzmann distribution is cumbersome. Therefore, some mathematical constructs are introduced that make finding the Boltzmann distribution easier. These constructs are called *free energies*, which come in different flavors.

---

[3] Here, we set $\beta = \frac{1}{T}$ for simplicity and without loss of generality. In physics, $\beta = \frac{1}{kT}$ where $k$ is Boltzmann's constant.

The *Helmholtz free energy* $F$ is defined as:

$$F = -\frac{1}{\beta} \log Z$$

As the partition function $Z$ is a function of $p(s_k)$, the free energy function thus assigns a value to a probability distribution over all possible microstates. Note that the energy function $E$ mentioned above assigns an energy value to each microstate, as opposed to a probability distribution over all microstates.

The Helmholtz free energy has the useful property that the Boltzmann distribution is recovered by minimizing $F$, for the energies $E(s_k)$ given, in function of $p(s_k)$. The free energy $F$ can be expressed in function of the internal energy $U$ and the entropy $S$:

$$U = \sum_k p(s_k) E(s_k)$$

$$S = -\sum_k p(s_k) \log p(s_k)$$

$$F = U - TS$$

The Helmholtz free energy can be used to obtain much useful information on a system. For example, the internal energy $U$ can be obtained in the following way:

$$-\beta \frac{\partial F}{\partial \beta} = -\frac{\partial \log(Z)}{\partial \beta}$$

$$= \frac{\partial \log \sum_k \exp(-\beta E(s_k))}{\partial \beta}$$

$$= U$$

A very useful concept for inference purposes is the *variational free energy* $G$ [786]. The variational free energy is used to construct approximations, when working with the Helmholtz free energy directly is intractable. This energy can be considered as a variant of the Helmholtz free energy $F$ in the presence of some constraints on the form of the probability distribution, which make the problem more tractable. In the absence of any constraints, $G$ becomes equal to $F$. Minimizing the variational free energy corresponds to finding the probability distribution $b(\mathbf{x})$ that respects the constraints, and is 'close' to the unconstrained probability distribution $p(\mathbf{x})$, where $\mathbf{x} = x_1, \ldots, x_N$ is discrete and $N$-dimensional. In practice, $b(\mathbf{x})$ is chosen so that it is a tractable approximation of $p(\mathbf{x})$. For example, one can use the *mean field* approach which insists on a fully factorized joint probability distribution for $b(\mathbf{x})$:

$$p(\mathbf{x}) \approx b(\mathbf{x}) = \prod_{n=1}^{N} b_n(x_n)$$

# 1 An Overview of Bayesian Inference and Graphical Models

To formulate this problem, we make use of the KL divergence. The goal is to find $b(\mathbf{x})$ that both respects the given constraints *and* minimizes the divergence with $p(\mathbf{x})$. The resulting divergence can be written in terms of entropies and energies in the following way:

$$
\begin{aligned}
\mathrm{KL}[b \parallel p] &= \sum_{\mathbf{x}} b(\mathbf{x}) \log \frac{b(\mathbf{x})}{p(\mathbf{x})} \\
&= \sum_{\mathbf{x}} b(\mathbf{x}) \log b(\mathbf{x}) - \sum_{\mathbf{x}} b(\mathbf{x}) \log p(\mathbf{x}) \\
&= \sum_{\mathbf{x}} b(\mathbf{x}) \log b(\mathbf{x}) + \sum_{\mathbf{x}} b(\mathbf{x}) \beta E(\mathbf{x}) + \log Z \\
&= -S_b + \beta U_b - \beta F_p
\end{aligned}
$$

where $E(\mathbf{x})$ is the energy of $\mathbf{x}$; $S_b$ and $U_b$ are the *variational entropy* and the *variational internal energy*; and $F_p$ is the Helmholtz free energy. For $S_b$, $U_b$ and $F_p$, the subscripts indicate which probability distribution is involved. The *variational free energy* $G_b$ is then equal to:

$$
G_b = U_b - T S_b
$$

After some re-arrangement and keeping in mind that $\beta = \frac{1}{T}$, we get the following expression for $G_b$:

$$
G_b = F_p + T \cdot \mathrm{KL}[b \parallel p]
$$

Hence, $G_b$ is equal to $F_p$ plus the temperature weighed KL divergence between the two probability distributions. In the absence of any constraints, the KL divergence is zero when $b(\mathbf{x})$ is exactly equal to $p(\mathbf{x})$, and $G_b$ becomes equal to $F_p$. In the presence of constraints, minimizing the variational free energy $G_b$ will minimize the free energy while respecting the constraints. In practice, because of the constraints, minimizing $G_b$ is easier than minimizing $F_p$ directly. The variational free energy is now widely used for approximate inference in otherwise intractable graphical models [785].

## 1.9 Graphical Models

### *1.9.1 Introduction*

Graphical models [62, 126, 345, 568] have their roots in a research field that used to be called *artificial intelligence* – often abbreviated to AI. One of the main outcomes of artificial intelligence were so-called *expert systems*: these systems combined a knowledge base from a particular branch of human activity with methods to perform inference. Expert systems for the diagnosis of disease are a classic example. Here, the knowledge base is a set of diseases and their associated symptoms, and the inference problem is the actual diagnosis of a patient's disease based on his particular symptoms. Such systems were developed in academia in the 1960s, and were commercialized and applied to practical problems in the 1980s.

Expert systems were initially based on rules that make use of *logical deduction*:

If A is true, therefore, B is true
  If B is false, therefore, A is false

However, it quickly became clear that systems solely based on deduction were severely limited. For example, a single unexpected or absent symptom can corrupt disease diagnosis completely. In addition, it is difficult to deal with unobserved variables – for example, a patient's smoking status is unknown – and the set of rules quickly becomes huge for even simple applications.

Therefore, attention shifted towards expert systems based on *logical induction*:

B is true, therefore, A becomes more plausible
  A is false, therefore, B becomes less plausible

Such expert systems were meant to reason in the face of uncertainty, by assigning degrees of plausibility or certainty factors to various statements. For that, one needs an algebra or calculus to calculate the plausibility of statements based on those of other statements. Various approaches were tried, including *fuzzy logic* and *belief functions*, all of which turned out to exhibit serious inconsistencies and shortcomings. In retrospects, it is of course clear that such reasoning should be performed according to the rules of the Bayesian probability calculus, but this was seen as problematic by the AI community. The first reason was that joint probability distributions over many variables were seen as intractable. The second reason was more subtle: at that time the ruling paradigm in statistics was the frequentist interpretation, which claims it is meaningless to assign probabilities to hypotheses.

Eventually both objections were overcome, thanks to pioneering work done by researchers such as Finn V. Jensen, Steffen Lauritzen, Judea Pearl and David Spiegelhalter [126, 345, 568]. The first problem was solved by making use of conditional independencies in joint probability distributions. The second perceived problem slowly faded away by the increasing acceptance of the Bayesian interpretation of probability. Finally, the availability of cheap and powerful computers brought many computationally intensive Bayesian procedures within practical reach.

## 1.9.2 Graphical Models: Main Ideas

Graphical models represent a set of random variables and their conditional independencies [62, 568]. This representation consists of a graph in which the nodes are the variables and the edges encode the conditional independencies. The goal is to solve problems of inference efficiently by taking the structure of the graph into account, either using systematic algebraic manipulation or using MCMC methods. Graphical models mostly come in two different flavors: Bayesian networks (BNs) [568] and Markov random fields (MRFs) [54]. In the former case, the graph is directed, while in the latter case the graph is undirected. Some independence structures can be encoded in a BN but not in a MRF, and vice versa. Some less common graphical models also exist that contain both directed and undirected edges. All graphical models can be understood in one general framework: the *factor graph*, which provides a unifying representation for both BNs and MRFs [402].

From a Bayesian point of view, there is no distinction between unknown "parameters" and unknown "variables". Indeed, MCMC methods, which are discussed at length in Chap. 2, provide a unified framework to perform Bayesian inference in graphical models. However, from a practical point of view, it is often useful to distinguish between parameters and latent variables. Therefore, in this chapter, we give an overview of some common methods to perform *inference* of latent variables and *learning* of parameters in graphical models, including maximum likelihood estimation. For a thorough treatment of MCMC methods, we refer to Chap. 2.

### 1.9.3 Bayesian Networks

#### 1.9.3.1 General Properties

As mentioned above, the graph of a Bayesian network is directed, that is, its edges are arrows. In addition, cycles are not allowed: it is not allowed to encounter the same node twice if one traverses the graph in the direction of the arrows. A directed arrow points from a *parent node* to a *child node*. Let's consider a general joint probability distribution $p(a, b, c)$ over three variables $a, b, c$. Using the product rule, this joint probability distribution can be factorized in different ways, of which we consider two possibilities:

$$p(a, b, c) = p(a \mid b, c) p(b \mid c) p(c)$$
$$= p(b \mid a, c) p(a \mid c) p(c)$$

Each factorization gives rise to a Bayesian network, as shown in Fig. 1.5. This example illustrates how a BN encodes a joint probability distribution. The joint

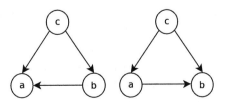

**Fig. 1.5** Two fully connected Bayesian networks that represent the joint probability distribution over three variables $a, b$ and $c$

probability distribution encoded in a BN is the product of factors (one for each variable), that each consist of the conditional probability distribution of the variable given its parents. More formally:

$$p(x_1, \ldots, x_N) = \prod_{n=1}^{N} p(x_n \mid \mathrm{pa}(x_n))$$

where $\mathrm{pa}(x_n)$ denotes the parents of node $x_n$. The conditional probability distributions are most often categorical distributions for discrete nodes and Gaussian nodes for continuous variables, but many other distributions are of course possible. In practice, a categorical distribution is specified as a *conditional probability table* (CPT), which tabulates the probability of a child's values, given the values of the parents. For example, consider a discrete node which can adopt two values that has two parents which can adopt three and four values, respectively. The resulting CPT will be a $4 \times 3 \times 2$ table, with 12 unique parameters because the probabilities need to sum to one. The case of a Gaussian node with one discrete parent is typically identical to the mixture model described in Sect. 1.3.4.

The example also clarifies that different BNs can give rise to the same probability distribution: in Fig. 1.5, both BNs correspond to different factorizations of the same joint probability distribution. In our example, the graph is fully connected: all nodes are connected to each other. The strength of BNs lies in the fact that one can leave out connections, which induces conditional independencies in the joint probability distribution. For example, if $a$ is independent of $b$ given $c$ then:

$$p(a \mid b, c) = p(a \mid c)$$
$$p(a, b \mid c) = p(a \mid c)p(b \mid c)$$

The absence of an arrow in a BN between two nodes $a, b$ *guarantees* that there is a third variable, or a set of variables, that renders $a$ and $b$ conditionally independent. Hence, a BN is a carrier of *conditional independence relations* among a set of variables. In other words, it defines the set of possible factorizations for a joint probability distribution. We'll use the shorthand notation $a \perp\!\!\!\perp b \mid c$ for conditional independence.

One of the appealing properties of BNs is that it is easy to examine these independence relationships by inspecting the graph [62]. This includes conditional independencies that are not limited to individual nodes. Consider three non-intersecting sets of nodes $A$, $B$ and $C$. The independence statement $A \perp\!\!\!\perp B \mid C$ is true if all paths from any node in $A$ to any node in $B$ are *blocked*: $A$ is then said to be *D-separated* from $B$ by $C$. A path is blocked if:

- Two arrows in the path meet head-to-tail at a node in $C$.
- Two arrows in the path meet tail-to-tail at a node in $C$.
- Two arrows meet head-to-head at a node that is not in $C$, and that does not have any descendants in $C$.

# 1 An Overview of Bayesian Inference and Graphical Models

**Fig. 1.6** A BN that is not fully connected, consisting of five nodes

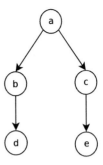

In general, the more independencies a BN specifies, the more tractable inference becomes. Consider the BN in Fig. 1.6. The graph of this BN is clearly not fully connected, and it is easy to see that this can speed up inference related calculations tremendously. According to the rules specified above, $d$ and $e$ are conditionally independent given $a$, as two arrows in the single path between $d$ and $e$ meet tail-to-tail in $a$. Suppose we want to calculate $p(d, e \mid a)$, and that each node can adopt values between 1 and 1,000. A simple brute force calculation, where the nodes $b, c$ are summed away, results in:

$$p(d, e \mid a) = \frac{p(a, d, e)}{p(a)} = \sum_b \sum_c \frac{p(a, b, c, d, e)}{p(a)}$$

This sum contains $10^6$ terms, if both $b$ and $c$ can adopt 1,000 values. However, when we make use of the conditional independencies as encoded in the BN, it is easy to see that this computation can be performed much more efficiently. First, note that:

$$p(a, b, c, d, e) = p(a) p(b \mid a) p(d \mid b) p(c \mid a) p(e \mid c)$$

From this factorization, it becomes clear that the sum can be written as:

$$p(d, e \mid a) = \left( \sum_b p(b \mid a) p(d \mid b) \right) \left( \sum_c p(c \mid a) p(e \mid c) \right)$$

Hence, the number of terms that need to be calculated – in this case 2,000 – is dramatically lower. The strength of BNs lies in the fact that it is possible to construct algorithms that perform these clever ways of inference in an automated way. This will be discussed in Sect. 1.9.6.1.

### 1.9.3.2 Dynamic Bayesian Networks

Dynamic Bayesian networks (DBNs) are Bayesian networks that represent sequential variables [219, 220]. The name "dynamic" is misleading, as it refers to the fact that these sequential variables often have a temporal character, such as for example in speech signals. However, many sequential variables, such as protein

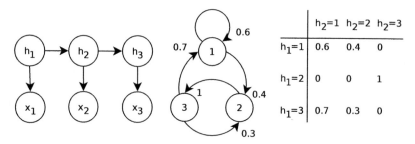

**Fig. 1.7** Bayesian network diagram versus an HMM state diagram. (*left*) An HMM with three slices shown as a Bayesian network diagram. (*middle*) An example of a possible state diagram for an HMM where the hidden nodes can adopt three possible values (states 1, 2 and 3). The *arrows* denote non-zero transition probabilities, which are shown next to the arrows. (*right*) The transition matrix associated with the shown state diagram. This conditional probability table (CPT) specifies $p(h_{n-1} \mid h_n)$ for all $n > 1$. Zeros in the matrix correspond to missing arrows in the state diagram

sequences, do not have a temporal character. Thus, there is nothing "dynamic" about a DBN itself, and the name "sequential Bayesian network" would have been more appropriate.

Formally, DBNs are identical to ordinary BNs. One of the simplest DBNs is the hidden Markov model (HMM) [164,594]. An example of an HMM with length three is shown in Fig. 1.7. Each sequence position, which corresponds to one *slice* in the DBN, is represented by one hidden or latent node $h$ that has one observed node $x$ as child. The use of a Markov chain of hidden nodes is a statistical "trick" that creates conditional dependencies *along the whole length of the observed sequence* – and not just between consecutive slices as is often erroneously claimed [757,794,795]. This can be clearly observed when considering the marginal probability of the observed sequence $\mathbf{x}$, with length $N$, according to an HMM:

$$p(\mathbf{x}) = \sum_{\mathbf{h}} p(\mathbf{x}, \mathbf{h}) = \sum_{\mathbf{h}} p(x_1 \mid h_1) p(h_1) \prod_{n=2}^{N} p(x_n \mid h_n) p(h_n \mid h_{n-1})$$

However, the elements in the sequence *do* become conditionally independent *given* the values of $\mathbf{h}$. The simple architecture of an HMM results in a flexible and powerful model that still remains computationally tractable, and has many applications [594], notably in bioinformatics [164].

The structure of a DBN is such that it can be "unrolled" when faced with sequences of different lengths. Formally, this corresponds to creating additional nodes and edges. The parameters associated with these nodes and edges are identical to those in the previous slices. Hence, a DBN can be fully specified by two slices: an initial slice at the first position and the slice at the second position. Sequences of any length can then be modelled by adding additional slices that are identical to the slice at position one. In the HMM example shown in Fig. 1.7 and for a sequence of length $N$, this corresponds to:

$$p(h_2 = a \mid h_1 = b) = p(h_3 = a \mid h_2 = b) = \ldots = p(h_N = a \mid h_{N-1} = b)$$

# 1 An Overview of Bayesian Inference and Graphical Models

for all possible states $a$ and $b$. Why are two slices needed, instead of one? Consider node $h_1$ at the start of the HMM. This node has no parents, and thus its conditional probability distribution will have a different number of parameters than the – shared – probability distribution of the consecutive hidden nodes, which do have a parent. Hence, we need to specify a starting slice and one consecutive slice to specify the model fully.

The conditional probability distribution $p(h_{n+1} \mid h_n)$ can be interpreted as the probability of evolving from one state at position $n$ to another state at position $n + 1$. If the sequence position $n$ is interpreted as an indicator of time, it is thus natural to represent $p(h_{n+1} \mid h_n)$ as a *state diagram*. Such a diagram is shown in Fig. 1.7. Here, the nodes do not represent random variables, but *states*; that is, values adopted by the hidden nodes; arrows do not represent conditional independencies, but connect states for which $p(h_{n+1} \mid h_n)$ is larger than zero. Figure 1.7 also gives the corresponding conditional probability table $p(h_{n+1} \mid h_n)$ for the given state diagram: a zero in the table corresponds to an arrow that is absent. Given the superficial similarity of a DBN graph and the HMM state diagram, it is important to understand the distinction.

Another interesting conceptual difference is that "structure learning" in HMMs corresponds to inference of the conditional probability distribution $p(h_{n+1} \mid h_n)$ in DBNs. Typically, the state diagram of an HMM, which specifies the possible state transitions but not their probabilities, is decided on before parameter estimation. Inference of the state diagram itself, as opposed to inference of the transition probabilities, is called *structure learning* in the HMM community; see for example [770] for structure learning in secondary structure prediction. However, in DBNs, the possible transitions, which correspond to non-zero entries in the conditional probability distribution $p(h_{n+1} \mid h_n)$, are typically inferred from the data during parameter estimation. In HMMs, parameter estimation and structure learning are often considered as distinct tasks, and the state diagram is often decided using "prior insights". It is of course possible to include prior information on the properties of $p(h_{n+1} \mid h_n)$ for DBNs, up to the point of specifying the possible "state transitions", just as is typical for HMMs.

Numerous extensions to the basic HMM architecture are possible, including higher order Markov models, where hidden nodes at slice $n + 1$ are not only connected to those in slice $n$, but also to one or more previous slices. Such models quickly become intractable [219,220]. The great advantage of considering HMMs as Bayesian networks is a unified view of standard HMMs and their numerous variants. There is no need to develop custom inference algorithms for every variant under this view. Inference and learning in DBNs and HMMs will be briefly discussed in Sects. 1.9.6 and 1.9.7.

### 1.9.3.3 Ancestral Sampling

Often, it is needed to generate samples from a probability distribution, for example in stochastic learning procedures. One of the attractive aspects of BNs is that they are *generative models*; they provide a full joint probability distribution over all variables, and can thus be used to generate "artificial data". In addition, several important sampling operations are trivial in BNs.

Generating samples from the joint probability distribution encoded in the BN can be done using *ancestral sampling* [62]. In order to do this, one first orders all the nodes such that there are no arrows from any node to any lower numbered node. In such an ordering, any node will always have a higher index than its parents. Sampling is initiated starting at the node with the lowest index (which has no parents), and proceeds in a sequential manner to the nodes with higher indices. At any point in this procedure, the values of the parents of the node to be sampled are available. When the node with the highest index is reached, we have obtained a sample from the joint probability distribution. Typically, the nodes with low indices are latent variables, while the nodes with higher indices are observed.

Ancestral sampling can easily be illustrated by sampling in a Gaussian mixture model. In such a model, the hidden node $h$ is assigned index zero, and the observed node $o$ is assigned index one. Ancestral sampling thus starts by sampling a value for the hidden node from $p(h)$: a random number $r$ in $[0, 1]$ is generated, and a value $h$ for the hidden node is sampled according to:

$$h = \arg\min_{h'} \left\{ r \leq \sum_{i=1}^{h'} p(i) \right\} \tag{1.8}$$

This amounts to sampling from a multinomial distribution. For most BNs with nodes that represent discrete variables, this will be the most common node type and associated sampling method. Next, we sample a value for the observed node $o$, conditional upon the sampled values of its parents. In a Gaussian mixture model, this marginal distribution is equal to a Gaussian distribution with mean $\mu_h$ and standard deviation $\sigma_h$ specified by $h$. Hence, we sample a value $y$ from the Gaussian distribution $\mathcal{N}(y \mid \mu_h, \sigma_h)$, and we are done. The pair $(y, h)$ is a sample from the joint probability distribution represented by the Gaussian mixture model.

Ancestral sampling can also be used when some nodes are observed, as long as all observed nodes either have no parents, or only observed parents. The observed nodes are simply clamped to their observed value instead of resampled. However, if some observed nodes have one or more unobserved parents, sampling of the parent nodes needs to take into account the values of the observed children, which cannot be done with ancestral sampling. In such cases, one can resort to Monte Carlo sampling techniques [62, 223] such as Gibbs sampling [212], as discussed in Sect. 1.9.6.2 and Chap. 2.

### 1.9.4 Markov Random Fields

Markov random fields or Markov networks [54, 62] are graphical models in which the graph is undirected: there is no direction associated with the edges. In general,

**Fig. 1.8** A simple Markov random field with two maximal cliques: $(a, b)$ and $(a, c, d)$

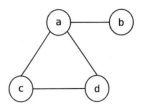

such models are much harder to work with than BNs, and problems of inference easily become intractable. In MRF, the joint probability distribution is a normalized product of potential functions, which assign a positive value to a set of nodes. Each potential function $\phi$ is associated with a maximal clique in the graph; a clique is a subgraph of the MRF in which all nodes are connected to each other, and a maximal clique ceases to be a clique if any node is added. For the MRF in Fig. 1.8, the joint probability distribution is given by:

$$p(a,b,c,d) = \frac{1}{Z} \phi_{acd}(a,c,d) \phi_{ab}(a,b)$$

The normalization factor $Z$ is called the partition function, and is equal to:

$$Z = \sum_a \sum_b \sum_c \sum_d \phi_{acd}(a,c,d) \phi_{ab}(a,b)$$

Many models that arise in statistical physics, such as the Ising model, can be readily interpreted as MRFs.

### 1.9.5 Factor Graphs

Factor graphs provide a general framework for both Bayesian networks and Markov random fields [62, 402]. Factor graphs consists of two nodes types: *factor nodes* and *variable nodes*. The factor nodes represent (positive) functions that act on a subset of the variable nodes. The edges in the graph denote which factors act on which variables. The joint probability distribution encoded in a factor graph is given by:

$$p(\mathbf{x}) = \frac{1}{Z} \prod_k f_k(\mathbf{x}_k)$$

As usual, $Z$ is a partition function which ensures proper normalization:

$$Z = \sum_{\mathbf{x}} \prod_k f_k(\mathbf{x}_k)$$

**Fig. 1.9** A factor graph with three variable nodes shown in *white*, and three factor nodes shown in *grey*

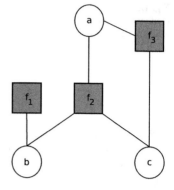

**Fig. 1.10** A simple BN with three nodes (*left*), and its corresponding representation as a factor graph (*right*). The factor nodes are shown in *grey*. Factor $f_1$ corresponds to $p(a)$; factor $f_2$ to $p(b \mid a)$ and factor $f_3$ to $p(c \mid a)$

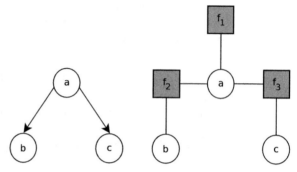

For example, the factor graph in Fig. 1.9 represents the following joint probability distribution:

$$p(a,b,c) = \frac{1}{Z} f_1(b) f_2(a,b,c) f_3(a,c)$$

To represent a BN as a factor graph, each node in the BN is represented by a variable node (just as in the case of the MRF). Then, each conditional distribution in the BN gives rise to one factor node, which connects each child node with its parents (Fig. 1.10). A MRF can be represented as a factor graph by using one variable node in the factor graph for each node in the MRF, and using one factor node in the factor graph for each clique potential in the MRF. Each factor node is then connected to the variables in the clique.

### 1.9.6 Inference for Graphical Models

The goal of inference is often – but not always – to estimate $p(\mathbf{h} \mid \mathbf{x}, \boldsymbol{\theta})$; the distribution of the hidden nodes $\mathbf{h}$ given the observed nodes $\mathbf{x}$ and the model parameters $\boldsymbol{\theta}$. In other cases, one wants to find the hidden node values that maximize $p(\mathbf{h} \mid \mathbf{x}, \boldsymbol{\theta})$. In some cases, the goal of inference is to obtain quantities such as entropy or mutual information, which belong to the realm of information

# 1 An Overview of Bayesian Inference and Graphical Models 43

theory. From a bird's eye perspective, inference in graphical models comes in two flavors: deterministic message passing algorithms and stochastic Monte Carlo methods. The latter are always approximative, while the former can be exact or approximative. Message-passing algorithms occur in Chap. 11 in the context of side chain modelling.

### 1.9.6.1 Message-Passing Algorithms

Message-passing algorithms are best understood in terms of factor graphs [62, 402]. Conceptually, they formalize the smart factoring that we illustrated in Sect. 1.9.3.1: message passing algorithms in factor graphs automate and generalize these ideas. As mentioned before, factor graphs provide a unified view on all graphical models, including Bayesian networks and Markov random fields. Many algorithms that were developed independently in statistical physics, machine learning, signal processing and communication theory can be understood in terms of message passing in factor graphs. These algorithms include the forward-backward and Viterbi algorithms in HMMs, the Kalman filter, and belief propagation in Bayesian networks [62, 402, 782–786].

In message passing algorithms applied to factor graphs, variable nodes $V$ send messages to factor nodes $F$, and vice versa. Each message that is sent is a vector over all the possible states of the variable node. Messages from a factor node $F$ to a variable node $V$ contain information on the probabilities of the states of $V$ according to $F$. Messages from a variable node $V$ to a factor node $F$ contain information on the probabilities of the states of $V$ based on all its neighboring nodes except $F$. The *sum-product algorithm* is a message passing algorithm that infers the local marginals of the hidden nodes; the *max-sum algorithm* finds the values of the hidden nodes that result in the highest probability. The application of the sum-product algorithm to Bayesian networks results in the belief propagation algorithm, originally proposed by Pearl [568]. If the graphical model is a tree or a polytree, these message passing algorithms produces exact results.

Applying the sum-product algorithm to HMMs results in the forward-backward algorithm [164, 594]. As the name of this classic algorithm already indicates, messages are first propagated from start to end (forward pass), and then in the opposite direction (backward pass). The application of the max-sum algorithm to HMMs results in the Viterbi algorithm. The appealing aspect of viewing these algorithms in the light of message passing in factor graphs is that all kind of extensions and modifications of HMMs can be treated in the same conceptual framework.

If the graphical model contains cycles, one has two options. The first option is to convert the factor graph to an equivalent factor graph without cycles, and then to apply standard belief propagation. The modified graph is called a *junction tree*, and the resulting belief propagation algorithm is called *junction tree inference* [126, 353]. In the junction tree, each node is a clique in the original graph. This approach easily gets intractable, as the algorithm is exponential in the number of nodes in a clique, and determining the optimal junction tree itself is *NP*-hard.

The second option is to turn to approximate message passing algorithms, which are called *generalized belief propagation* algorithms [786]. In this algorithm, regions of nodes send messages to neighboring regions of nodes. The choice of the regions is critical for the quality of the approximation and the computational efficiency. In the *Bethe approximation* [60], the regions contain at most two nodes; this approximation becomes exact when the graph is a tree. In the *Kikuchi approximations* [372], the regions contain more than two nodes. Interestingly, these approximations can be understood in terms of free energies and entropies; they were first developed in the field of statistical physics [60, 372]: generalized belief propagation can be viewed as a message passing implementation of the variational approach to inference that we mentioned in Sect. 1.8.3.

### 1.9.6.2 Monte Carlo Methods

Belief propagation algorithms are complex, especially when the BN contains nodes that are not discrete or Gaussian. An attractive alternative is to turn to sampling methods, which are typically easy to implement and general due to the independence relations that are encoded in the graph structure of a graphical model [62, 223]. We will briefly discuss Gibbs sampling in BNs.

*Gibbs sampling* [212] in BNs makes use of *Markov blanket sampling* [62]. The goal of Markov blanket sampling is to sample a value for one node, conditional upon the values of all the other nodes. An important application of this type of sampling is to generate samples when some node values are observed. In that case, ancestral sampling typically does not apply as it assumes that all observed nodes have observed or no parents. In Gibbs sampling, we sample a value for node $x_i$, conditional upon the values of all other nodes, $\{x_1, \ldots, x_N\} \setminus x_i$:

$$x_i \sim p(x_i \mid \{x_1, \ldots, x_N\} \setminus x_i)$$

where the sum runs over all possible values of $x_i$. For a BN with $N$ discrete nodes, this distribution can be written as:

$$
\begin{aligned}
p(x_i \mid \{x_1, \ldots, x_N\} \setminus x_i) &= \frac{p(x_1, \ldots, x_N)}{\sum_{x_i} p(x_1, \ldots, x_N)} \\
&= \frac{\prod_{n=1}^{N} p(x_n \mid \mathrm{pa}(x_n))}{\sum_{x_i} \prod_{n=1}^{N} p(x_n \mid \mathrm{pa}(x_n))}
\end{aligned}
$$

It can easily be seen that all factors that do not contain $x_i$ can be taken out of the sum, and cancel in the numerator and the denominator. The factors that are remaining are $p(x_i \mid \mathrm{pa}(x_i))$ and any factor in which the node $x_i$ itself is the parent of a node. In simple words, Gibbs sampling of a node depends on a node's parents, a node's children and the parents of the node's children. This set of nodes is called the

**Fig. 1.11** A Bayesian network with seven nodes. The nodes in the Markov blanket of node $b$ are shown in *gray*

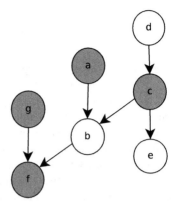

*Markov blanket* of a node. For example, the Markov blanket of node $b$ in Fig. 1.11 is $\{a, c, f, g\}$. Here, $a$ and $c$ are parents; $f$ is a child; and $g$ is a parent of the child. Gibbs sampling for node $b$ is thus done by sampling from the following distribution:

$$b \sim p(b \mid a, c) p(f \mid b, g)$$

For a discrete BN, one simply loops through all values of node $b = \{1, \ldots, B\}$ and calculates the value of the expression above. One then obtains a vector of probabilities $\{p(b = 1), \ldots, p(b = B)\}$ by normalization. Finally, a value for node $b$ is obtained by sampling from the normalized probabilities according to Eq. 1.8.

The disadvantage is that sampling methods can take a long time to converge, and that the diagnosis of convergence itself is difficult. Popular sampling methods include Gibbs sampling, Metropolis-Hastings sampling and importance sampling [223]. These and other much more refined Monte Carlo methods, are explained in detail in Chap. 2.

## 1.9.7 Learning for Graphical Models

So far we have assumed that the parameters of the graphical model are known, and that we are interested in inferring the values of the hidden variables, or in integrating out hidden nuisance variables to obtain the probability of the observed variables. Naturally, in most cases we also need to estimate the values of the parameters themselves. We point out again that from a Bayesian point of view, there is no distinction between inference of hidden node values and inference of the parameters of the conditional probability distributions, the potentials or the factors. However, in practice, these two tasks often give rise to different algorithms and methods [62, 219, 223].

In most cases of parameter learning, only a subset of nodes is observed, which complicates parameter estimation. Specifically, in parameter estimation we are interested in:

$$p(\boldsymbol{\theta} \mid \mathbf{d}) = \sum_{\mathbf{h}} p(\boldsymbol{\theta}, \mathbf{h} \mid \mathbf{d})$$

The sum becomes an integral in the continuous case. The fact that we need to integrate the hidden variables away complicates things. It is impossible to give an overview of all possible learning methods for graphical models; for many models, parameter estimation is a topic of intense research. In this section, we will discuss two of the most popular learning algorithms for Bayesian networks: ML estimation using Expectation Maximization (EM) [144, 219, 353] and approximative Bayesian estimation using Gibbs sampling [62, 212, 223].

Expectation maximization corresponds to simple maximum likelihood estimation of the parameters, in the presence of hidden nodes.

$$\boldsymbol{\theta}_{ML} = \arg\max_{\boldsymbol{\theta}} \sum_{\mathbf{h}} p(\mathbf{h}, \mathbf{d} \mid \boldsymbol{\theta})$$

Conceptually, EM is easy to understand. In a first step, the expectation step (E-step), the values of the hidden nodes are inferred. In the maximization step (M-step), the values of the parameters are estimated by making use of the inferred hidden node values, as obtained in the E-step. The E and M steps are repeated iteratively until convergence. Theoretically, the EM algorithm is guaranteed to converge to a local maximum or saddlepoint of the likelihood. More formally, in the E-step one constructs the probability distribution over the hidden nodes, conditional upon the observed nodes and the current parameters, which is $p(\mathbf{h} \mid \mathbf{d}, \boldsymbol{\theta}_{old})$. In the M-step, a new set of parameters $\boldsymbol{\theta}_{new}$ is chosen that maximizes the expected value of the log-likelihood under this probability distribution:

$$\boldsymbol{\theta}_{new} = \arg\max_{\boldsymbol{\theta}} \sum_{\mathbf{h}} p(\mathbf{h} \mid \mathbf{d}, \boldsymbol{\theta}_{old}) \log p(\mathbf{h}, \mathbf{d} \mid \boldsymbol{\theta})$$

The E-step can be performed in several ways. In tractable models such as the HMM, it is possible to calculate the probability distribution over the hidden nodes, which is done using the well known *forward-backward algorithm* [164, 594]. The resulting distribution is then used to update the parameters of the HMM in the M-step. For HMMs, the EM algorithm is called the *Baum-Welch algorithm* [62, 164, 594].

For many Bayesian networks, the probability distribution of the hidden variables can only be approximated, as deterministic methods quickly become intractable. In general Bayesian networks, one can approximate this distribution by sampling, as described in Sect. 1.9.6.2. The E-step thus consists of Gibbs sampling, which corresponds to blanket sampling in BNs. The sampled values are then used in the M-step. Such an EM algorithm is called a Monte Carlo EM algorithm [223]. Typically, one approximates the distribution over the hidden nodes using a large amounts of samples.

1 An Overview of Bayesian Inference and Graphical Models

However, it is also possible to draw a single value for each hidden, thus effectively creating a 'completed' dataset, where all the values of the hidden nodes are explicitly filled in. Such a version of the EM algorithm is called stochastic EM (S-EM) [223]. The S-EM method has many advantages. Classic EM and Monte Carlo EM algorithms are notorious for getting stuck in local maxima or saddle points; the S-EM method is much less prone to this problem [544]. In addition, S-EM is extremely computationally efficient, as one only needs to sample a single value for each hidden node.

Once the completed dataset is obtained, parameter estimation becomes trivial: the parameters of the conditional probability distributions associated with the nodes and their parents are simply estimated by ML or MAP estimation. In the case of a CPT, for example, all occurrences of a node's value together with its parent's values in the completed data are collected in a suitable table of counts, which is subsequently normalized in a table of probabilities. We will encounter the S-EM algorithm again in Chap. 10, where it is used to estimate the parameters of a probabilistic model of the local structure of proteins.

The EM algorithm provides a ML estimate of the parameters, and makes the – rather artificial – distinction between parameters and hidden variables. An analytic Bayesian treatment is intractable for most models, but good approximations can be obtained using stochastic methods based on sampling, typically using Markov chain Monte Carlo methods [223]. Note that sampling is also used in the S-EM algorithm, but in that case one still ends up with a ML point estimate of the parameters. Here we want to approximate the probability distribution $p(\boldsymbol{\theta} \mid \mathbf{d})$ by a set of samples. A classic way to do this is by Gibbs sampling [62, 212, 223]. Here one creates a BN where the prior distributions are explicitly represented as nodes with given parameters. Gibbs sampling, and other Markov chain Monte Carlo methods that can be used for the same purpose, are discussed at length in Chap. 2.

## 1.10 Conclusions

The Bayesian probability calculus provides a gold standard for inference; it is based on a firm axiomatic foundation, and can be applied to a wide variety of problems, without the need to develop *ad hoc* methods [342]. Unfortunately, a fully Bayesian analysis is often intractable or impractical, for example for speed reasons. In that case, approximations to a full Bayesian analysis – such as ML, MAP, empirical Bayes, pseudolikelihood or even moment estimators – often offer useful and tractable solutions. However, in such cases one should keep in mind that one is using an approximation to a rigorous treatment of the problem; these approximations might fail spectacularly for some specific cases [342,509]. Bayesian probability theory, and its application to learning and inference in graphical models, has close ties to statistical physics and information theory; in subsequent chapters we will encounter numerous examples of this fascinating interplay.

## 1.11 Further Reading

This introductory chapter provides a bird's eye view on Bayesian inference and graphical models, and points out some interesting references for further reading. We end with mentioning some books that we found particularly useful. A short, accessible introduction to Bayesian statistics can be found in Lee's *Bayesian statistics* [428]. Bernardo and Smith's *Bayesian theory* [48], and Robert's *The Bayesian choice* [606] provide more in-depth discussion; the latter from the point of view of decision theory. Jaynes' classic book *Probability theory: The logic of science* [342] provides a thorough defense of the Bayesian view, and presents many examples of the paradoxes that arise when alternative views are adopted. Lemm's *Bayesian field theory* [433] gives a physicist's view on Bayesian statistics. Bishop's *Machine learning and pattern recognition* [62] presents a timely, Bayesian view on machine learning methods, including graphical models and belief propagation. Introductory books on Bayesian networks include Pearl's seminal *Probabilistic reasoning in intelligent systems* [568], Neapolitan's *Learning Bayesian networks* [538] and Cowell, Dawid, Lauritzen and Spiegelhalter's *Probabilistic networks and expert systems* [126]. Jordan's *Learning in graphical models* [353] is one of the few books that addresses approximate algorithms. Zuckerman's *Statistical physics of biomolecules* [801] and Dill and Bromberg's *Molecular driving forces* [150] are excellent introductions to statistical physics, with a firm basis in probability theory and a focus on biomolecules.

Finally, the fascinating history of the rise, fall and resurrection of the Bayesian view of probability is presented in a very accessible manner in McGrayne's *The theory that would not die* [494], while Stigler's *The history of statistics: The measurement of uncertainty before 1900* [692] provides a scholarly view on its emergence.

# Chapter 2
# Monte Carlo Methods for Inference in High-Dimensional Systems

**Jesper Ferkinghoff-Borg**

## 2.1 Introduction

Modern Monte Carlo methods have their roots in the 1940s when Fermi, Ulam, von Neumann, Metropolis and others began to use random numbers to examine various problems in physics from a stochastic perspective [118, 413]. Since then, these methods have established themselves as powerful and indispensable tools in most branches of science. In general, the MC-method represents a particular type of numerical scheme based on random numbers to calculate properties of probabilistic models, which cannot be addressed by analytical means. Its wide-spread use derives from its versatility and ease of implementation and its scope of application has extended considerably due to the dramatic increase within the last 2–3 decades in accessible computer power. In this chapter we shall mainly focus on the *Markov Chain Monte Carlo* method (MCMC) as a tool for inference in high-dimensional probability models, with special attention to the simulation of bio-macromolecules.

In Sect. 2.2.1 we briefly review the conceptual relations between probabilities, partition functions, Bayes factors, density of states and statistical ensembles. We outline central inference problems in Sect. 2.2.2, which will be the focus of the remaining chapter. This concerns calculation of *expectation values*, *marginal distributions* and *ratios of partition functions*. In particular, we discuss why these calculations are not tractable by analytical means in high-dimensional systems.

Basic concepts from sampling theory are delineated in Sect. 2.3. Section 2.4 focuses on the MCMC-method with emphasis on the central *detailed balance* equation, the *Metropolis-Hastings* algorithm [293, 501] (Sect. 2.4.1), *Gibbs sampling* [212] (Sect. 2.4.2) and specific concerns regarding the use of the MCMC-method for continuous degrees of freedom (Sect. 2.4.3). In Sect. 2.5 we discuss how the MCMC-simulation can be used to address the inference problems outlined in

---

J. Ferkinghoff-Borg (✉)
Department of Electrical Engineering, Technical University of Denmark, Lyngby, Denmark
e-mail: jfb@elektro.dtu.dk

T. Hamelryck et al. (eds.), *Bayesian Methods in Structural Bioinformatics*,
Statistics for Biology and Health, DOI 10.1007/978-3-642-27225-7_2,
© Springer-Verlag Berlin Heidelberg 2012

Sect. 2.2.2. We conclude the section by reviewing inherent deficiencies of the standard MCMC-approach, in particular with respect to the limited information provided by the sampling as well as its potential failure to ensure ergodicity. These deficiencies makes the standard MCMC-approach inapplicable for inference in more complex model systems, one example being bio-macromolecules.

The last four sections are devoted to various recent methodologies aimed at alleviating the aforementioned drawbacks. In Sect. 2.6, we discuss the ergodicity and convergence of the Markov chain from a general perspective and present a number of system-specific improvements to enhance sampling efficiency. Section 2.7 presents a general approach known as *extended ensemble* MCMC [328]. This includes the parallel tempering method [218, 323, 373, 483, 708] (Sect. 2.7.1), the simulated tempering method [455, 484] (Sect. 2.7.2), the multicanonical ensemble [41] (Sect. 2.7.3) and the $1/k$-ensemble [303] (Sect. 2.7.4). While these ensembles differ with respect to their statistical properties and domain of application (Sect. 2.7.5) they all provide means to circumvent the problems pertaining to the standard MCMC-algorithm. However, unlike the standard MCMC-method extended ensemble algorithms rely on a number of parameters which are not a priori known. As further discussed in Sect. 2.7.6, inference based on these ensembles is vulnerable to erroneous choices of these parameters. A central aspect is therefore to *learn* the appropriate parameters of the ensemble in concern.

In Sect. 2.8 we review a number of recent methods to accomplish this task. In particular, we discuss two widely popular non-Markovian learning algorithms, the *Wang-Landau* method [754] (Sect. 2.8.4) and *metadynamics* (Sect. 2.8.5) [411]. Here, detailed-balance is partly violated to speed up the parameter tuning. In Sect. 2.8.6 we shall detail a method (Muninn) developed in our own group, which aims at providing fast parameter convergence while preserving detailed balance [180, 196].

In Sect. 2.9 we return to the three inference problems outlined in Sects. 2.2.2 and 2.5. We discuss how sampling in extended ensembles provides a solution to these problems, using a technique known as *reweighting* [702]. While MCMC-methods play a central role in Bayesian inference [62], it is a curiosity that these methods themselves have not yet been subject to a full Bayesian treatment. In the concluding section we discuss possible entrance points to such a generalization.

## 2.2 Probabilities and Partition Functions

Let $\Omega$ be the domain of a single- or multi-valued ($D$ dimensional) stochastic variable, $x = (x_1, \cdots, x_D)^T$, distributed according to some given probability density function, $p(x)$. In a general statistical setting, we can typically only evaluate $p(x)$ up to a normalization constant, $Z$, also known as the *partition function*. Accordingly, for a continuous state space the distribution takes the form

$$p(x) = \frac{\omega(x)}{Z}, \quad Z = \int_\Omega \omega(x) dx, \tag{2.1}$$

where the *weights*, $\omega(x)$, are assumed to be known explicitly. In graphical models in general and in Random Markov fields in particular $\omega$ would represent the product of the potential functions defining the model [62]. Part IV of this book include several examples of such type of models for protein and RNA structures.

In statistical physics, $\omega$ is a particular function of the *total energy*, $E(x)$, of the system and possibly other extensive variables as well, see Table 2.1. With respect to the table, the main focus will be on the *microcanonical* and *canonical ensemble*. For sampling and inference in other ensembles we refer to the excellent overviews given in the textbooks [5, 198, 413]. The interest in the first ensemble derives from the fact that all thermodynamic potentials can be calculated once the number of states (or density of states $g$ in the continuous case), $\Gamma$, is known. As we shall see, many advanced Monte Carlo techniques also make direct use of $\Gamma$ to enhance the efficiency of the sampling. The interest in the second ensemble is due to the fact that any probability distribution can be brought to the canonical form by a suitable (re-) definition of the energy function. For a continuous systems this ensemble is defined by the probability density distribution

$$p_\beta(x) = \frac{\exp(-\beta E(x))}{Z_\beta}, \quad Z_\beta = \int_\Omega \exp(-\beta E(x))dx, \quad (2.2)$$

where $\beta = \frac{1}{kT}$ is the inverse temperature times the Boltzmann constant. For notational convenience we have suppressed the dependence of $V$ and $N$ in Eq. 2.2, compared to the expression given in Table 2.1. The weights in the canonical ensemble, $\omega_\beta(x) = \exp(-\beta E(x))$, are referred to as the *Boltzmann weights*. Note that the other statistical ensembles in Table 2.1 can be brought on the canonical form by a change of energy function. In the isobaric-isothermal ensemble, for example, the *enthalpy function*, $H = E + pV$, leads to the Boltzmann weights, $\omega_\beta(x) = \exp(-\beta H(x))$.

Comparing Eq. 2.2 to the general form, Eq. 2.1, one observes that by simply associating an "energy", $E$, to a state $x$ according to the probability weights,

$$E(x) = -\ln[\omega(x)], \quad (2.3)$$

and subsequently setting $\beta = 1$, the two distributions become identical. In keeping with this construction, we will in the following refer to both Eqs. 2.2 and 2.1 as the canonical distribution, in distinction to various modified or extended ensembles to be discussed later. Similarly, we shall refer to "energies" and "temperatures" without limiting ourselves to distributions of thermal nature.

## 2.2.1 Reference Distributions and Measures

As indicated in Table 2.1, the partition functions, $Z$, play a central role in thermal physics due to their relation to thermodynamic potentials. In Bayesian modeling, $Z$ corresponds to the *evidence* of the model, a quantity which forms an essential part of

**Table 2.1** The most commonly encountered distributions in statistical mechanics, assuming a discrete state space. Here, $x$ represents a specific microstate of the physical system, and $\Omega$ is defined from the thermodynamical variables specifying the boundary conditions. The minimum set of relevant conjugated pairs of thermodynamic variables typically considered is $\{(E, \beta), (V, p), (N, \mu)\}$, where $E$ is the energy, $\beta = \frac{1}{kT}$ is the inverse temperature times the Boltzmann constant $k$, $V$ is the volume, $p$ is the pressure, $N$ is the particle number and $\mu$ is the chemical potential. Note the special role played by the *number of states*, $\Gamma(E, V, N)$, the knowledge of which allows the calculation of all thermodynamic potentials. As further discussed in the text, the continuous version of these ensembles is obtained by replacing $\Gamma$ with the *density of states* $g$ and summation with integrals. Consequently, $p(x)$ becomes a probability density in $\Omega$

|  Microcanonical ensemble | |
| --- | --- |
| Definition | Fixed $E, V, N$ |
| Domain | $\Omega = \{x \mid E(x) = E, V(x) = V, N(x) = N\}$ |
| Partition function | $\Gamma(E, V, N) = \sum_{x \in \Omega} 1$ |
| Potential | *Entropy:* $S(E, V, N) = k \ln(\Gamma)$ |
| Distribution | $p(x) = \frac{1}{\Gamma}$ |

|  Canonical ensemble | |
| --- | --- |
| Definition | Fixed $\beta, V, N$ |
| Domain | $\Omega = \{x \mid V(x) = V, N(x) = N\}$ |
| Partition function | $Z(\beta, V, N) = \sum_{x \in \Omega} \exp(-\beta E(x))$ $= \sum_{E} \Gamma(E, V, N) \exp(-\beta E)$ |
| Potential | *Helmholtz free energy:* $F(\beta, V, N) = -\beta^{-1} \ln(Z)$ |
| Distribution | $p(x) = \frac{\exp(-\beta E(x))}{Z}$ |

|  Grand canonical ensemble | |
| --- | --- |
| Definition | Fixed $\beta, V, \mu$ |
| Domain | $\Omega = \{x \mid V(x) = V\}$ |
| Partition function | $\Xi(\beta, V, \mu) = \sum_{x \in \Omega} \exp(-\beta E(x) + \beta \mu N(x))$ $= \sum_{E, N} \Gamma(E, V, N) \exp(-\beta E + \beta \mu N)$ |
| Potential | $p(\mu, \beta)V = -\beta^{-1} \ln(\Xi)$ |
| Distribution | $p(x) = \frac{\exp(-\beta E(x) + \beta \mu N(x))}{\Xi}$ |

|  Isobaric-isothermal ensemble | |
| --- | --- |
| Definition | Fixed $\beta, p, N$ |
| Domain | $\Omega = \{x \mid N(x) = N\}$ |
| Partition function | $Y(\beta, p, N) = \sum_{x \in \Omega} \exp(-\beta E(x) - \beta p V(x))$ $= \sum_{E, V} \Gamma(E, V, N) \exp(-\beta E - \beta p V)$ |
| Potential | *Gibbs free energy:* $G(\beta, p, N) = -\beta^{-1} \ln(Y)$ |
| Distribution | $p(x) = \frac{\exp(-\beta E(x) - \beta p V())}{Y}$ |

Bayesian model comparison. However, these important properties of $Z$ seem to be of little practical use in keeping with the fact that partition functions are almost always intractable in high dimensional systems. Fortunately, for most statistical

## 2 Monte Carlo Methods for Inference in High-Dimensional Systems

considerations it is not $Z$ per se that is of interest but rather the ratio, $Z/Z_\pi$, to some other partition function $Z_\pi$, and these ratios *are* tractable – as we shall see. Here, $\pi$ is a particular probability distribution defined on the same domain $\Omega$

$$\pi(x) = \frac{\omega_\pi(x)}{Z_\pi}, \quad Z_\pi = \int_\Omega \omega_\pi(x)dx. \tag{2.4}$$

Bayesian model comparison, for instance, involves two distributions, $p$ and $\pi$, and their relative evidence is precisely given by the Bayes factor $Z/Z_\pi$. The evidence function of a single model can also be made tractable by expressing it as a partition function ratio, $Z/Z_\pi$, provided that the normalization of the likelihood function is known. This is accomplished by identifying $\pi(x)$ with the *prior*-distribution over the model parameters $x$ and $p(x)$ with the *posterior* distribution, $p(x) = p(x|\mathcal{D})$, where $\mathcal{D}$ are the acquired data. The posterior probability weights $\omega$ become

$$\omega(x|\mathcal{D}) = p(\mathcal{D}|x)\omega_\pi(x),$$

where $p(\mathcal{D}|x)$ is the likelihood. The evidence, $p_\pi(\mathcal{D})$, is now obtained as

$$p_\pi(\mathcal{D}) \doteq \int_\Omega p(\mathcal{D}|x)\pi(x)dx \tag{2.5}$$

$$= Z_\pi^{-1} \int_\Omega \omega(x|\mathcal{D})dx = \frac{Z}{Z_\pi}.$$

In the following we shall in general refer to $\pi$ as the *reference distribution* or the *reference state* in distinction to the *target distribution p*.

A natural common reference state for thermal partition functions is the uniform distribution, $\omega_\pi(x) = 1$ corresponding to the Boltzmann weights with $\beta = 0$. The partition function of this state equals the total state space volume, $\Gamma_{tot}$, a quantity which often can be evaluated in a given statistical application:

$$Z_{\pi=1} = \int_\Omega dx = \Gamma_{tot}. \tag{2.6}$$

In fact, it is a fundamental prerequisite in statistical physics to have a unique way of counting the number of states in $\Omega$. While this quantity is naturally defined for discrete systems, its translation to continuous ensembles relies on a proper choice of the integration measure. Setting $\omega_\pi = 1$ is equivalent to the notion that the normal differential volume-element $dx = \prod_i dx_i$ is the appropriate measure for the number of states. A more rigorous translation would require $\omega_\pi dx$ to be dimensionless. For example, in classical statistical mechanics, the integration measure for a set of $D$ distinguishable atoms with positions $r_i$ and linear momenta $p_i$ is given by the dimensionless quantity $\omega_\pi \prod_{i=1}^{D} dr_i dp_i$, where $\omega_\pi = h^{-3D}$ and $h$ is Planck's constant, see also Chap. 3. For most thermodynamic considerations we can safely ignore this proportionality constant. In Sect. 2.4.3 we shall briefly discuss cases where the manifold $\Omega$ is non-Euclidean, implying that the proper integration is non-uniform, $\omega_\pi \neq$ const.

Provided the correctness of the uniform measure, $\Gamma_{tot}$ can be identified with the total number of states in $\Omega$. Similarly, the *density of states* $g(E)$ is given as

$$g(E) = \int_\Omega \delta(E - E(\mathbf{x})) d\mathbf{x}, \tag{2.7}$$

where $\delta(\cdot)$ is the Dirac delta-function and $E(\mathbf{x})$ is the physical energy function. The analogue of the discrete micro-canonical partition function, $\Gamma(E)$ (see Table 2.1), then becomes the *statistical weight* [440], $\Gamma(\mathcal{E})$, defined as

$$\Gamma(\mathcal{E}) = \int_\mathcal{E} g(E') dE' \approx g(E) \Delta E, \quad \mathcal{E} = \left[ E - \frac{\Delta E}{2}, E + \frac{\Delta E}{2} \right), \tag{2.8}$$

where the last expression holds true when $\Delta E$ is chosen sufficiently small.

In order to treat inference in thermal and non-thermal statistical models at a unified level even when the latter models involve non-uniform reference distributions, it will prove convenient to generalize the definition of the non-thermal "energy" associated with a state $\mathbf{x}$, Eq. 2.3, according to

$$E(\mathbf{x}) = -(\ln[\omega(\mathbf{x})] - \ln[\omega_\pi(\mathbf{x})]). \tag{2.9}$$

Note that when $\pi$ is identified with the prior distribution in a Bayes model, Eq. 2.9 is the "energy" associated with the likelihood-function. Similarly, if a uniform reference distribution is used, Eq. 2.9 reduces to the original definition, Eq. 2.3.

### 2.2.2 Inference and the Curse of Dimensionality

In any statistical problem, a set of functions $\boldsymbol{f}(\mathbf{x}) = (f_1(\mathbf{x}), \cdots, f_d(\mathbf{x}))^T : \Omega \rightarrow \mathbb{R}^d$ will be given, for which expectation values, joint or marginal distributions are of particular interest. Often the $f_i$'s will be function of all of most of components of $\mathbf{x}$, in which case they are also referred to as *collective variables*. In statistical physics, relevant collective variables would most certainly entail the total energy along with specific order parameters, structural descriptors or reaction coordinates. Examples of collective variables used protein structure characterization are compactness [182, 264], hydrogen-bond network [249, 488], number of native contacts [550] and contact order [581]. Other examples of collective variables or structural features of interest can be found in Chaps. 3 and 5, referred to as $\boldsymbol{\lambda}$. In the following we shall outline three central inference problems, which will be the focus of this chapter. The first and foremost statistical problem is to calculate the *expectation value*, $\mathbb{E}_p[\boldsymbol{f}(\mathbf{x})]$, of $\boldsymbol{f}(\mathbf{x})$ – here taken to be single-valued – with respect to $p$:

2 Monte Carlo Methods for Inference in High-Dimensional Systems 55

$$\mathbb{E}_p\left[f(\boldsymbol{x})\right] = \int_\Omega f(\boldsymbol{x})p(\boldsymbol{x})\mathrm{d}\boldsymbol{x}. \tag{2.10}$$

The second, a more elaborate inference problem is to determine the full *marginal distribution*, $p_f(y)$, defined as

$$p_f(y) = \int_\Omega \delta\left(f(\boldsymbol{x}) - y\right) p(\boldsymbol{x})\mathrm{d}\boldsymbol{x}. \tag{2.11}$$

Since $\mathbb{E}_p\left[f(\boldsymbol{x})\right] = \int_\mathbb{R} p_f(y)y\mathrm{d}y$, it is straight-forward to calculate the expectation value, once $p_f(y)$ is known. For a multi-valued function $\boldsymbol{f} = (f_1, \cdots, f_d)^T$ the corresponding formula becomes

$$p_f(\boldsymbol{y}) = \int_\Omega \prod_{i=1}^{d} \delta\left(f_i(\boldsymbol{x}) - y_i\right) p(\boldsymbol{x})\mathrm{d}\boldsymbol{x}. \tag{2.12}$$

The final inference problem we wish to address is how to estimate the *ratio of partition functions*, $Z/Z_\pi$, as presented in Sect. 2.2.1.

Common to all three problems outlined above is that the calculation involves an integral over $\Omega$. If the dimension, $D$, of $\Omega$ is low, analytical approaches or direct numerical integration schemes can be used. However, in most cases of interest $D$ is large, which leaves these approaches unfeasible. While approximate inference is partly tractable with deterministic approaches such as variational Bayes and belief propagation [62] as further discussed in Chap. 11, the Monte Carlo based sampling approach offer a particular versatile and powerful computational alternative to solve inference problems for high-dimensional probability distributions.

## 2.3 Importance Sampling

As an introduction to Monte Carlo methods in general and the Markov Chain Monte Carlo methods (MCMC) in particular, let us first focus on Eq. 2.10 for the average of a given function, $f$. The simplest possible stochastic procedure of evaluating the integral over $\Omega$ involves an unbiased choice of points, $\{\boldsymbol{x}_i\}_{i=1}^{N}$ in the state space and use

$$\mathbb{E}_p\left[f(\boldsymbol{x})\right] \approx \frac{1}{N}\sum_{i=1}^{N} p(\boldsymbol{x}_i)f(\boldsymbol{x}_i). \tag{2.13}$$

This procedure is known as *random sampling* or *simple sampling*. However, this procedure is highly inefficient since $p(\boldsymbol{x})$ typically vary many orders of magnitude in $\Omega$, so there is no guarantee that the region of importance for the average will be sampled at all. Furthermore, it is often the case that $p(\boldsymbol{x})$ can only be evaluated up to a normalization constant, cf. Eq. 2.1, leaving the expression on the r.h.s.

indeterminate. The basis of *importance sampling* is to impose a certain bias in the sampling method, so that states are approximately chosen according to $p(x)$. There are in principle two ways of realizing such sampling. The first method relies on the the explicit use of a *proposal distribution* $q(x)$ which approximates $p(x)$ and which is easy to draw samples from. The second method constructs $q(x) \approx p(x)$ implicitly, using an iterative procedure.

### 2.3.1 Sampling from a Target-Approximated Distribution

The basic considerations of any importance sampling method can be elucidated by focusing on the estimator of expectation values, when samples are drawn according to some explicit proposal distribution $q(x)$. For the sake of generality, we assume that $q(x)$ also can be evaluated up to a constant only, so $q(x) = \frac{\omega_q(x)}{Z_q}$. The expression for the expectation value of $f(x)$ then becomes

$$\mathbb{E}_p\left[f(x)\right] = \int_\Omega f(x)p(x)\mathrm{d}x \tag{2.14}$$

$$= \frac{Z_q}{Z}\int_\Omega f(x)\frac{\omega(x)}{\omega_q(x)}q(x)\mathrm{d}x \tag{2.15}$$

$$\approx \frac{Z_q}{Z}\frac{1}{N}\sum_{i=1}^N r_i f(x_i), \tag{2.16}$$

where $r_i = \frac{\omega(x_i)}{\omega_q(x_i)}$ are known as the *importance weights*. The ratio $\frac{Z}{Z_q}$ can be evaluated as

$$\frac{Z}{Z_q} = \frac{1}{Z_q}\int_\Omega \omega(x)\mathrm{d}x = \int_\Omega \frac{\omega(x)}{\omega_q(x)}q(x)\mathrm{d}x \approx \frac{1}{N}\sum_{i=1}^N r_i.$$

Consequently,

$$\mathbb{E}_p\left[f(x)\right] \approx \sum_{i=1}^N \tilde{r}_i q(x_i), \quad \tilde{r}_i = \frac{r_i}{\sum_j r_j} = \frac{\omega(x_i)/\omega_q(x_i)}{\sum_j \omega(x_j)/\omega_q(x_j)}.$$

Note that in the case when $q(x) = p(x)$, Eq. 2.16 reduces to a simple arithmetic average

$$\mathbb{E}_p\left[f(x)\right] \approx \frac{1}{N}\sum_{i=1}^N f(x_i). \tag{2.17}$$

The efficiency of the importance sampling relies crucially on how well $q(x)$ approximates $p(x)$. Often $q(x)$ will be much broader than $p(x)$ and hence the set of importance weight $\{r_i\}$ will be dominated by a few weights only. Consequently, the effective sample size can be much smaller than the apparent sample size $N$. The problem is even more severe if $q(x)$ is small in regions where $p(x)$ is large. In that case the apparent variances of $r_i$ and $r_i q(x_i)$ may be small even though the expectation is severely wrong. As for random sampling it has the potential to produce results that are arbitrarily in error with no diagnostic indication [62].

## 2.3.2 Population Monte Carlo

The merit the importance sampling method in its iterative form is the ease by which in principle any distribution, $p(x)$ can be sampled. These algorithms can broadly be ordered in two categories: the *Population Monte Carlo* algorithms and the *Markov Chain Monte Carlo* (MCMC) algorithms. This distinction is not strict, however, as the individual iteration steps of population MC-methods may involve a Markov chain type of sampling, and conversely MCMC-methods may involve the use of a "population" of chains.

We will not be concerned with the population Monte Carlo algorithms any further, but simply give a list of relevant references. The population Monte Carlo method is called by a lot of different terms depending of the specific context: "quantum Monte Carlo" [634], "projection Monte Carlo" [104], "transfer-matrix Monte Carlo" [545, 648] or "sequential Monte Carlo" ([155] and references herein). The method has also been developed for polymer models by different groups ([210, 250, 251, 553]). Most notably is the PERM-algorithm developed by Grassberger et al. [250, 251], which has been shown to be very efficient to compute finite temperature properties of lattice polymer models [194]. Finally, the population Monte Carlo method has interesting analogies to Genetic Algorithms [304].

One recent variant of the Population Monte Carlo method which deserves special attention is the *nested sampling* by Skilling [666, 669]. This sampling, developed in the field of applied probability and inference, provides an elegant and straightforward approximation for the partition function and calculation of expectation values becomes a simple post-processing step. Therefore, the inference problems outlined in Sect. 2.2.2 can all be addressed by the nested sampling approach. In this respect, it offers the same merits as the extended ensemble techniques to be discussed in Sects. 2.7 and 2.8. Furthermore, nested sampling involves only few free parameters the settings of which seems considerably less involved than the parameter setting in extended ensembles [562]. On the other hand, nested sampling relies on the construction of a sequence of states with strictly decreasing energies. While the method compares favorably to the popular parallel-tempering extended ensemble for Lennard-Jones cluster simulations [562] and has also found use in the field of astrophysics [181], it it yet not clear if the restriction of this sequence construction can be effectively compensated by the choice of the population size for systems with more complicated energy landscapes. We shall briefly return to the method in Sect. 2.7.4 and this discussion in Sect. 2.10.

## 2.4 Markov Chain Monte Carlo

The principle of the Markov chain Monte Carlo (MCMC) algorithm is to represent the probability distribution, $p(x)$, by a chain of states $\{x_t\}_t$. Formally, a (time-independent) Markov chain is a sequence of stochastic variables/vectors $\{X_t\}_t$, for which the probability distribution of the $t$'th variable only depends on the preceding one [77]. While the MCMC-method can be applied equally well to continuous state spaces (see Sect. 2.4.3), we shall in the following assume $\Omega$ to be discrete for notational convenience. Consequently, the stochastic process in $\Omega$ is fully specified by a fixed matrix, $W$, of transition probabilities. Here, each matrix element, $W(x'|x)$, represents the conditional probability that the next state becomes $x'$ given that the current state is $x$. This matrix must satisfy

$$\sum_{x'} W(x'|x) = 1, \quad W(x'|x) \geq 0. \tag{2.18}$$

Let $p_0(x)$ be the probability distribution of the initial state. Starting the Markov chain from a unique configuration, $x_0$, the probability distribution will be 1 for this particular state and 0 otherwise. The distribution after $t$ steps will then be given by,

$$p_t(x) = \sum_{x'} W^t(x'|x) p_0(x'), \tag{2.19}$$

where $W^t$ represents the $t$'th power of $W$ viewed as a matrix in the space of configurations. The so-called Perron-Frobenius theorem of linear analysis guarantees that an unique invariant limit distribution exists,

$$p_\infty(x) = \lim_{t \to \infty} p_t(x) = \lim_{t \to \infty} \sum_{x'} W^t(x'|x) p_0(x'), \tag{2.20}$$

provided that two conditions are satisfied:

1. $W$ is irreducible, i.e. $W^t(x'|x) > 0$ for all states $x$ and $x'$ and some $t > 0$. In words, it is possible with nonzero probability to move from $x$ to $x'$ in a finite number of steps. This feature of the transition matrix is usually referred to as *ergodicity*
2. $W$ is *aperiodic*, i.e. $W^t(x|x') > 0$ for all $t > t_{\min}$ and all $x$. If this condition fails, one can have probability distributions that oscillate between two or more forms, without a unique $p_\infty(x)$.

2 Monte Carlo Methods for Inference in High-Dimensional Systems 59

The Perron-Frobenius theorem states that when the normalization (2.18) and these two conditions are fulfilled, the maximum eigenvalue of $W$ will be 1 with $p_\infty(x)$ as the corresponding unique eigenvector;

$$\sum_{x'} W(x|x')p_\infty(x') = p_\infty(x).$$ (2.21)

The power of the Markov Chain Monte Carlo method is that it offers a simple yet general recipe to generate samples from any distribution $q(x)$, as defined through the limiting behavior of the Markov chain, $q(x) = p_\infty(x)$. As before, we will focus on the case where the aim is to draw samples from a predefined probabilistic model, $p(x)$, i.e. we set $q(x) = p(x)$. From Eqs. 2.21 and 2.18 it follows that a sufficient requirement to insure that the sampled distribution will converge to the equilibrium distribution, $p_\infty(x) = p(x)$, is that the transition matrix elements satisfy the *detailed balance* equation,

$$p(x)W(x'|x) = p(x')W(x|x').$$ (2.22)

## 2.4.1 Metropolis-Hastings Algorithm

It is computationally instructive to factorize the transition matrix elements into two parts;

$$W(x'|x) = q(x'|x)a(x'|x).$$ (2.23)

Here, $q(x'|x)$ is a selection or proposal function – a conditional probability of attempting to go to state $x'$ given that the current state is $x$ [755]. The fulfillment of detailed balance is ensured by a proper choice of the second term, $a(x'|x)$ of Eq. 2.23 which is the acceptance probability for $x'$ once this state has been selected. This requires that

$$\frac{a(x'|x)}{a(x|x')} = \frac{p(x')q(x|x')}{p(x)q(x'|x)} = \frac{\omega(x')q(x|x')}{\omega(x)q(x'|x)}.$$ (2.24)

Note that the normalization constant $Z$ of the probability distribution, $p(x) = \omega(x)/Z$ cancels out in the expression of the acceptance ratio. This is particular convenient as a large variety of models involve intractable normalization constants, including factor graphs, random Markov fields or thermal distributions in statistical physics (Table 2.1).

There is a considerable freedom in the choice of acceptance rates, $a$. The standard Metropolis-Hastings algorithm [293, 501] is to use

$$a(x'|x) = \min\left\{1, \frac{\omega(x')q(x|x')}{\omega(x)q(x'|x)}\right\}.$$ (2.25)

The chance of staying in the same state, $W(x|x)$, is automatically defined by the rejection probability, $W(x|x) = 1 - \sum_{x' \neq x} q(x'|x)a(x'|x)$. According to Eq. 2.24 any pair of valid acceptance probabilities must have the same ratio as any other valid pair. Therefore they can be obtained by multiplying the Metropolis-Hastings probabilities by a quantity less than one. This implies a larger rejection rate, which harms the asymptotic variance of the chain. In this sense (although not by all measures) using Eq. 2.25 is optimal [530, 573].

Although the MCMC-procedure guarantees convergence to $p(x)$ when the condition of detailed balance is satisfied, it is important to pay attention to the degree of *correlations* between the generated states in any particular implementation of the method. Typically, a certain number, $t_{skip}$ of the initial generated states, $\{x_t\}_{t=0}^{t_{skip}}$, are discarded from the statistics in order to minimize the transient dependence of the sampling on the (arbitrary) choice of $x_0$. This stage of the sampling is known as the "*equilibration*" or "*burn-in*" stage. If $T$ denotes the total number of generated states a standard MCMC-algorithm using the Metropolis-Hastings acceptance criteria then proceeds through the following steps:

---
**Metropolis-Hastings sampling**

---
0.    Choose $T$ and $t_{skip} < T$. Generate $x_{t=0}$.
1.    **for** $t = 1 \ldots T$
2.       Propose $x' \sim q(\cdot|x_{t-1})$
3.       Compute $a = \min\left\{1, \frac{\omega(x')q(x_{t-1}|x')}{\omega(x)q(x'|x_{t-1})}\right\}$.
4.       Draw $r$ uniformly in $[0; 1)$.
5.       Set $x_t = \begin{cases} x' & \text{if } r < a \\ x_{t-1} & \text{otherwise} \end{cases}$
6.       If $t > t_{skip}$ add $x_t$ to statistics.
7.    **end for**

---

A considerable element of trial-and-error pertains to finding a proper choice of $t_{skip}$ and $T$ which we shall return to in Sects. 2.6, 2.7.6 and 2.8. Specific examples of the Metropolis-Hastings sampling procedure are given in Chaps. 3, 5, 6, and 12.

When the proposal function is *symmetric*, $q(x|x') = q(x'|x)$, a property sometimes referred to as *microscopic reversibility* [551], the acceptance rate takes the particular simple form,

$$a(x'|x) = \min\left\{1, \frac{\omega(x')}{\omega(x)}\right\}, \quad \text{when } q(x|x') = q(x'|x). \tag{2.26}$$

For example, the acceptance-ratio in the canonical ensemble, Eq. 2.2, for a symmetric proposal distribution becomes

$$a(x'|x) = \min\{1, \exp(-\beta \Delta E)\}, \quad \Delta E = E(x') - E(x). \tag{2.27}$$

This is the original Metropolis algorithm [501].

## 2.4.2 Gibbs Sampling

According to Eq. 2.25 a high acceptance can be achieved when the proposal function is chosen so $\frac{q(x|x')}{q(x'|x)} \approx \frac{p(x)}{p(x')}$. Although the MCMC-procedure is brought in to play only in situations where one cannot directly sample from the joint distribution $p(x)$ of some multivariate quantity, $x = (x_1, \cdots, x_D)^T$, it is in many applications possible to sample from the conditional distributions, $p(x_i | x_{\backslash i})$, where $x_{\backslash i}$ denotes the set of all variables except the $i$'th, $x_{\backslash i} = (x_1, x_2, \cdots, x_{i-1}, x_{i+1}, \cdots, x_D)^T$. Since

$$p(x) = p(x_i | x_{\backslash i}) p(x_{\backslash i}),$$

the use of the $i$'th conditional distribution as proposal distribution leads to the Metropolis-Hastings acceptance rate of $a = 1$. In effect, by choosing $i = 1, \cdots, D$ in turn (or selecting $i \in \{1, \cdots, D\}$ at random) and sample a new $x = (x_i, x_{\backslash i})$ according to $x_i \sim p(x_i | x_{\backslash i})$ a Markov chain will be generated that converges to $p(x)$. This sampling procedure is known as *Gibbs sampling* [212]. Examples of the use of Gibbs sampling in the context of inferential protein structure determination and in directional statistics can be found in Chaps. 12 and 6, respectively.

## 2.4.3 Continuous State Space

In the digression of the MCMC-method outlined above, we have assumed the state space $\Omega$ to be discrete. However, the MCMC-method can straightforwardly be applied to systems with continuous degrees of freedom as well. Formally, the probabilities $p(x)$ are simply replaced with probability densities and matrix-multiplications with kernel-integrations. Thus, Eq. 2.25, the Metropolis-Hasting procedure (Sect. 2.4.1) and Gibbs sampling (Sect. 2.4.2), remains unaltered, provided that the sampling in $\Omega$ do not involve particular geometrical considerations.

Some applications, however, do involve non-trivial geometries, meaning that the volume elements in $\Omega$ will depend on the values assumed by $x = (x_1, \cdots, x_D)^T$. Examples of such spaces are given in Chaps. 9, 8, 6, 7 and 10, including Euler angles, spheres, projective planes and tori. Here, $\prod_i dx_i$, is *not* the natural integration measure and should be replaced with $J(x) \prod_i dx_i$, where $\omega_\pi(x) = J(x)$ is the Jacobian of the map from $\Omega$ to the embedding Euclidean space. Configurational-dependent volume factors are also known to appear in molecular dynamics from the integral over the conjugate momenta, if $x$ is chosen different from the atomic positions, as further discussed in Chap. 3.

Non-trivial volumetric factors may also come into play if some geometrical constraints between the $D$ components of $x$ and $x'$ are present in the proposal distribution, $q(x'|x)$. *Concerted rotations algorithms* in proteins, RNA and other polymeric systems constitute a distinct example of this problem. In these methods, a stretch of monomers is resampled while the ends remain in place [57, 72, 153,

309, 505, 730]. This condition, known as the *loop-closure* condition [231], imposes particular constraints in the degrees of freedom which lead to a non-trivial volume factors, a point first realized by Dodd et al. [153].

Whether the state-space itself $\Omega$ involves a particular non-trivial metric (case I) or such metric arises from the geometrical constraints involved in a particular proposal function (case II), the relative change of volume elements can be determined by the ratio of the associated Jacobians. The requirement of detailed balance, Eq. 2.22, implies that the Metropolis-Hastings acceptance becomes

$$a(x'|x) = \min\left\{1, \frac{\omega(x')q(x|x')J(x')}{\omega(x)q(x'|x)J(x)}\right\}, \qquad (2.28)$$

where $J(x)$ and $J(x')$ are the Jacobians evaluated in $x$ and $x'$ respectively. In the following we shall assume that the target weights $\omega(x)$ includes the proper geometrical distribution $\omega_\pi(x) = J(x)$ (case I). This implies that the general expression for the thermal probability weights becomes

$$\omega_\beta(x) = \exp(-\beta E(x))\omega_\pi(x). \qquad (2.29)$$

Similarly, we assume that the Jacobian of the proposal distribution (case II) is "absorbed" into $q(x)$. This will allow us – in both cases – to express the acceptance rate on the standard form, Eq. 2.25 and use the Euclidean measure, $\prod_i dx_i$, in expressions defined by integrals over $\Omega$. We refer to the excellent review by Vitalis and Pappu [741] and references therein for more details on this subject.

## 2.5 Estimators

In this section we shall discuss how MCMC-sampling can be used to address the three inference problems outlined in Sect. 2.2.2. We conclude the section by discussing some inherent deficiencies of the MCMC-algorithm which makes it inapplicable for inference in more complex and/or high-dimensional systems. These deficiencies are the main motivation for studying various improvements and extensions to the algorithm which will be the subject of the subsequent sections.

### 2.5.1 Expectation Values

First, the expectation value, Eq. 2.10, of any function $f(x)$ can be estimated as the arithmetic time average of the Markov chain, once this has converged to its equilibrium distribution, $p_t(x) \approx p_\infty(x) = p(x)$. Specifically, since we assume $p_t(x) \approx p(x)$ when $t > t_{\text{skip}}$ the MCMC-estimate, $\mathbb{E}_p[f(x)]$ of the expectation value of $f(x)$ follows Eq. 2.17,

$$\hat{\mathbb{E}}_p[f(x)] = \bar{f}(x) = \frac{1}{N}\sum_t f(x_t). \tag{2.30}$$

Here, $N = T - t_{\text{skip}}$ is the total number of states used for the statistics and the bar over a quantity indicates an MCMC "time"-average. The summation is over recorded states only, $\sum_t = \sum_{t=t_{\text{skip}}+1}^T$.

### 2.5.2 Histogram Method

When the aim is to estimate the full probability distribution of $f$, Eq. 2.11, rather than the expectation value alone, some extra considerations must be taken. Kernel density methods constitute one obvious approach for estimating $p_f(y)$. For simplicity and in keeping with most applications of the MCMC-method we shall make use of histogram based estimators. According to Eq. 2.11 the probability, $p(f(x) \in \mathcal{Y})$, of obtaining a value of $f$ in some set $\mathcal{Y} = [\tilde{y}_-, \tilde{y}_+)$ is given by

$$p(f(x) \in \mathcal{Y}) = \int_{\mathcal{Y}} p_f(y)\mathrm{d}y = \int_{\Omega} \chi_{\mathcal{Y}}(f(x))p(x)\mathrm{d}x = \mathbb{E}_p\left[\chi_{\mathcal{Y}}(f(x))\right],$$
$$\tag{2.31}$$

where $\chi_{\mathcal{Y}}(\cdot)$ is the indicator function on $\mathcal{Y}$,

$$\chi_{\mathcal{Y}}(y) = \begin{cases} 1 \text{ if } y \in \mathcal{Y} \\ 0 \text{ otherwise} \end{cases}.$$

Since $p(f(x) \in \mathcal{Y})$ can be expressed as an expectation value, Eq. 2.31, an estimator is given by

$$\hat{p}(f(x) \in \mathcal{Y}) = \overline{\chi_{\mathcal{Y}}(f)} = \frac{1}{N}\sum_t \chi_{\mathcal{Y}}(f(x_t)).$$

A Taylor expansion of Eq. 2.31 in $\Delta y = \tilde{y}_+ - \tilde{y}_-$ around $y = \frac{\tilde{y}_- + \tilde{y}_+}{2}$ yields

$$p(f(x) \in \mathcal{Y}) = p_f(y)\Delta y + \frac{1}{2}p'_f(y)\Delta y^2 + \mathcal{O}(\Delta y^3).$$

Consequently, by choosing $\Delta y \ll \left|\frac{p(y)}{p'(y)}\right| = \left|\frac{\mathrm{d}\ln(p_f(y))}{\mathrm{d}y}\right|^{-1}$, the probability density $p_f(y)$ can be estimated as

$$\hat{p}_f(y)\Delta y = \frac{1}{N}\sum_t \chi_{\mathcal{Y}}(f(x_t)), \quad \Delta y \ll \left|\frac{\mathrm{d}\ln p_f(y)}{\mathrm{d}y}\right|^{-1}. \tag{2.32}$$

The *histogram method* for estimating the full density function $p_f(y)$ arises naturally by partitioning the image of $f$. Specifically, let $\{\mathcal{Y}_i\}_{i=1}^L$, be such a partition, where $\mathcal{Y}_i = [\tilde{y}_i, \tilde{y}_{i+1})$, $y_i = \frac{\tilde{y}_i + \tilde{y}_{i+1}}{2}$ and $\Delta y_i = \tilde{y}_{i+1} - \tilde{y}_i$ for $\tilde{y}_1 < \tilde{y}_1 < \ldots < \tilde{y}_{L+1}$. Then

$$\hat{p}_f(y_i) = \frac{n(y_i)}{N|\Delta y_i|}, \qquad (2.33)$$

where $n(y_i)$ is the total number of states in the sampling belonging to the $i$'th bin. Although Eq. 2.32 suggests that bin-sizes should be chosen according to the variation of $\ln[p_f(y)]$ it is quite common in the histogram method to set the bin-widths constant.

Note, that the histogram method easily generalizes to multi-valued functions, $f = (f_1, \cdots, f_d)$, by considering $\mathcal{Y}_i$ to be a box with edge lengths $\Delta y_i = (\Delta y_{i1}, \cdots, \Delta y_{id})$ and volume $|\Delta y_i| = \prod_{j=1}^d |\Delta y_{ij}|$. This approach could in principle be used to estimate the the probability density, $p_f(y)$, in the multi-variate case, Eq. 2.12. In practice, however, kernel-methods offer a better alternative for estimating probability densities, particular in higher dimensions, $d > 1$ [62].

### 2.5.3 Ratio of Partition Functions

When the aim of the MCMC-simulation is to estimate ratios of partition functions, $Z/Z_\pi$, it is convenient to express the target distribution, Eq. 2.1 as

$$p(x) = \frac{\omega(x)}{Z} = \frac{\omega_E(E(x))\omega_\pi(x)}{Z}, \qquad (2.34)$$

$$Z = \int_\Omega \omega_E(E(x))\omega_\pi(x)\mathrm{d}x, \qquad (2.35)$$

where $E$ refer to the physical energy function in the thermal case, and to Eq. 2.9 in the non-thermal case. Correspondingly, $\omega_E(E)$ is used as a general notation for the weight associated with $E$. For a thermal distribution $\omega_E(E) = \omega_\beta(E) = \exp(-\beta E)$, where $\beta$ can attain any value, whereas for a non-thermal distribution $\omega_E(E) = \exp(-E)$ (see Sect. 2.2). This notation will allow us to treat the estimation of partition functions in the same manner for both thermal and non-thermal problems.

Now by setting $f(x) = E(x)$ in Eq. 2.11, we obtain

$$p_E(E) = \int_\Omega \delta(E(x) - E)p(x)\mathrm{d}x = \frac{g_\pi(E)\omega_E(E)}{Z}, \qquad (2.36)$$

$$Z = \int_\mathbb{R} g_\pi(E)\omega_E(E)\mathrm{d}E,$$

where $g_\pi$ is given by a marginalization of the reference state

## 2 Monte Carlo Methods for Inference in High-Dimensional Systems

$$g_\pi(E) = \int_\Omega \delta(E(x) - E)\omega_\pi(x)\mathrm{d}x. \tag{2.37}$$

When $\omega_\pi(x) = 1$, (or $\omega_\pi(x) = J(x)$ for non-Euclidian geometries), $g_\pi(E)$ takes the physical meaning of the density of states, Eq. 2.7. In particular we obtain for the canonical ensemble

$$p_\beta(E) = \frac{g_\pi(E)\exp(-\beta E)}{Z_\beta}, \quad Z_\beta = \int_\mathbb{R} g_\pi(E)\exp(-\beta E)\mathrm{d}E. \tag{2.38}$$

Similarly, we can introduce the statistical weights $\Gamma_\pi(\mathcal{E}_e)$ for the reference state, defined as

$$\Gamma_\pi(\mathcal{E}_e) = \int_{\mathcal{E}_e} g_\pi(E)\mathrm{d}E \approx g_\pi(E_e)\Delta E_e, \tag{2.39}$$

where $e$ refers to the $e$'th bin of the partition of the energies $E(x)$

$$\{\mathcal{E}_e\}_e = \left\{\left[E_e - \frac{\Delta E_e}{2}, E_e + \frac{\Delta E_e}{2}\right)\right\}_e.$$

Equation 2.39 reduces to the statistical weights for the thermal distribution, when $\omega_\pi$ is the uniform (or geometrical) measure. The reference partition function can be expressed in terms of $\Gamma_\pi(\mathcal{E}_e)$ as

$$Z_\pi = \int \omega_\pi(x)\mathrm{d}x = \int_\mathbb{R} g_\pi(E)\mathrm{d}E \approx \sum_e \Gamma_\pi(\mathcal{E}_e). \tag{2.40}$$

According to Eqs. 2.33, 2.36 and 2.39 an estimate of $g_\pi$ and $\Gamma_\pi$ is given by

$$\frac{\hat{g}_\pi(E_e)\Delta E_e}{\hat{Z}} = \frac{\hat{\Gamma}_\pi(\mathcal{E}_e)}{\hat{Z}} = \frac{n(E_e)}{N\omega_E(E_e)}, \tag{2.41}$$

where the estimated partition function is

$$\hat{Z} = \sum_e \omega_E(E_e)\hat{\Gamma}_\pi(\mathcal{E}_e). \tag{2.42}$$

In the context of statistical physics, estimating $\Gamma(\mathcal{E})$ from Eq. 2.41 is known as the *single histogram method* [702]. Since multiplying the estimates $\{\hat{\Gamma}_\pi(\mathcal{E})\}_e$, with an arbitrary constant $\{\hat{\Gamma}_\pi(\mathcal{E}_e)\}_e \to \{c\hat{\Gamma}_\pi(\mathcal{E}_e)\}_e$ leads to the same rescaling of the partition function, $\hat{Z} \to c\hat{Z}$, the statistical weights can only be estimated in a relative sense. This reflects the fact, that MCMC-sampling only allows ratios of partition functions to be estimated. In practice, an absolute scale is most simply obtained by prescribing a particular value to either $Z$ or $\Gamma(\mathcal{E}_{ref})$, where $\mathcal{E}_{ref}$ is some selected reference bin. In the latter case, $Z$ is then estimated from Eq. 2.42. The corresponding value for the reference partition function is subsequently found by inserting the estimated statistical weights into Eq. 2.40. This procedure will lead to a unique estimate of $Z/Z_\pi$.

### 2.5.4 Limitations of the Standard MCMC-Estimators

While the Metropolis-Hasting importance sampling has been the workhorse in statistics for the past half-century it suffers from a number of limitations. First of all, it is a requirement for all estimators discussed above that the Markov chain has converged, $p_t(x) \approx p(x)$. However, it may be difficult to assess the convergence of the chain due to the *high degree of correlation* between the sampled states $x_t$. A point we shall return to in the next Section. Secondly, the sampling typically only provides limited information of the marginal distributions, Eqs. 2.33 or 2.36. Essentially, from a given total simulation time $T$, one can at best hope to sample regions $\mathcal{Y}_i$ (or $\mathcal{E}_e$) with the standard MCMC-algorithm for which $p(f(x) \in \mathcal{Y}_i) \gtrsim 1/T$. However, in many statistical problems it is in fact the properties in the low-probability regions that are of particular interest [411].

The *limited support* of the estimated distribution $\hat{p}_E(E)$ has important consequences for the use of the MCMC-technique to calculate partition functions. Assuming that the uncertainty, $\delta$, of the fraction of observed states $\frac{n(E_e)}{N}$ in Eq. 2.41 is of the order $\delta \sim 1/T$ when $n(E_e) \lesssim 1$, the corresponding uncertainty of $\frac{\hat{\Gamma}_\pi}{\hat{Z}}$ becomes

$$\sigma \left[ \frac{\hat{\Gamma}_\pi(\mathcal{E}_e)}{\hat{Z}} \right] \simeq \frac{1}{T \omega_E(E_e)}.$$

Consequently, for energy-intervals $\mathcal{E}_e$ where $n(E_e) \lesssim 1$ the estimate of the statistical weight can become arbitrarily inaccurate, depending on the value of $T \omega_E(E_e)$. If the bulk of the probability mass for the reference distribution, $\pi$, falls outside the support of $\hat{p}_E$, the standard MCMC-sampling cannot provide a reliable estimate of $Z/Z_\pi$. For thermal distributions this implies that only ratios $Z_\beta/Z_{\beta'}$ can be determined when $\beta'$ is close to $\beta$ [34]. In the following sections we will discuss various approaches to improve the properties of the Monte Carlo sampling to provide more accurate estimators of $\mathbb{E}[f(x)]$, $p_f(y)$ or $Z/Z_\pi$.

## 2.6 Ergodicity and Convergence

As previously mentioned, the direct application of the MCMC-algorithm is often hampered by the high degree of correlations between the generated states, a property also referred to as *poor mixing* or *slow convergence* of the Markov chain. Poor mixing reduces the effective number of samples and may easily lead to results which

# 2 Monte Carlo Methods for Inference in High-Dimensional Systems

are erroneously sensitive to the arbitrary initialization, $x_{t=0}$, thereby compromising the ergodicity of the Markov chain.

The problem of poor mixing or slow convergence is typically associated with multi-modal probability distributions, where the transition between different modes involves intermediate states with very low probability. This situation is generic for models with frustrated or *rough energy landscapes* [36,375,503] or physical systems in vicinity of *phase transitions* [42,413,523]. Hetero-polymeric systems generally belong to the first class [288, 557, 644] and many bio-molecules also display cooperative behavior akin to first order type of phase-transitions, including proteins [592, 593] and RNA [13]. Slow convergence is also invariably related to low-temperature sampling, where even small energy increments have low acceptance probabilities.

Formally, the convergence time, $\tau_{conv}$, is given by the next highest eigenvalue, $\lambda_2 < \lambda_1 = 1$, of the transition matrix $W$, as $\tau_{conv} \simeq (\ln \lambda_2)^{-1}$. This quantity is important as it dictates both the proper choice of the burn-in time, $t_{skip} \simeq \tau_{conv}$, and total simulation time, $T \gg \tau_{conv}$. However, even for the simplest non-trivial problems, the size of the transition matrix is far too large to compute the eigenvalue explicitly. On a practical basis, the key is to identify a set of *slowly varying* variables, for which variance- or covariance-based measures can be applied. Typically, this set overlaps with the quantities of interest to the MCMC-application in the first place (see Sect. 2.2.2), and we shall for simplicity use the same notation for these variables, $f = \{f_i\}_{i=1}^d$. For physical systems, slowly varying quantities invariably involve specific order parameters as well as the total energy itself [413,523,741].

The first and foremost test should always be the reproducibility of the sampling with respect $\{f_i\}_i$, using a sufficient number of identical replicas with different starting conditions [741]. Convergence can then be assessed by the magnitude of the standard deviations of the ensemble averages of $\{f_i\}_i$. Further type of statistical error analysis involve measures of time correlation functions for $\{f_i\}_i$ [802], block averaging [198], jackknife analysis [508] or tools from exploratory data analysis and probability theory [125]. Recent work for convergence assessment in the the field of bio-macromolecule simulations has been dominated by identifying relevant collective variables using either principal component analysis [302], clustering techniques [200] or other reduced-dimensional quantities [679,741].

A number of methods have been proposed to improve the ergodicity and convergence properties of the MCMC sampling technique. To classify these methods we shall follow Y. Iba [328] and distinguish between three aspects of the MCMC-method: the actual sampling *algorithm*, the employed *MCMC-kinetics* and the chosen *ensemble*. Whereas the first method essentially is a computational trick to bypass the generation of a long sequence of rejected states, the two latter methods aim at reducing the actual convergence time by modifying the transition matrix, either through an extension of the proposal distribution or by replacing the target distribution/ensemble with an artificial one.

In this section, we shall briefly discuss various improvements to the MCMC-algorithm and its kinetics. We emphasize that these two first methods do not overcome the fundamental deficiency of the canonical sampling regarding the limited support of the density estimators, $\hat{p}_f(y)$ and $\hat{p}_E(E)$, discussed in Sect. 2.5.4. The realization of the ensemble-approach will be reviewed in Sects. 2.7 and 2.8.

### 2.6.1 Event-Driven Simulation

The common strategy to improve the Markov chain algorithm itself, is to perform an integration of the MC-time. This can be particular effective in cases where the average acceptance rate $\bar{a}$ is very low, which is a problem typically encountered in the Metropolis-Hastings algorithm at low temperatures. The idea of the MC-time integration is roughly, that instead of rejecting $\sim 1/\bar{a}$ moves on average one can increase the MC-time with $1/\bar{a}$ and enforce a move to one of the possible trial states according to the conditional probability distribution. This method, which is limited to discrete models, is known as event-driven simulation [421] or the N-fold way [71,548]. The event-driven method provide a solution to the problem of trapping in local energy minima, in cases where these minima only comprise one or a few number of states, and where the proposal distribution is sufficient local to make the successive rescaling of the MC-time feasible. However, this approach is more involved for continuous degrees of freedom and cannot circumvent the long correlation times associated with cooperative transitions.

### 2.6.2 Efficient Proposal Distributions

The MCMC-kinetics is encoded in the proposal distribution $q(x'|x)$. Typically, $q(x'|x)$ will be composed of a mixture of some basic proposal-functions or move-types $q_i(x'|x)$,

$$q(x'|x) = \sum_i \tilde{q}_i(x)q_i(x'|x),$$

where $\tilde{q}_i(x)$ is the probability of choosing move-type i and $\sum_i \tilde{q}_i(x) = 1$. Detailed balance is most easily ensured by requiring that it is satisfied for each move-type individually. Consequently, if move-type $i$ has been chosen then the Metropolis-Hastings acceptance probability for the trial state, $x'$ is

$$a(x'|x) = \min\left\{1, \frac{\omega(x')\tilde{q}_i(x')q_i(x'|x)}{\omega(x)\tilde{q}_i(x)q_i(x|x')}\right\}.$$

When the $\tilde{q}_i$'s are constant (independent of $x$) this expression reduces to its usual form, Eq. 2.24, for each move-type. In general, a compromise needs to be made between the characteristic step-size of the proposal and the acceptance rate. If the step-size is too large compared to the typical scale by which $p(x)$ varies, the acceptance rate will become be exceedingly small, but too small moves will make the exploration inefficient. Both cases compromise convergence [62].

A successful strategy is often to mix local and non-local types of proposals. The best example of this approach is in MCMC-applications to spin-systems. The cluster-flip algorithm, developed in the seminal works of R.H. Swendsen and J.-S. Wang [704] and U. Wolff [769], was used in conjunction with single spin flip and proven extremely valuable in reducing critical slowing down near second order phase transitions.

Another important example is the pivot-algorithm for polymer systems [462] which has been used with great success to calculate properties of self avoiding random walks [734]. A Pivot-move is defined by a random choice of a monomer as a pivot and by a random rotation or reflection of one segment of the chain with the pivot as origin. The pivot-move is also efficient in sampling protein configurations in the unfolded state but must be supplemented with local type of moves to probe the conformational ensemble in the dense environment of the native state. These type of proposal distributions are commonly referred to as *concerted rotation algorithms* (see also Sect. 2.4.3). For instance, Favrin et al. have suggested a semi-local variant of the pivot-algorithm for protein simulations [176]. This work has given inspiration to the fully local variant suggested by Ulmschneider et al. [730] and which provides improved sampling performance compared to earlier concerted rotation algorithms with strict locality [730]. However, a key bottleneck of all existing methodologies with strict locality is that some of the degrees of freedom of the polypeptide chain are not controlled by the proposal function but post-determined by the condition for chain closure. Based on an exact analytical solution to the chain-closure problem we have recently devised a new type of MCMC algorithm which improves sampling efficiency by eliminating this problem [72]. Further examples of approximate distributions for protein and RNA structures that serve as effective proposal distributions in a MCMC-sampling framework are discussed in Chaps. 10 and 5.

In general, the extension of the proposal distribution needs to provide both a reasonable constancy of the acceptance rate, Eq. 2.25, as well as ensuring that the gain in sampling efficiency is not lost in the computer time to carry out the move. In contrast to the method discussed in the next section, efficient choices of $q(x|x')$ will always be system-specific.

## 2.7 Extended Ensembles

In the past few decades, a variety of MCMC-methods known as *extended ensembles* [328] have been suggested to improve the ergodicity of the chain and alleviate the problem of poor mixing. In these methods, the target (or canonical) distribution, $p(x)$, is replaced with an "artificial" distribution constructed either as an extension or by compositions of $p(x)$. The underlying idea of these constructions is to build a "bridge" from the part of the probability distribution, where the Markov chain suffers from slow relaxation (e.g. low temperatures/energies) to the part, where the sampling is free from such problems (e.g. high temperatures/energies) [328]. This construction can be viewed as a refinement of the *chaining technique* [62, 535]. While extended ensemble methods have previously been prevalent only in the field of statistical physics, they are slowly gaining influence in other fields, such as statistics and bioinformatics [197, 450, 530]. For an excellent and in-depth review of extended ensembles and related methods in the machine-learning context, we refer to the dissertation by Murray [530].

An attractive feature of the extended ensemble approach, is that they not only ensure a more efficient sampling but also provide estimators for the tails of the marginal distributions, $p_f(y)$ or $p_E(E)$, which are inaccessible to the standard MCMC-approach (Sect. 2.5.4). Consequently, extended ensembles are particular suited for calculating key multivariate integrals including evidence or partition functions. To appreciate this aspect of the extended ensembles in a general statistical context, we shall in the following retain the explicit use of the reference weights, $\omega_\pi$. In a physical context and when $\Omega$ is considered Euclidean, $\omega_\pi = 1$ and can therefore be ignored.

The merit of the extended ensembles is the generality of the approach. They can in principle be combined with any legitimate proposal distribution and they can be applied to any system. However, unlike a simple canonical sample the extended ensembles all introduce a set of parameters, which are not a priori known. The central strategy in the methods is then to learn (or tune) the parameters of the algorithm by a step-by-step manner in preliminary runs which typically involve a considerable amount of trial-and-error [179, 285]. This stage is termed the *learning* stage [328] which we shall discuss in some details in Sect. 2.8. After the tuning has converged, a long run is performed where the quantities of interest are sampled. This stage is called the *sampling* or production stage. For all ensembles, it is then straight-forward to reconstruct the desired statistics for the original target distribution, $p(x)$. This reconstruction technique which we shall return to in Sect. 2.9, is known as *reweighting* [702]. It should be emphasized, that the statistical weights $\Gamma_\pi$ plays a key role in all extended ensembles, since both the unknown parameters as well as the reweighting can be expressed in terms of $\Gamma_\pi$.

It is convenient to distinguish between two types of extended ensembles, *tempering* and *generalized* ensembles [285]. In tempering based ensemble extensions, the new target distribution is constructed from a pre-defined set of inverse temperatures $\{\beta_r\}_r$, including the original $\beta$. These inverse temperatures enter in a Boltzmann-type of expression for a corresponding set of weights, $\{\omega_r(x)\}_r$, where $\omega_r(x) \propto \exp(-\beta_r E(x))\omega_\pi(x)$. In generalized ensembles (GE), the Boltzmann form is abandoned altogether and $\omega_{GE}(E)$ can be any function designed to satisfy some desired property of the GE-target distribution. Parallel and simulated tempering presented in Sects. 2.7.1 and 2.7.2 belong – as their names indicate – to the first type of category, whereas the multicanonical and $1/k$-ensemble presented in Sects. 2.7.3 and 2.7.4 belong to the second category. As clarified in Sect. 2.7.5 the tempering-based methods have a more limited domain of application compared to generalized ensembles, since in the former case certain restrictions are imposed on the allowed

# 2 Monte Carlo Methods for Inference in High-Dimensional Systems

functional form of the density of states $g_\pi$. We conclude the section by discussing general considerations regarding parameter tuning and convergence properties of extended ensembles.

## 2.7.1 Parallel Tempering

The parallel tempering algorithm [483] has been independently discovered by several different group of authors in the period 1990–1996 and as a result bears a variety of different names (Exchange MC-algorithm [323], Metropolis-Coupled Chain Algorithm [218], Time-homogeneous Parallel Annealing [373], Multiple Markov Chain algorithm [708]).

In parallel tempering (PT) the configuration space $\Omega_{PT}$ is composed of $R$ *replica* of the original configuration space $\Omega_{PT} = \Omega^R$, so a $PT$-state $x_{PT}$ is a family of replica states, $x_{PT} = \{x_r\}_{r=1}^R$. Each replica state $x_r$ is sampled according to a canonical ensemble with its own inverse temperature $\beta_r$. Consequently, the target probability distribution is a product of canonical distributions:

$$p_{PT}(x_{PT}) = \prod_{r=1}^R p_{\beta_r}(x_r) = \prod_r \frac{\exp(-\beta_r E(x_r))\omega_\pi(x_r)}{Z(\beta_r)}. \qquad (2.43)$$

The idea of construction the system as a series of replica at different temperatures is to incorporate global types of move in the proposal distribution. Besides the conventional proposal function, $q(x_r'|x_r)$, applied to each replica individually, a putative *replica-exchange* move $q_{rs}$ is introduced between pairs of replica $r$ and $s$. In this step, candidates of new configurations $x_r'$ and $x_s'$ are defined by the exchange of configurations of the two replica, $x_r' = x_s$ and $x_s' = x_r$. If the acceptance probability, $a_{rs} = a(\{x_s, x_r\}|\{x_r, x_s\})$, is on the Metropolis form, $a_{rs} = \min\{1, \tilde{a}_{rs}\}$, it will be given by

$$\tilde{a}_{rs} = \frac{p_{\beta_r}(x_s)p_{\beta_s}(x_r)}{p_{\beta_r}(x_r)p_{\beta_s}(x_s)} = \exp[(\beta_r - \beta_s)(E(x_r) - E(x_s))]. \qquad (2.44)$$

Consequently, the exchange rate between replica $r$ and $s$ becomes $W_{rs}^{re} = q_{rs}a_{rs}$. The simultaneous probability distribution $p_{PT}$ will be invariant with this choice, so detailed balance is automatically satisfied. The temperatures of the two replica $r$ and $s$ have to be close to each other to insure non-negligible acceptance rates. In a typical application, the predefined set, $\{\beta_r\}$, will span from high to low temperatures, and only neighboring temperature-pairs will serve as candidates for a replica exchange. This construction is illustrated in Fig. 2.1.

The averages taken over each factor $p_{\beta_r}(x_r)$ reproduces the canonical averages at inverse temperature $\beta_r$, because the replica-exchange move does not change the simultaneous distribution $p_{PT}(x_{PT})$. At the same time, the states of the replicas are

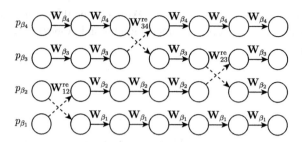

**Fig. 2.1** An illustration of parallel tempering. The *columns of circles* represents states in the Markov chain (with state space $\Omega^4$). Each *row of circles* has stationary distribution $p_{\beta_r}$ for $r \in \{1, 2, 3, 4\}$. The *horizontal arrows* represent the conventional transition matrix, $W_{\beta_r}$, that makes independent transitions at inverse temperature $\beta_r$. The *dotted arrows* represent the replica exchange transition matrix, $W_{rs}^{\text{re}}$, that exchanges the states of chains with adjacent inverse temperatures $r$ and $s$ (Figure by Jes Frellsen [195], adapted from Iain Murray [530])

effectively propagated from high to lower temperatures and the mixing of Markov chain is facilitated by the fast relaxation at higher temperatures. Note that $p_{\beta=0}(x)$ is identical to the reference distribution $p_{\beta=0}(x) = \pi$. Consequently, by setting $\beta_{r=1} = 0$ and $\beta_{r=R} = \beta$ in the replica set, where $\beta$ is the inverse temperature of the original target distribution ($\beta = 1$ in the non-thermal case), the parallel tempering sampling makes it feasible to estimate $Z/Z_\pi$ using reweighting, see Sect. 2.9.

The PT-method can be considered as a descendant of the simulated annealing algorithm [375]. The term "annealing" indicates that the simulations are started at high temperatures, which are then gradually decreased to zero. The object of this method is typically not statistical inference but only to search for the ground state of complex problems. Simulated annealing is a widely used heuristic, and is one of the key algorithms used in protein structure prediction programs, such as Rosetta from the Baker Laboratory [426]. While the annealing is useful for escaping from shallow local minima it does not allow the escape from deep meta-stable states. Such jumping is in principle facilitated by the PT-method by allowing the temperature to change "up and down" alternately.

In the field of biopolymer simulations the parallel tempering/replica exchange method has become very popular in conjunction with molecular dynamics (MD) algorithms [694], the direct integration of Newtons equations of motion. In these approaches the sampling of each individual replica is simply given by the MD-trajectory, whereas replica exchange is performed according to Eq. 2.44. The replica exchange method has also been used successfully in the context of inferential structure determination by Habeck and collaborators, as detailed in Chap. 12.

The challenge in the learning stage of parallel tempering algorithm is to set the number of inverse temperatures $R$ and spacing between them. The standard requirement is that the exchange-rate between all adjacent pairs of replica happens with a uniform and non-negligible probability. It can be shown [323, 324, 328] that this is satisfied when the spacing between adjacent inverse temperatures $|\beta_{r+1} - \beta_r|$ is selected approximately as $Q(\beta_r)^{-1}$, where

# 2 Monte Carlo Methods for Inference in High-Dimensional Systems

$$Q(\beta) \propto \sqrt{\mathbb{E}_\beta[E^2] - \mathbb{E}_\beta[E]^2} = \sqrt{\sigma_\beta^2(E)} \tag{2.45}$$

Here, $\mathbb{E}_\beta[E]$ denotes the expectation value of $E$ with respect to the Boltzmann distribution, Eq. 2.2. Note that $\sigma_\beta^2(E)$ is related to the *heat capacity* $C$ as $C = k\beta^2\sigma_\beta^2(E)$, see Eq. 2.72, and may also be expressed as

$$\sigma_\beta^2(E) = -\mathbb{E}_\beta\left[\frac{d^2 \ln[p_\beta(x)]}{d\beta^2}\right] = \mathcal{I}(\beta),$$

where $\mathcal{I}(\beta)$ is the *Fisher information*, see Chap. 1. Since $E$ typically scale with system size or number of degrees of freedom, $D$, as $E \propto D$ so will the variance, $\sigma_\beta^2(E) \propto D$. Consequently, the required number of replica scales with system size as

$$R \simeq \int_{\beta_{\min}}^{\beta_{\max}} Q(\beta)d\beta \propto \sqrt{D}$$

However, as $\sigma_\beta^2(E)$ is not a priori known, the number of inverse temperatures and their internal spacing have to be estimated, typically through an iterative approach [322, 362]. Further aspects and perspectives of the PT-method can be found in the recent review by Earl and Deem [166].

## 2.7.2 Simulated Tempering

An algorithm closely related to the parallel tempering is the simulated tempering (ST) [371, 484] or expanded ensemble [455, 745]. Its use for biomolecular simulations was pioneered by Irbäck and collaborators [329]. In this approach, the temperature is treated as a dynamical variable, so the state space $\Omega_{ST}$ is constructed as a direct product of the original state space $\Omega$ and a predefined set of inverse temperatures, $\Omega_{ST} = \Omega \times \{\beta_r\}_{r=1}^R$. The target distribution in this extended space takes the form,

$$p_{ST}(x, \beta_r) = \frac{1}{Z_{ST}} \exp\left(-\beta_r E(x) + \eta(\beta_r)\right)\omega_\pi(x), \tag{2.46}$$

$$Z_{ST} = \sum_{r=1}^R e^{\eta(\beta_r)} Z_{\beta_r},$$

where $\eta$ is a weight function that controls the distribution in $\{\beta_r\}_r$. The Markov chain is constructed by simulating the system with the ordinary proposal function, $q(x'|x)$, combined with a probability $q(s|r)$ of selecting the temperature move, $\beta_r \to \beta_s$, where $\beta_r$ is the current inverse temperature. According to Eq. 2.46 the latter proposal would have the Metropolis acceptance probability

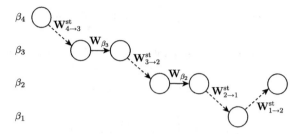

**Fig. 2.2** An illustration of simulated tempering. The *circles* represent states in the Markov chain. The *rows* represent different values of the inverse temperature. The *horizontal arrows* represent the conventional transition matrix, $W_{\beta_r}$ at inverse temperature $\beta_r$. The *dotted arrows* represent the transition matrix, $W^{st}_{r \to s}$, that changes the inverse temperature to an adjacent level (Figure by Jes Frellsen [195], adapted from Iain Murray [530])

$$a(s, x | r, x) = \min\left\{1, \exp\left(-(\beta_s - \beta_r) E(x) + \eta(\beta_s) - \eta(\beta_r)\right)\right\}.$$

Consequently, the temperature transition rate becomes $W^{st}_{r \to s}(x) = q(s|r) a(s, x | r, x)$. The ST-construction is illustrated in Fig. 2.2.

In simulated tempering the marginal distribution of the inverse temperature $\beta_r$ is given by

$$p_{ST}(\beta_r) = \frac{Z_{\beta_r} \exp(\eta(\beta_r))}{Z_{ST}}. \tag{2.47}$$

Consequently, to get a uniform sampling in the temperature space the weight-function $\eta(\beta_r)$ should be chosen as

$$\eta(\beta_r) \propto -\ln[Z_{\beta_r}]. \tag{2.48}$$

The unbiased choice, $\eta(\beta_i) = const.$, would on the other hand trap the system in one of the extreme ends of the temperature region. Accordingly, the learning stage of simulated tempering involves choosing the set of inverse temperatures, $\{\beta_r\}_{r=1}^{R}$, and finding the weight function $\eta$. The optimal spacing of the inverse temperatures can be iteratively estimated in a similar way as described for parallel tempering [371]. Equivalently, $\eta$ can be estimated iteratively [371], which means estimating the partition functions, $\{Z_{\beta_r}\}$, when $\eta$ is chosen according to Eq. 2.48. Thus, by setting $\beta_{r=1} = 0$ and $\beta_{r=R} = \beta$ – the target inverse temperature – the ratio $Z/Z_\pi = \exp(\eta_1 - \eta_R)$ can be obtained. An alternative estimation approach is discussed in Sect. 2.9.

The ST-method is less memory demanding compared to PT, since one simulates only one replica at the time. However, the circumstance that the partition sums also needs to be iteratively estimated makes the PT-ensemble a better choice both in terms of convenience and robustness [328], provided that memory is not a limiting factor.

## 2.7.3 Multicanonical Ensemble

A very popular alternative to the parallel and simulated tempering is the *multicanonical ensemble* (MUCA) [40–42]. A more convenient formulation of the ensemble, which we shall adopt here, is given by J. Lee [427], under the name "entropic ensemble". Methods based on similar idea are also known as *Adaptive umbrella sampling* [23,313,504]. In their original forms, multicanonical ensembles deals with extensions in the space of total energy, while Adaptive umbrella sampling focuses on the extensions in the space of a reaction coordinate for a fixed temperature. The two methods correspond to different choices of $f$ used for the weight-parameterization, but from a conceptual point of view they are closely related. More recently, the methodology of the adaptive umbrella sampling has been generalized in the *metadynamics* algorithm proposed by Laio and Parrinello [411], which we will return to in Sect. 2.8.5.

As opposed to the ST- and PT-method, the state space itself is not extended in the multicanonical approach, $\Omega_{MUCA} = \Omega$. Instead, the multicanonical ensemble is a particular realization of a *generalized ensemble* (GE), where the use of Boltzmann weights are given up all-together and replaced by a a different weight function $\omega_{GE}$. As further discussed in Sects. 2.7.5 and 2.8.5, $\omega_{GE}$ can in principle be a function of any set of collective coordinates, $f = (f_1, \cdots, f_d)^T$. For sake of clarity we shall here assume $\omega_{GE}$ to be a function of energy $E$ only

$$\omega_{GE}(x) = \omega_{E,GE}\big(E(x)\big)\omega_\pi(x). \tag{2.49}$$

The procedure of sampling according to $\omega_{GE}$ follows directly from the Metropolis-Hastings expression for the acceptance probability

$$a(x'|x) = \min\left\{1, \frac{\omega_{GE}(x')q(x' \to x)}{\omega_{GE}(x)q(x \to x')}\right\}. \tag{2.50}$$

According to Eq. 2.36 the marginal distribution over $E$ for any arbitrary weight function $\omega_{E,GE}$ is given as

$$p_{E,GE}(E) = \frac{\omega_{E,GE}(E)g_\pi(E)}{Z_{GE}}, \qquad Z_{GE} = \int_{\mathbb{R}} g_\pi(E)\omega_{E,GE}(E)dE.$$

The question is how to choose $\omega_{E,GE}$ in an optimal way. The choice of Boltzmann weights generates a sampling, which only covers a rather narrow energy-region (Sect. 2.5.4). Instead, a uniform sampling in the energy-space would be realized by the function,

$$\omega_{E,MUCA}(E) = g_\pi(E)^{-1}. \tag{2.51}$$

These weights define the multicanonical ensemble.

The multicanonical ensemble was originally devised as a method to sample first order phase transitions [41], and in this respect it is superior to the "temperature based" approaches of the PT- and ST-ensemble. In the original work [41], the method was shown to replace the exponential slowing down related to a canonical sampling of a first order phase transition with a power like behavior. The disadvantage of the MUCA-ensemble resides in the extensive number of parameters which need to be estimated in the learning stage. In principle, the weights for each individual energy (or energy-bin) $\omega_{E,GE}(E)$ represent an independent tunable parameter. For the uniform reference state, the total entropy, $S = k \ln[\Gamma_{tot}]$, will scale with system size as $S \propto D$. Consequently, the required number of weight-parameters to resolve $\ln[g(E)]$ sufficiently fine also scales proportionally to $D$, as opposed to the $\sqrt{D}$ scaling for the required number of temperatures in the PT- and ST-ensemble. We shall return to this problem in Sect. 2.8 .

The multicanonical ensemble has been applied in a wide range of problems in statistical physics, including systems within condensed matter physics, in gauge theories and in different types of optimization problems (reviewed in [39, 328, 755]). U.H.E. Hansmann and Y. Okamoto pioneered its use in studies of proteins [284] and by now a quite extensive literature exists which is compiled in [288, 332].

### 2.7.4 1/k-ensemble

The MUCA-ensemble is diffusive in the energy space and may therefore spend unnecessary amount of time in high energy regions. In many statistical problems, one is often more interested in properties of the system around the ground state. However, an ensemble with too much weight at low energies may become fragmented into "pools" at the bottoms of "valleys" of the energy function, thus suffering from the same deficiencies as canonical ensemble at low temperatures, where ergodicity is easily lost. It has been argued that optimal compromise between ergodicity and low temperature/energy sampling is provided by the *1/k-ensemble* [303]. The $1/k$-ensemble is also a generalized ensemble, defined by the weight-function

$$\omega_{1/k}(x) = \omega_{E,1/k}(E(x))\omega_\pi(x),  \tag{2.52}$$

where

$$\omega_{E,1/k}(E) = \frac{1}{k_\pi(E)}, \quad k_\pi(E) = \int_{E'<E} g_\pi(E')\mathrm{d}E'.  \tag{2.53}$$

Consequently, the $1/k$-ensemble assigns larger weights to the low energy region as compared to the multicanonical ensemble. Furthermore, the $1/k$-ensemble is less sensitive to errors in an estimate of density of states, due to the integral form of the weights [303].

It can be shown that samples from the $1/k$-ensemble approximately have a uniform marginal distribution over $\ln[g_\pi(E)]$ [303]. This appealing feature is shared by the *nested sampling* method [666, 669], briefly presented in Sect. 2.3.2. In fact, nested sampling parameterizes the partition- or evidence function, Eq. 2.35, precisely according to the cumulant distribution $k_\pi(E)$. Specifically, the total cumulant reference (or prior) mass covering all energies below a given $E$ can be expressed as $X(E) = \frac{k_\pi(E)}{Z_\pi}$. Since this is a monotonously increasing function one may equally well parameterize $E$ according to $X$. Consequently, Eq. 2.35 can also be written as

$$Z = \int_0^1 \omega_E(E(X))\mathrm{d}X. \tag{2.54}$$

This reparameterization is used in a quite original way in the nested sampling method [669].

### 2.7.5 Extensions along Reaction Coordinates

So far we have discussed extended ensembles using the energy as the axis of extension. The energy, however, plays a special role as this quantity dictates the probability weights $\omega(x)$ of the target distribution, as discussed in Sect. 2.2. In the following we shall shortly outline how generalized ensembles can be applied to facilitate broad sampling along any choice of variable(s) $f$, here considered as a reaction coordinate. The digression will also serve to highlight the limitations of using tempering based methods for this task.

When the axis of extension is not some simple function of the energies, Eq. 2.49 is not the most convenient form of the GE-weights. Rather, we shall use

$$\omega_{GE}(x) = \omega_{f,GE}(f(x))\omega(x), \tag{2.55}$$

where $\omega(x)$ is the original target weights. The marginal probability distribution $p_{f,GE}(y)$ then becomes

$$
\begin{aligned}
p_{f,GE}(y) &= Z_{GE}^{-1} \int_\Omega \delta(f(x) - y)\omega_{GE}(x)\mathrm{d}x \\
&= \frac{Z}{Z_{GE}}\omega_{f,GE}(y)p_f(y),
\end{aligned}
\tag{2.56}
$$

where

$$Z_{GE} = Z \int_{\mathbb{R}} \omega_{f,GE}(y)p_f(y)\mathrm{d}y. \tag{2.57}$$

Consequently, the flat histogram/multicanonical ensemble is realized with the choice

$$\omega_{f,\mathrm{MUCA}}(y) = \frac{1}{p_f(y)}. \tag{2.58}$$

Equation 2.58 can be applied to any marginal distribution $p_f(y)$. Tempering based extensions on the other hand, are by and large limited to cases where $\ln[p_f(y)]$ is *concave*. This observation can be deduced from Eq. 2.56 which for a replica with inverse "temperature" $\lambda$, the Lagrange parameter conjugated to $y$, would read

$$p_{f,\lambda}(y) = \frac{Z}{Z_{f,\lambda}} \exp(-\lambda y) p_f(y), \quad Z_{f,\lambda} = Z \int_{\mathbb{R}} \exp(-\lambda y) p_f(y) \mathrm{d}y. \quad (2.59)$$

This probability distribution will have a local maximum at a $y$ satisfying

$$\frac{\mathrm{d}\ln[p_{f,\lambda}(y)]}{\mathrm{d}y} = -\lambda + \frac{\mathrm{d}\ln[p_f(y)]}{\mathrm{d}y} = 0.$$

If $\ln[p_f(y)]$ is not concave, this equation will have multiple solutions for some range of $\lambda$-values, implying that $p_{f,\lambda}(y)$ is not uni-modal. Such distributions are generally difficult to sample efficiently (Sect. 2.6). It is precisely the ability of the GE-techniques to handle a non-concave behavior of $\ln[g(E)]$ that allow them to alleviate the exponential slowing down associated with first order phase transitions. For a more detailed discussion of the caveats of tempering based methods to sample non-concave functions we refer to the excellent presentation by Skilling [669].

### 2.7.6 Sampling Times and Efficiency Optimization

While extended ensembles in general improves sampling efficiency, these methods do not alleviate the need for making critical assessment of the convergence of the simulation. The theoretical aspects related to this analysis are similar to those for standard MCMC-simulations, discussed in Sect. 2.6. In particular, convergence times, $\tau_{\mathrm{conv}}$, are still formally given by second largest eigenvalue of the (extended) transition matrix $W(x'|x)$ and dictates the proper choice of sampling time $T \gg \tau_{\mathrm{conv}}$.

Since the problems of determining eigenvalues from $W$ prevail in the extended ensembles, alternative methods are required to assess $\tau_{\mathrm{conv}}$ including those reviewed in Sect. 2.6. There are also a number of diagnostic tools specifically developed for extended ensembles. One commonly used test is to compare the observed distributions over energies or temperatures with the one expected from the ensemble choice. In MUCA-ensemble this translates into a test of flatness of the accumulated energy histogram [754]. Similarly, the time dependence on the replica exchange rates or temperature exchange rates in the PT- and ST-ensemble respectively, often serves as a sensitive probe of convergence [483]. A more accurate assessment is provided by estimating the "tunneling time" $\tau_{tun}$, defined as the average time to make a round trip from low to high energies/temperatures and back again [138, 534, 724]. This serves as a lower-bound proxy for $\tau_{\mathrm{conv}}$, but may in some cases actually differ quite significantly from the true convergence time [124]. The minimum requirement for equilibration is then that the total simulation time $T$ satisfies $T \gg \tau_{tun}$.

Clearly, both the learning stage and the sampling stage. of extended ensembles are sensitive to erroneously small sampling times. For instance, if the sampling time in each step of the iteration scheme in the learning stage is not sufficiently much longer than $\tau_{conv}$, one may observe "oscillatory" behavior of the weights [328, 443] and the iteration scheme will never converge. Even examples of "quasi"-convergence can be found in the literature (e.g. [287, 294]), where longer simulations (or different ensemble approach) lead to significantly different results (e.g. [286, 485]).

Due to its fewer number of tunable parameters, parallel tempering/replica-exchange is typically less vulnerable to erroneous choices of simulation times than other extended ensembles, such as the MUCA- or 1/k-ensemble. However, the works by several groups on the *in silico* folding of the 20 amino-acid mini-protein Trp-cage [539] using the replica-exchange molecular dynamics method (REMD), clearly demonstrates the difficulties in setting simulation parameters correctly for this ensemble as well [355, 580]. In particular, the problem of apparent simulation convergence has been emphasized in the recent studies by Garcia and collaborators [137, 563, 564]. Here, careful calculations show that the actual correlation time of the REMD-trajectories of Trp-Cage exceeds the total replica simulation times used in previous studies, implying that full convergence was not obtained in earlier works.

Since the convergence time $\tau_{conv}$ in extended ensembles is often much longer than their theoretical minimum [138, 283], recent advances in this field have been aimed at tuning parameters to directly minimize $\tau_{conv}$, viz. maximizing round trips. This key idea was first formulated by Trebst et al. [138, 724] in the context of the MUCA-ensemble. The modified generalized ensemble, which we shall refer to as the *tunneling-optimized ensemble*, was demonstrated to give a dramatic speed-up of the convergence on various Ising models compared to the MUCA-ensemble. The scheme was subsequently adapted to the parallel tempering ensemble and used to iteratively determine optimal temperature spacing [362].

## 2.8  Learning Aspects of Generalized Ensembles

As discussed in the previous section, generalized ensembles such as the multi-canonical and $1/k$-ensemble, are more complicated to implement compared to the tempering-based ensemble extensions, since the number of tunable parameters scale proportionally to the system size, $D$, rather than as $\sqrt{D}$. On the other hand, once these parameters have been obtained generalized ensembles have a larger range of applicability (Sect. 2.7.5). In this section we shall therefore focus on algorithms for obtaining the GE-weights. For simplicity, we will mostly consider the energy as the axis of extension although the discussion generalizes to other parameterizations, $f$, following the outline in Sect. 2.7.5. A specific example of a learning algorithm designed primarily for estimating the distribution, $p_f$, is given in Sect. 2.8.5.

The main complication of generalized ensembles is how to sample with weights defined as function of $g_\pi$ or $p_f$ without prior knowledge of these quantities. Common to all learning algorithms to be discussed below is to use an iterative approach, whereby the probability weights $\omega_{E,GE}$ ($\omega_{f,GE}$) for the next iteration is based on the best current estimates of $g_\pi = \Gamma_\pi / \Delta E \; (p_f)$. In Sect. 2.8.1 we present the most common approach, originally used in the simulation field, and which is based on the single histogram method (Sect. 2.5.2). Due to its poor convergence properties a number of alternative approaches have since then been suggested, including transition based methods (Sect. 2.8.2), hybrid methods (Sect. 2.8.3) and non-Markovian approaches (Sects. 2.8.4 and 2.8.5). In Sect. 2.8.6 we present a learning algorithm (Muninn) developed in our own group, which aims at combining the merits of non-Markovian approaches in terms of their adaptiveness and speed with the requirement of detailed-balance. Each section is concluded with references to model systems where the particular algorithm has been successfully applied.

### 2.8.1 Single Histogram Method

The simplest approach [38] to obtain the GE-weights is based on applying the single histogram method (see Sect. 2.5.3) iteratively, given some pre-partitioning $\{\mathcal{E}_e\}_{e=1}^L$ of the energy-space with constant bin-sizes $\Delta E_e = \Delta E$. Accordingly, let $n_k(E_e)$ be the histogram obtained using the weight scheme

$$\omega_k(x) = \omega_{E,k}(E_e)\omega_\pi(x),$$

where $e$ refers to the energy bin and $\omega_{E,k}(E_e)$ is the generalized ensemble weights for the $k$'th iteration. From Eq. 2.41 one obtains the following estimates of the statistical weights

$$\hat{\Gamma}_\pi(\mathcal{E}_e) \propto \frac{n_k(E_e)}{\omega_{E,k}(E_e)}. \tag{2.60}$$

In turn, this estimate can be used to define the GE-weights $\omega_{E,k+1}$ for the subsequent iteration according to the choice of ensemble. Typically, constant weights are used for the first iteration.

The standard iteration scheme, as defined by the single histogram method, entails a number of problems which were first systematically addressed in Refs. [672, 673]. First of all, it suffers from the *loss of statistics* inherited in the updating rule, as the information obtained from the previous $k - 1$ iterations is neglected [179, 180]. Secondly, the scheme requires a *careful choice of sampling time* $T_k$ for each iteration $k$. If $T_k$ is not much longer than the convergence time, $\tau_{\text{conv}}$ of the extended ensemble the iteration procedure will fail to converge as discussed in Sect. 2.7.5. However, choosing $T_k$ too high will compromise the speed of convergence. Thirdly, the scheme is *sensitive to errors* in $\hat{g}$ resulting from low statistics, which can cause the convergence process to become irregular [672]. Another aspect of this problem is how to assign weights to energies that have not yet been visited by the simulation. Finally, the choice of $\Delta E$ involves a compromise between the accuracy of the estimator $\hat{\Gamma}_\pi(\mathcal{E}_e) \approx \hat{g}_\pi(E_e)\Delta E$ and the resolution (viz. the efficiency) of the ensemble. Ideally, $\Delta E$ should be chosen as a function of $E$ to *ensure a uniform resolution* of $\ln(g_\pi)$ or $\ln(\omega_{E,GE})$, as indicated by the requirement for the bin-size in the histogram-method, Eq. 2.32. However, this approach is faced with the same problem as setting the GE-weights themselves, namely that $g_\pi$ is not known a priori. The consequence of these problems is that the sampling suffers from the so-called "scalability problem" [754], namely that systematic errors and substantial deviations will rapidly increase with system size.

A number of different methods have been proposed to overcome some of these problems pertaining to the learning aspects of generalized ensembles. Berg has devised a method to accumulate histogram based statistics in [37, 38]. This method focuses solely on relations between neighboring bins and in its current form it is limited to univariate weight parameterizations. In the following we discuss recent alternative approaches to facilitate the generalized ensemble sampling.

### 2.8.2 Transition Based Methods

As first realized by Oliveira et al. [140], it is possible to accumulate statistics from previous runs by using estimators based on the marginalized trial transition probabilities $q(E'|E)$ rather than histograms, provided that the proposal distribution is symmetric, $q(x'|x) = q(x|x')$. Here, $q(E'|E) = \int \delta(\mathcal{E}(x') - E')\delta(\mathcal{E}(x) - E)q(x'|x) \, dx' \, dx$. For uniformly discrete systems the number of states can then be estimated by solving the eigenvector problem $\sum_E q(E'|E)\hat{\Gamma}(E) = \hat{\Gamma}(E')$. This procedure forms the basis of the *broad histogram method* [140], where the eigenvector equation is approximated by neglecting elements away from the tridiagonal band of $q(E'|E)$, which in effect is similar to the approach of Berg [37]. A variant of the broad histogram method is given by the *transition matrix method* [755], where a more accurate solution-scheme to the eigenvector problem is used. Transition based methods have however generally been limited to discrete systems.

### 2.8.3 Hybrid Methods

Another way of accelerating the parameter estimation of generalized ensemble is to make initial use of the replica-exchange/PT method. This approach has been suggested by Mitsutake et al. and is known as the REMUCA- or MUCARE-method (from concatenation of replica-exchange and multicanonical) [513–515]. In the first stage a replica exchange run is performed using a predefined number of replica with associated temperatures. $\Gamma_\pi$ can subsequently be estimated from the replica exchange method and used as input for one or more single- or parallel multicanonical runs.

While a number of parameters still needs to be set appropriately prior to production run, including the temperature ladder of the replica method, the simulation time(s) in the learning stage and the energy resolution for the MUCA-ensemble, the hybrid approach suggested by Mitsutake et al. is a marked improvement compared to the single-histogram method, both in terms of convenience and robustness. However, the efficiency of the scheme relies on the initial applicability of the parallel tempering approach which may become problematic for systems displaying first order type of phase transitions [42].

### 2.8.4 Wang-Landau Method

More recently, *non-Markovian* (viz. history-dependent) histogram methods designed specifically to the multicanonical ensemble, have become very popular. Among others, these methods entail the random-walk algorithm by Wang and Landau (WL) [753, 754] and *metadynamics* by Laio and Parrinello [411]. Although the WL-algorithm is typically applied in the space of energies and the meta-dynamics is applied in the space of reaction coordinates, the two methods are closely related. Both methods change the weight of the visited energy (or reaction coordinate) at regular steps in the simulation by a constant modification factor $\gamma$ to enforce a flat sampling.

In the Wang-Landau (WL) algorithm, a flat histogram is generated by modifying the estimate of the density of states $\hat{g}_\pi(E_e)$ every time a given energy bin $e$ is visited in the simulation. Since the algorithm uses a fixed bin size, the distinction between $g_\pi$ and $\Gamma_\pi$ becomes obsolete in the following. The updating rule proposed in [753, 754] is on the form

$$\hat{g}_\pi(E_e) \rightarrow \gamma \times \hat{g}_\pi(E_e),$$

where $\gamma > 1$ is a "user-supplied" modification factor. Defining the probability weights according to the multicanonical ensemble

$$\omega_{\text{MUCA}}(x) = \omega_{E,\text{MUCA}}(E)\omega_\pi(x) = g_\pi^{-1}(E)\omega_\pi(x),$$

the Metropolis-Hastings acceptance rate in the simulation becomes

$$a(x'|x) = \min\left\{1, \frac{\hat{g}_\pi(E(x))\omega_\pi(x')q(x|x')}{\hat{g}_\pi(E(x'))\omega_\pi(x)q(x'|x)}\right\}.$$

Initially, $\hat{g}_\pi(E_e) \equiv 1$ and with $\gamma$ typically set to $\gamma_0 = \exp(1)$, the simulation will visit a wide range of energies quickly [754]. The modification procedure is repeated until the accumulated histogram satisfy some prescribed criteria of flatness which is periodically checked. Typically, the histogram is considered sufficiently flat when the minimum entry is higher than $\sim$80% of the mean value. At this point, the histogram is reset and the modification factor is reduced according to the recipe, $\gamma_{k+1} = \sqrt{\gamma_k}$. Note, that this procedure implicitly guarantees a proper choice of the simulation time at iteration $k$, $T_k$. The simulation is stopped when the modification factor is less than a predefined final value $\gamma_{final}$. For continuous state-spaces it is

2 Monte Carlo Methods for Inference in High-Dimensional Systems 83

customary to also supply the algorithm with a specific energy-window $[E_{\min}, E_{\max}]$ to which the sampling is constrained.

It is important to emphasize that detailed balance, Eq. 2.22 is only ensured in the limit $\gamma \to 1$. Indeed, the error is known to scale as $\epsilon \sim \sqrt{\ln(\gamma)}$ [796]. Consequently $\ln(\gamma_{final})$ is typically set very small, $\ln(\gamma_{final}) \simeq 10^{-6} - 10^{-8}$ [413, 753]. At this point, $\hat{g}_\pi$ readily provides an accurate estimate of the true density of states.

The elegance and apparent simplicity of WL-algorithms over other methods has lead to a large variety of applications, including spin systems [753, 754], polymer models [561, 596, 640, 746], polymer films [331], fluids [725, 774], bio-molecules [217, 599] and quantum problems [727]. However, despite being generally regarded as very powerful, the algorithm suffers from a number of drawbacks and performance limitations [138, 413, 796], especially when applied to continuous systems [590]. These drawbacks stem from the difficulty of choosing simulation parameters optimally, dealing with the boundaries of the accessible conformational space, as well as from the practical discretization of the energy range [590]. The problems pertaining to the constant binning in the WL-approach has also been highlighted in recent work on the thermodynamics of Lennard-Jones clusters using *nested sampling* [562] (Sect. 2.3.2).

## 2.8.5 Metadynamics

In the metadynamics algorithm the aim is to estimate the marginalized probability distribution, $p_f(y)$ in Eq. 2.11, for a given set of collective variables $f = (f_1, \cdots, f_d)^T$. When $f$ is chosen as the total energy itself ($f$ being one-dimensional) [507], the method will provide estimates for the density of states, $g(E)$ (Sect. 2.5.3). However, most applications of the metadynamics focus on variables relevant for chemical reactions or structural transitions within fixed temperature molecular dynamics simulations [411].

As for the Wang-Landau algorithm the metadynamics (M) uses a set of biasing weights $\omega_M(y)$ to modify the original distribution $p_f(y)$, into a flat distribution, cf. Eqs. 2.56 and 2.58

$$p_M(y) = \omega_M(y)p_f(y) = \text{const.} \tag{2.61}$$

In this sense, the method can be considered as a learning algorithm for the multicanonical ensemble. Once a flat distribution has be realized, the original probability distribution $p_f(y)$ can be obtained from the weights as $p_f(y) \propto \omega_M^{-1}(y)$.

Metadynamics is more conveniently discussed in terms of the *free energy surface*, $F(y)$, defined as

$$e^{-\beta(F(y)-F)} = p_f(y), \tag{2.62}$$

where $\beta$ is the inverse temperature of the original target distribution $p(x)$ and $F$ is the total free energy, here simply serving the role as a normalization constant. Similarly, Eq. 2.61 is expressed as

$$p_M(y) \propto e^{-\beta\left(F(y)+F_M(y)\right)},$$

where $F_M$ is the potential associated with the biasing weights of the metadynamics algorithm, $\exp(-\beta F_M(y)) = \omega_M(y)$. Metadynamics achieves a flat distribution in $f$ by adding a Gaussian term to the existing biasing potential $F_M$ at regular time intervals $\tau$. For the one-dimensional case ($f \to f$), the total biasing potential at time $t$ will then be given by

$$F_M(y,t) = h \sum_{t=\tau,2\tau,3\tau,\cdots} \exp\left(-\frac{\left(y - f(x_t)\right)^2}{2\delta y^2}\right),$$

where $h$ and $\delta y$ is the height and width of the Gaussians [411, 412]. In an MD-simulation the components of the forces coming from the biasing potential will discourage the system from revisiting the same spot. The same effect will take place in an MCMC-simulation using the the acceptance probability implied by the history-dependent weights, cf. Eq. 2.55

$$a(x'|x) = \min\left\{1, \frac{\omega(x')q(x|x')}{\omega(x)q(x'|x)} \exp\left(-\beta[F_M(f(x')) - F_M(f(x))]\right)\right\}.$$

This non-Markovian approach to ensure flat sampling is in close analogy to the approach of WL. The bin-size $\Delta$ for the WL-histogram is basically equivalent to the width of the Gaussian bias, $\Delta \sim \delta y$ and the WL-modification factor $\gamma$ roughly corresponds to the Gaussian height $h$ as $\ln(\gamma) \sim \frac{\delta y \beta h}{\tau}$. This equivalence is supported by an error-analysis of the metadynamics which shows that the (dimensionless) error scales as $\epsilon \sim \sqrt{\frac{C \delta y \beta w}{\tau}}$, corresponding to $\epsilon \sim \sqrt{\ln(\gamma)}$ in WL. Here, the prefactor $C$ is a system-dependent quantity which is function of the range and dimension of the $f$-space as well as the effective diffusion constant [412].

An important difference between the WL-algorithm and the usual implementation of the metadynamics is that in the latter case no attempt is made to reconcile with the requirement of detailed balance. Consequently, the parameters for the simulation protocol set an upper limit for the accuracy of the method, implying that the target distribution is only approximately reproduced even in the limit of long simulation times. However, for many applications this seems not to be an essential limitation of the method.

An interesting hybrid between the replica-exchange method and metadynamics has recently been proposed by Piana and Laio [576] called "bias-exchange metadynamics" (BE-META). In this work, each replica $r$ performs metadynamics along an individual collective variable $f_r$. As for the replica-exchange method an extra move is introduced to increase sampling efficiency, whereby two replicas $r$ and $s$ are exchanged with a probability $a(\{x_s, x_r\}|\{x_r, x_s\})$. This probability is obtained from the standard replica-exchange move, Eq. 2.44 by replacing the Boltzmann weights $\omega_{\beta_r}$ and $\omega_{\beta_s}$ with $\omega_{M_r}$ and $\omega_{M_s}$, so

$$a(\{x_s, x_r\}|\{x_r, x_s\})$$
$$= \min \left\{ 1, \exp\left( \beta \left[ F_{M,r}(x_r) + F_{M,s}(x_s) - F_{M,s}(x_r) - F_{M,r}(x_s) \right] \right) \right\},$$

where $F_{M,i}(x_j)$ is short-hand notation for $F_{M,i}(f_i(x_j))$.

Metadynamics or BE-META has been used very successfully on several systems including protein folding and structure characterization [122, 123, 576, 649], substrate binding [579], crystal structure prediction [489], ion permeation [791], quantum problems [129] and other systems. As a method for reconstruction density of states, results from simulations on Ising systems [507] seems to suggest that metadynamics in its current form is less efficient than other existing extended ensemble algorithms [180]. The appropriate values of simulation parameters also depend on system properties and are therefore not trivial to select [412, 791].

### 2.8.6 Muninn: An Automated Method Based on Multi-histogram Equations

Muninn is a histogram based method that aims at combining the merits of the non-Markovian approaches in terms of their adaptiveness with the requirement of detailed-balance. As opposed to other GE-learning algorithms, the estimates of the statistical weights $\Gamma_\pi$ are based on a maximum likelihood approach which in principle allows one to combine all the information obtained from previous iterations. As detailed below, this approach also makes it possible to adapt the histogram bins according to the variations of $\ln(g_\pi)$ or $\ln(\omega_{E,GE})$ to ensure a uniform ensemble resolution [196]. Here, $\omega_{E,GE}$ refers to the target weights as defined by the multicanonical, $1/k$ or any other generalized ensemble.

#### 2.8.6.1 Generalized Multihistogram Equations

Let the energy space be partitioned into $L$ bins $\{\mathcal{E}_e\}_{e=1}^L$ with variable bin-sizes $\Delta E_e$ and let
$$\omega_i(x) = \omega_{E,i}(E(x))\omega_\pi(x)$$

be the weights for the $i$'th simulation, where $\omega_{E,i}$ are the GE-weights of the energies. Furthermore, assume that $M$ independent MCMC simulations have been carried out with weights $\{\omega_i\}_{i=1}^M$, leading to the set of histograms $\{n_i\}_{i=1}^M$. It follows from Eqs. 2.36 and 2.39 that the probability of visiting $\mathcal{E}_e$ in simulation $i$ within a restricted energy domain $\mathcal{S}_i$ is approximately given by

$$p_i(E \in \mathcal{E}_e | \Gamma_\pi) = \frac{\Gamma_\pi(\mathcal{E}_e)\omega_{E,i}(E_e)}{Z_i}, \quad Z_i = \sum_e \chi_i(E_e)\Gamma_\pi(\mathcal{E}_e)\omega_{E,i}(E_e),$$

where $\Gamma_\pi(\mathcal{E}_e) \approx g_\pi(E_e)\Delta E_e$ are the statistical weights associated with the reference state and $\chi_i$ is the indicator function for $\mathcal{S}_i$

$$\chi_i(E_e) = \begin{cases} 1 \text{ if } E_e \in \mathcal{S}_i \\ 0 \text{ otherwise.} \end{cases}$$

Given that the $i$'th simulation is in equilibrium within the region $\mathcal{S}_i$, the histogram $n_i$ will be a member of the multinomial probability distribution

$$p_i(n_i|\Gamma_\pi) = N_i! \prod_{E_e \in \mathcal{S}_i} \frac{p_i(E_e|\Gamma_\pi)^{n_i(E_e)}}{n_i(E_e)!},$$

where $N_i = \sum_e \chi_i(E_e)n_i(E_e)$ is the total number of counts within the region $\mathcal{S}_i$ for the $i$'th simulation. Note that $n_i$ denotes all bins of the $i$'th histogram, while $n_i(E_e)$ only denotes the $e$'th bin. The probability for observing the full set of histograms is then given by

$$p(\{n_i\}_{i=1}^M|\Gamma_\pi) = \prod_{i=1}^M p_i(n_i|\Gamma_\pi). \tag{2.63}$$

Here, we can consider $\mathcal{D} = \{n_i\}_{i=1}^M$ as the data and $\Gamma_\pi$ to be the unknown model parameters. In the Bayesian formalism, a posterior distribution of the model is then constructed from the data as

$$p(\Gamma_\pi|\mathcal{D}) \propto p(\mathcal{D}|\Gamma_\pi)p(\Gamma_\pi), \tag{2.64}$$

where $p(\mathcal{D}|\Gamma_\pi)$ represents the *likelihood*, Eq. 2.63, and $p(\Gamma_\pi)$ represents the *prior* distribution for $\Gamma_\pi$. Choosing a constant prior, the most likely statistical weight function, $\hat{\Gamma}_\pi$ will by obtained by maximizing Eq. 2.63 with respect each $\Gamma_\pi(\mathcal{E}_e)$. It can be shown [179, 180, 196] that this maximum likelihood estimator can be expressed as

$$\hat{\Gamma}_\pi(\mathcal{E}_e) = \frac{\sum_{i=1}^M n_i(E_e)}{\sum_{i=1}^M \chi_i(E_e)N_i \omega_{E,i}(E_e)\hat{Z}_i^{-1}}. \tag{2.65}$$

The partition sums $\hat{Z}_i$ must be estimated self-consistently from Eq. 2.65. This estimation problem can be formulated as the root of $M$-nonlinear equations in $\{\hat{Z}_i\}_{i=1}^M$ which in turn can be determined efficiently using the iterative Newton–Raphson method [180, 196]. This solution is then inserted into Eq. 2.65 to obtain the statistical weight estimates. While the multihistogram equations improves the estimates for the statistical weights compared the single histogram method, Eq. 2.41, these estimators are still only defined up to an arbitrary scale. Consequently, one must rely on some choice of normalization procedure as discussed in Sect. 2.5.3.

Equation 2.65 can be viewed as a generalization of the multi-histogram equations derived by Ferrenberg and Swendsen and used for combining simulations at different temperatures [703]. We shall therefore refer to Eq. 2.65 as the generalized

2 Monte Carlo Methods for Inference in High-Dimensional Systems    87

multihistogram equations (GMH). The multihistogram equations arise by identifying $M$ with the number of temperature replica, $M = R$, setting $\chi_i = 1$ and $\omega_{E,i}(E) = \exp(-\beta_i E)$

$$\hat{\Gamma}_\pi(\mathcal{E}_e) = \frac{\sum_{i=1}^R n_i(E_e)}{\sum_{i=1}^R N_i \exp(-\beta_i E_e)\hat{Z}_i^{-1}}, \quad \hat{Z}_i = \sum_e \hat{\Gamma}_\pi(\mathcal{E}_e) \exp(-\beta_i E_e). \quad (2.66)$$

In fact, the GMH-equations can also be recast into the weighted histogram analysis method (WHAM) developed by Kumar and coworkers [409] and which constitutes a powerful and popular approach for reconstructing free energy profiles along reaction coordinates from a set of equilibrium simulations. These equations arise by repeating the probabilistic reasoning above for the estimation of $p_f(y \in \mathcal{Y}_e)$ from $M$ equilibrium distributions with weights $\omega_i(x) = \omega_{f,i}(f(x))\omega(x)$, cf. Eq. 2.55:

$$\hat{p}_f(\mathcal{Y}_e) = \frac{\sum_{i=1}^M n_i(y_e)}{\sum_{i=1}^M N_i \omega_{f,i}(y_e)\hat{Z}_i^{-1}}, \quad \hat{Z}_i = \sum_e \hat{p}_f(\mathcal{Y}_e)\omega_{f,i}(y_e). \quad (2.67)$$

From a mathematical point of view the derivation of Eq. 2.65 only differs from that of the multihistogram and WHAM-equations by the use of probability arguments instead of a variance minimization technique. We shall retain the probabilistic formulation as it generalizes more naturally into a Bayesian inference framework. Note that the GMH/WHAM equations can be used straight-forwardly to multidimensional problems as well.

The reason for introducing he restricted regions, $\mathcal{S}_i$, can deduced from Eq. 2.65: due to the presence of the $\chi_i$ functions, the counts for simulation $i$ will only influence the overall estimate of the statistical weights at a given energy, if this energy belongs to $\mathcal{S}_i$. In the absence of a regularizing prior, these restrictions allows us to account for the sensitivity of the estimators associated with low statistics in an easy manner. Assuming that the histogram are sequentially ordered with the newest histogram first, the definition $\mathcal{S}_i = \{\mathcal{E}_e | \sum_{j=i}^M n_i(E_e) > \kappa\}$ will ensure that bins are only included in the support of $i$ if the accumulated statistics at this point in the iteration scheme is larger than a given cut-off value, $\kappa \simeq 20 - 30$. Muninn is insensitive to this choice provided $\kappa \gg 1$.

### 2.8.6.2 Iteration Scheme

The GMH-equations can be used to define an efficient iteration scheme to obtain the generalized ensemble weights and the statistical weights. Applying Eq. 2.65 on the statistics obtained in the last $M \gg 1$ iterations will provide a much more accurate estimate of $\Gamma_\pi$ – and thus the GE-weights for the next iteration – than the single ($M = 1$) histogram method. This makes three improvements possible. First of all, it allows the weights to be changed more adaptively (using shorter simulation runs) in the spirit of the non-Markovian approaches, without compromising the requirement of detailed balance. Such an update is facilitated by the following simple exponential scheme for the sampling times [180]

$$T_{k+1} = \begin{cases} \gamma T_k & \text{if } \mathcal{S}_k \subseteq \mathcal{S} \\ T_k & \text{otherwise} \end{cases}, \quad \text{and} \quad \mathcal{S} = \bigcup_{i=1}^{M} \mathcal{S}_i = \left\{ \mathcal{E}_e \mid \sum_{i=1}^{M} n_i(E_e) \geq \kappa \right\}.$$

Here, $T_k$ is the simulation time at iteration $k$, $\gamma > 1$ and $\mathcal{S}$ represents the domain where reliable estimates are readily available. $T_{k=1}$ can be set to any some small constant times the total number of degrees of freedom, say $T_{k=1} \simeq 100 \times D$ [180]. This schemes ensures fast weight-changes while guaranteeing that the total statistics accumulated within the last $M$ histograms scales proportionally with the total simulation time in the limit of many iterations. This proportionality constant is given by $1 - \gamma^{-M}$ [196] which implies that iteration scheme preserves $1 - 2^{-2} = 75\%$ of the statistics when default values, $\gamma = 2^{1/10}$ and $M = 20$, are used. Consequently, the total simulation time $T$ will eventually satisfy $T \gg \tau_{\mathrm{conv}}$, and the scaling of the accumulated statistics ensures that estimation errors will decay as $\epsilon \propto T^{-1/2}$ [196].

### 2.8.6.3 Resolution Scheme

The second improvement related to Eq. 2.65 is that one can obtain reliable estimates of the slope $\alpha$ of the target log-weights $\ln[\omega_{GE}]$ by using the weights $\omega_k$ obtained from the latest iteration,

$$\alpha(E_e) = \frac{d \ln[\omega_{E,GE}(E_e)]}{dE} \approx \frac{d \ln[\omega_{E,k}(E_e)]}{dE}, \quad \forall E_e \in \mathcal{S}.$$

This slope function can be applied both to define weights for unobserved energies $E \notin \mathcal{S}$ as well as to *ensure an uniform resolution of* $\ln[\omega_{E,GE}]$. Indeed, as detailed in [196] the bin-width in an unobserved region, $\mathcal{E}$, can be defined as $\Delta E = \frac{r}{|\alpha(E_e)|}$, where the slope is evaluated at the energy $E_e \in \mathcal{S}$ closest to $\mathcal{E}$ and where $r$ represents the chosen resolution. This definition ensures that the change of log-weights across one bin is approximately constant *for all energies*

$$\left| \ln[\omega_{E,GE}(E_e)] - \ln[\omega_{E,GE}(E_{e+1})] \right| \simeq r, \quad \forall E_e \in \mathcal{S}.$$

In other words, setting $r \ll 1$ the binning procedure will automatically ensure that the bin-widths for the histogram technique are chosen appropriately according to the variation of the target weights $\omega_{E,GE}$, as required by Eq. 2.32. This procedure alleviates the deficiencies associated with a constant binning [562, 590], an advantage that Muninn shares with the nested sampling approach [562] (Sect. 2.3.2).

Finally, the "running" estimates of $\alpha(E_e)$ allows one to restrict the generalized ensemble sample to a particular temperature window of interest [196], rather than to a particular energy window of interest. For continuous systems it is most often

2 Monte Carlo Methods for Inference in High-Dimensional Systems

easier to express the region of state space of relevance in terms of the temperatures rather than the energies.

#### 2.8.6.4 Summary

In summary, we have discussed how the main difficulties pertaining to the learning stage of generalized ensembles, as outlined in Sect. 2.8.1, are addressed in Muninn. This includes estimation technique, simulation time, binning procedure and assignment of weights to unobserved regions. The original algorithm was benchmarked on spin-systems, where a marked improvement of the accuracy was demonstrated compared to existing techniques, including transition-based methods and the WL-algorithm, particularly for larger systems [180]. The method has subsequently been applied to folding and design of simplified protein models [6–8, 276, 726], all-atom protein and RNA structure determination from probabilistic models [69, 70, 197] as well as from data obtained by small angle X-ray scattering [693] and nuclear magnetic resonance [552]. A recent study clearly demonstrates that the method also compares favorably to the WL-algorithm on protein folding studies and other hard optimization problems [196], where order(s) of magnitude faster convergence is observed.

While Muninn has not been applied for sampling along reaction coordinates, we would expect the algorithm to be equally efficient for these problems due to the statistical equivalence between estimating $\Gamma_\pi$ and $p_f(y)$. The code is freely available at http://muninn.sourceforge.net.

## 2.9 Inference in Extended Ensembles

We shall conclude this section by returning to the three main inference problems outlined in Sects. 2.2.2 and 2.5, namely how to estimate expectation values, marginal density distributions and partition functions from the sampling.

The basic premise of extended ensembles is that the sampling along the axis of extension satisfies the dual purpose of enhancing the mixing of the Markov chain as well as allowing the calculation of particular multivariate integrals of interest. Assuming that the appropriate parameters of the extended ensembles have been learned, the inference problems outlined above can be solved based on sufficient samples, $T \gg \tau_{\text{conv}}$, from the equilibrium distribution.

### 2.9.1 Marginal Distributions

Under these above-mentioned premises, the marginal distribution, Eq. 2.11, is most efficiently estimated using a generalized ensemble/flat-histogram sampling along the coordinate $y = f(x)$. The metadynamics algorithm has been designed especially for this purpose and has been shown to be applicable also for higher-variate distributions [576]. Other generalized-ensemble algorithms including those discussed in Sect. 2.8 could in principle be applied for this problem as well, although it is still unclear how these various algorithms compare to each other in terms of accuracy, generality and efficiency. As discussed in Sect. 2.7.5, tempering based methods are not suited for this task.

### 2.9.2 Partition Functions and Thermodynamics

The calculation of partition function ratios, $Z/Z_\pi$ can be carried out choosing $E$ as the axis of extension, where $E$ either refers to the physical energy function or to Eq. 2.9 for non-thermal problems. The energy extension is required in cases where $Z/Z_\pi$ is not of the order of unity, otherwise the standard MCMC-approach will suffice [34]. More precisely, the estimation requires that the range of observed energies $\mathcal{E}$ covers the bulk part of the probability mass for both $p(x)$ and $\pi(x)$. The ratio of the partition functions can generally be obtained from the estimates of the statistical weights $\Gamma_\pi(\mathcal{E}_e)$, where $\{\mathcal{E}_e\}_{e=1}^L$ is some fine-grained partitioning of $\mathcal{E}$. In the parallel- or simulated tempering approach, $\Gamma_\pi$ can be found from the multi-histogram equations (see Eq. 2.66) [703]. In the generalized ensemble approach, $\Gamma_\pi$ is given by the particular learning algorithm employed, as discussed in Sect. 2.8. While $\Gamma_\pi$ can only be estimated up to a constant factor, combining Eqs. 2.42 and 2.40 will provide a unique estimate of the ratio $Z/Z_\pi$:

$$\frac{\hat{Z}}{\hat{Z}_\pi} = \frac{\sum_e \omega_E(E_e)\hat{\Gamma}_\pi(\mathcal{E}_e)}{\sum_e \hat{\Gamma}_\pi(\mathcal{E}_e)}. \tag{2.68}$$

The weights, $\omega_E$, for the target distribution are defined from its associated inverse temperature $\omega_E(E) = \exp(-\beta E)$, where $\beta = 1$ in the non-thermal case as discussed in Sect. 2.2. As shown by Eq. 2.36 it now becomes straightforward to calculate the marginalized probability distribution over energies for *any* inverse temperature $\tilde{\beta}$, by reweighting

$$p_{\tilde{\beta}}(\mathcal{E}_e) = \frac{\Gamma_\pi(\mathcal{E}_e)\exp(-\tilde{\beta}E_e)}{Z_{\tilde{\beta}}}, \tag{2.69}$$

where

$$Z_{\tilde{\beta}} = \sum_e \Gamma_\pi(\mathcal{E}_e)\exp(-\tilde{\beta}E_e). \tag{2.70}$$

The only two prerequisites for this reweighting scheme is that the variation of the probability weights $\omega_E(E) = \exp(-\tilde{\beta}E)$ is small within each energy-bin $\mathcal{E}_e$, and that the bulk of the probability mass of $p_{\tilde{\beta}}$ is within the observed energy region $\mathcal{E}$.

In the tempering based methods, the latter requirement implies that $\beta_R \leq \tilde{\beta} \leq \beta_0$, where $\beta_R$ and $\beta_0$ are the highest and lowest value of the range of $\beta$-values used in the simulation. Note that setting $\tilde{\beta} = 0$ leads to the reference distribution and $\tilde{\beta} = \beta$ leads to the target distribution.

Equations 2.69 and 2.70 are particular convenient for calculating thermodynamic quantities at any temperature $T = (k_B \beta)^{-1}$, where $\beta$ in the following is considered as variable ($\tilde{\beta} \rightarrow \beta$). First, the expectation value of any moment of the energies, $E^m$, as function of $\beta$ can be obtained from Eq. 2.69

$$\mathbb{E}_\beta[E^m] = \sum_e E_e^m p_\beta(\mathcal{E}_e). \tag{2.71}$$

Consequently, the *heat capacity* $C(\beta)$ is directly calculable

$$C(\beta) \doteq \frac{d\mathbb{E}_\beta[E]}{dT} = -k\beta^2 \frac{d\mathbb{E}_\beta[E]}{d\beta}$$

$$= k\beta^2 \left( \mathbb{E}_\beta[E^2] - \mathbb{E}_\beta[E]^2 \right). \tag{2.72}$$

Furthermore, *Helmholtz free energy* $F(\beta)$ (see Table 2.1) and the *thermodynamic entropy* $S(\beta)$ can be evaluated from the knowledge of the partition function and by using the relation $F = \mathbb{E}_\beta[E] - TS$

$$\beta F(\beta) = -\ln(Z_\beta), \tag{2.73}$$

$$S(\beta) = k\beta \left( \mathbb{E}_\beta[E] - F(\beta) \right). \tag{2.74}$$

### 2.9.3 Expectation Values and Microcanonical Averages

The final question to address is how to estimate expectation values $\mathbb{E}_p[f(x)] = \mathbb{E}_\beta[f(x)]$ from an extended ensemble simulation. Irrespective of the type of extension, $\mathbb{E}_p[f(x)]$ can be reconstructed by calculating the average of $f$ for each bin of the weight-scheme. In the energy-based extension this average is given by

$$\mathbb{E}_{p(\cdot|\mathcal{E}_e)}[f(x)] = \Gamma_\pi(\mathcal{E}_e)^{-1} \int_\Omega \chi_{\mathcal{E}_e}(E(x)) f(x) \omega_\pi(x) dx, \tag{2.75}$$

where $\chi_{\mathcal{E}_e}$ is the indicator function on $\mathcal{E}_e$. Equation 2.75, which in a physical context represents the *microcanonical average*, can be estimated from the sampled states $\{x_t\}_t$, as

$$\hat{\mathbb{E}}_{p(\cdot|\mathcal{E}_e)}[f(x)] = n(E_e)^{-1} \sum_t \chi_{\mathcal{E}_e}(E(x_t)) f(x_t), \tag{2.76}$$

where

$$n(E_e) = \sum_t \chi_{\mathcal{E}_e}\big(E(\boldsymbol{x}_t)\big) \tag{2.77}$$

is the total number of observed states belonging to $\mathcal{E}_e$. Since $p(\boldsymbol{x}) = p(\boldsymbol{x}|\mathcal{E}_e)p(\mathcal{E}_e)$ the expectation value of $f$ with respect to $p(\boldsymbol{x})$ can be expressed as

$$\mathbb{E}_x[f(\boldsymbol{x})] = \sum_e \mathbb{E}_{p(\cdot|\mathcal{E}_e)}[f(\boldsymbol{x})]p(\mathcal{E}_e) = \frac{1}{Z_\beta}\sum_e \mathbb{E}_{p(\cdot|\mathcal{E}_e)}[f(\boldsymbol{x})]\Gamma_\pi(\mathcal{E}_e)\exp(-\beta E_e). \tag{2.78}$$

Consequently, an estimate of $\mathbb{E}_{p(x)}[f(\boldsymbol{x})]$ is obtained from $\hat{\Gamma}_\pi$ and Eq. 2.76.

## 2.10 Summary and Discussion

In this chapter we have focused on the MCMC-method in general and the extended ensemble approach in particular as a tool for inference in high-dimensional model systems, described by some given (target) probability distribution $p(\boldsymbol{x})$. By introducing the general notion of a reference state (Sect. 2.2.1) we have aimed at presenting the use of these methods for inference in Bayesian models and thermal systems at a unified level. This reference state $\pi(\boldsymbol{x})$ can be either an alternative Bayesian model describing the same data, the prior distribution for a single Bayesian model or simply the geometrical measure of the manifold $\Omega$, usually assumed to be Euclidian ($\omega_\pi = 1$). When the "energy" of a state is defined according to Eq. 2.9 for non-thermal models, the reference state will in all cases be associated with the inverse temperature $\tilde{\beta} = 0$ and the target distribution with the inverse temperature $\tilde{\beta} = \beta$. Using energy as the axis of extension the extended ensemble approach facilitates a sampling which smoothly interpolates from $\tilde{\beta} = 0$ to $\tilde{\beta} = \beta$ while at the same time improving the mixing of the Markov chain (Sect. 2.7). This construction allows the calculation of partition function ratios used for evidence estimates, model averaging or thermodynamic potentials (Sect. 2.9), quantities which are usually not tractable by the standard MCMC-procedure (Sects. 2.5 and 2.6). The proper sample weights for the target distribution $p$ and the estimation of expectation values becomes a simple postprocessing step (Sect. 2.9).

One noticeable drawback of the extended ensemble approach is the use of a number of parameters which needs to be *learned* before reliable inference can be made from the sampling. The popularity of the parallel tempering/replica exchange ensemble derives from the fact that this step is less involved than in the generalized ensemble approaches (Sect. 2.8). On the other hand, the Wang-Landau and metadynamics algorithm (Sects. 2.8.4 and 2.8.5) used for the multicanonical ensemble constitute a significant step forward with respect to the automatization of this parameter learning, an advantage we believe Muninn shares with these non-Markovian approaches.

The nested sampling MC-method proposed by Skilling [666, 669] (Sect. 2.3.2) provides an interesting alternative to extended ensembles, because it only involves

## 2 Monte Carlo Methods for Inference in High-Dimensional Systems

one essential parameter; the population size $K$. However, since nested sampling uses a sequence of strictly decreasing energies the ergodicity or accuracy of the sampling has to be ensured by a sufficiently large choice of $K$. In this respect it is very different from the extended ensemble approach where ergodicity is established by "seamless" transitions up and down the energy/temperature axis. How these two different approaches generally compare awaits further studies.

As a conclusive remark, it is an interesting curiosity that while Monte Carlo methods continues to play a central role in Bayesian modeling and inference [62], the method itself has never been subject to a full Bayesian treatment [272]. Rather than providing beliefs about quantities from the computations performed, MC-algorithms always lead to "frequentist" statistical estimators [272, 530, 598]. Rasmussen and Ghahramani have proposed a Bayesian MC methodology for calculating expectation values and partition function ratios using Gaussian Processes as functional priors. This approach leads to superior performance compared to standard MC-estimators for small sample sizes [598]. However, in its current form the method does not extend easily to large samples for computational reasons [598]. Skilling has argued that the nested sampling method is inherently Bayesian, because the reparameterization of the partition function, Eq. 2.54, imposes a definitive functional form on the estimated values $\hat{Z}$ during sampling which in turn induces an unambiguous distribution for the estimates $p(\hat{Z})$ [669]. One may tentatively argue that a general Bayesian treatment should include a *prior* distribution over $Z$ or – alternatively – over the statistical weights $p(\Gamma_\pi)$. For a further discussion on the Bayesian aspects of the nested sampling, we refer to Murrays dissertation [530].

One natural entrance point to a Bayesian Monte Carlo methodology in the context of extended ensembles is given by the maximum-likelihood approach used in Muninn. Indeed, Eq. 2.64 is directly amenable to a full Bayesian treatment, where inference of the statistical weights are based on the posterior distribution $p(\Gamma_\pi | \mathcal{D})$, rather than on the likelihood function $p(\mathcal{D} | \Gamma_\pi)$. In practice, this requires the formulation of a suitable mathematical framework, where general smoothness, scaling or other regularizing properties of $\Gamma_\pi$ can be incorporated in the prior $p(\Gamma_\pi)$ while at same time allowing the posterior knowledge to be summarized in a tractable manner.

# Part II
# Energy Functions for Protein Structure Prediction

# Chapter 3
# On the Physical Relevance and Statistical Interpretation of Knowledge-Based Potentials

**Mikael Borg, Thomas Hamelryck, and Jesper Ferkinghoff-Borg**

## 3.1 Introduction

Most approaches in protein structure prediction rely on some kind of energy or scoring function in order to single out the native structure from a set of candidate structures, or for use in Monte Carlo simulations. In some cases, the energy function is based on the underlying physics, including electrostatic and van der Waals interactions. However, an accurate physics-based description would require quantum mechanical calculations which are computationally too demanding for large macromolecules like proteins. Approximate force fields used in molecular dynamics simulations are typically tuned to reproduce the results from quantum chemical calculations on small systems [354, 458, 459]. Furthermore, as physics-based force fields are approximations of potential energies, they can be used to calculate thermodynamic quantities.

Another approach is to construct an approximation of the free energy using the information from experimentally determined protein structures. The resulting energy functions are often called statistical potentials or *knowledge-based potentials* (KBPs).

The concept of knowledge based potentials has a long history. For example, Pohl derived torsional angle potentials and Tanaka and Scheraga extracted inter-action parameters from experimentally determined protein structures in the 1970s [587, 706]. Other early efforts include work by Warme and Morgan [758, 759] and by Lifson and Levitt [441]. Currently, a common knowledge-based strategy

---

M. Borg (✉) · T. Hamelryck
The Bioinformatics Centre, University of Copenhagen, Copenhagen, Denmark
e-mail: borg@binf.ku.dk; thamelry@binf.ku.dk

J. Ferkinghoff-Borg
Department of Electrical Engineering, Technical University of Denmark, Lyngby, Denmark
e-mail: jfb@elektro.dtu.dk

T. Hamelryck et al. (eds.), *Bayesian Methods in Structural Bioinformatics*,
Statistics for Biology and Health, DOI 10.1007/978-3-642-27225-7_3,
© Springer-Verlag Berlin Heidelberg 2012

is to optimize a function that discriminates native protein structures from decoy structures [452,512,719]. Sippl pioneered the use of the knowledge-based approach to construct energy functions based on pairwise distances [661]. He was also the first to point out analogies between KBPs and the *potential of mean force* in liquid systems [390,665].

Today, knowledge-based potentials are for example commonly used for detecting errors in experimentally determined protein structures [663], computational studies of protein interactions [47, 115, 234, 346, 419] and protein engineering [65, 617].

In spite of their widespread use, KBPs lack a rigorous statistical foundation which makes it difficult to determine in what situations and for what purposes they are applicable. While KBPs are often viewed as approximations to physical *free energies* or potentials of mean force (PMFs), they do not readily have a thermodynamic interpretation. In particular, the formal resemblance between KBPs and PMFs only arises from the common use of a *reference state*, the definition of which, however, has until now been quite elusive in the knowledge-based context [32, 390, 715].

There are several excellent reviews on the subject [17, 88, 232, 383, 390, 423, 438, 541, 589, 670, 798]. Here, we focus on the theoretical background and on the importance of the reference state when deriving these type of potentials. In particular, we discuss a recently introduced probabilistic explanation of KBPs that sheds a surprising new light on the role and definition of the reference state [276].

Since KBPs are often presented using terminology borrowed from statistical mechanics, we will discuss some basic concepts from this field in Sect. 3.2. In particular we will give an introduction to thermodynamic free energies in Sect. 3.2.1 and the concept of the potential of mean force in Sect. 3.2.2. As the two concepts are often used interchangeably in the field, their differences are clarified in Sect. 3.2.3, the content of which also lays the ground for how KBPs are constructed in general.

Section 3.3 focuses on the use of knowledge based potentials for proteins. Since the construction of a KBP depends on both the chosen parameterization of protein structures as well as the coarse-grained structural properties extracted from these we shall shortly review these two aspects in Sects. 3.3.1 and 3.3.2. Section 3.3.3 discusses how KBPs are constructed from the observed statistics of these properties. As elucidated in Sect. 3.3.4, all these potentials involve a definition of the reference state which refers to some hypothetical state of the protein system. In Sect. 3.3.6 we present various approaches to ensure that the derived KBPs are self-consistent. Self-consistent means that protein conformations obtained using a force field faithfully reproduce the statistics used in constructing that force field. We conclude Sect. 3.3 by analyzing to which extent KBPs can be related to physical free energies at all.

In Sect. 3.4 we discuss the definition of KBPs in a probabilistic framework. A first, purely qualitative explanation was given by Simons et al. [658, 659]. In addition to being qualitative, this explanation also relies on the incorrect assumption of *pairwise decomposability*. This assumption implies that the joint probability distribution of all pairwise distances in a protein is equal to the product of the marginal probabilities of the individual distances.

3 On the Physical Relevance and Statistical Interpretation 99

In the concluding section, we present our recent work that uses simple Bayesian reasoning to provide a self-consistent, quantitative and rigorous definition of KBPs and the reference state, without introducing unwarranted assumptions such as pairwise decomposability. Our work also elucidates how a proper definition of the reference state relates to the intended application of the KBP, and extends the scope of KBPs beyond pairwise distances. As summarized in Sect. 3.5, KBPs are in general not related to physical potentials in any ordinary sense. However, this does not preclude the possibility of defining and using them in a statistically rigorous manner.

## 3.2 Theoretical Background

In this section a brief introduction to statistical physics and liquid state theory is given, as most KBPs for protein structure prediction are based on concepts from these areas. We summarize the theoretical background, and introduce the two main concepts, the *free energy* and the *potential of mean force*, and show how they can be related to the probability that a system adopts a specific configuration. The main results are given by Eq. 3.9 relating physical free energies to the entropy and average potential energy of a thermal system; Eq. 3.18 defining the potential of mean force for liquid systems and Eq. 3.33 relating free energy differences to probability ratios. The latter expression forms the basis for the construction of all KBPs.

### 3.2.1 Free Energy and the Boltzmann Distribution

A sample of protein molecules in thermal equilibrium can be viewed as a canonical ensemble of individual molecules, provided that the sample is dilute enough for the molecules to be non-interacting. Each protein molecule can be considered an isolated member of the ensemble that make up the sample. Thus, the protein molecules are distributed over different states or conformations according to the *Boltzmann distribution*. The Boltzmann distribution can be derived as the most probable distribution of states under the assumption that all degenerate conformations (conformations with the same energy) are equally probable, with the constraint that the total energy of the system is constant.[1] The probability that a molecule with $N$ atoms adopts a specific conformation $\mathbf{r} = (\mathbf{r}_1, \mathbf{r}_2, \ldots \mathbf{r}_N)$ and conjugate momenta $\mathbf{p} = (\mathbf{p}_1, \mathbf{p}_2, \ldots \mathbf{p}_N)$ is given by

$$p_\beta(\mathbf{r}, \mathbf{p}) d\mathbf{r} d\mathbf{p} = \frac{e^{-\beta H(\mathbf{r}, \mathbf{p})}}{Z_{rp}} d\mathbf{r} d\mathbf{p} \tag{3.1}$$

---

[1] See [33] for a good presentation of Boltzmann statistics and the relation to information theory.

where

$$H(\mathbf{r}, \mathbf{p}) = E(\mathbf{r}) + \sum_i \frac{\mathbf{p}_i^2}{2m_i} \tag{3.2}$$

is the Hamiltonian, $E(\mathbf{r})$ is the conformational energy, $m_i$ is the mass of the $i$th atom and $\beta$ is the inverse of the temperature of the system times the Boltzmann constant, $\beta = (k_B T)^{-1}$. The partition function, $Z_{rp}$ is obtained by integrating over all possible positions and momenta using the semi-classical measure [414]

$$\frac{d\mathbf{r}d\mathbf{p}}{h^{3N}} = \prod_i \frac{d\mathbf{r}_i d\mathbf{p}_i}{h^3} \tag{3.3}$$

resulting in

$$Z_{rp} = \int e^{-\beta H(\mathbf{r},\mathbf{p})} \prod_i \frac{d\mathbf{r}_i d\mathbf{p}_i}{h^3}. \tag{3.4}$$

Here, $h$ is the Planck constant. Since this integral factorizes we can write

$$Z_{rp} = Z_r \times Z_p, \quad \text{and}$$

$$p_\beta(\mathbf{r}, \mathbf{p}) = p_\beta(\mathbf{r}) p_\beta(\mathbf{p}) = \frac{e^{-\beta E(\mathbf{r})}}{Z_r} \frac{e^{-\beta \sum_i \frac{\mathbf{p}_i^2}{2m_i}}}{Z_p}, \tag{3.5}$$

where,

$$Z_p = \int e^{-\beta \sum_i \frac{\mathbf{p}_i^2}{2m_i}} \prod_i \frac{d\mathbf{p}_i}{h^3} = \prod_i \frac{(2\pi m_i k_b T)^{3/2}}{h^3} \quad \text{and}$$

$$Z_r = \int_\Omega e^{-\beta E(\mathbf{r})} d\mathbf{r}. \tag{3.6}$$

$\Omega$ represents the full configuration space. In other words, $Z_p$ is independent of $\mathbf{r}$, when the configuration space is parameterized using atomic positions. In the following we set $Z = Z_r$ for notational convenience.

The thermodynamic potential associated with $Z$ is the Helmholtz free energy

$$F = -k_B T \ln Z. \tag{3.7}$$

This potential is a scalar function which represents the thermodynamic state of the system, as further discussed in Chap. 2. The entropy of the ensemble is defined as

$$S \equiv -k_B \mathbb{E}_\beta[\ln p\,(\mathbf{r})] = -k_B \int_\Omega \frac{e^{-\beta E(\mathbf{r})}}{Z} \left(-\beta E(\mathbf{r}) - \ln Z\right) d\mathbf{r}$$

$$= T^{-1} \left(\mathbb{E}_\beta[E] - F\right), \tag{3.8}$$

where $\mathbb{E}_\beta[\cdot]$ represents the expectation value of a quantity with respect to the configurational Boltzmann distribution. Consequently, the free energy can be viewed as a sum of contributions from the average energy and entropy,

$$F = \mathbb{E}_\beta[E] - TS. \tag{3.9}$$

In the widely accepted Anfinsen's hypothesis [11, 12], a protein sample under folding conditions is assumed to reach thermal equilibrium, and the state with highest probability is the folded state, or *native* state. Protein structure prediction can thus be viewed as a search for the state with the highest probability. Free energies are in general difficult to compute, as information of the entire distribution of states is required. The development of new methods for calculating or estimating free energies from Monte Carlo or molecular dynamics simulations is an active field of research [386, 388, 728].

### 3.2.2 Potential of Mean Force in Liquids

The potential of mean force was introduced in the twentieth century in theoretical studies of fluids [376]. Here, we follow the presentation given by McQuarrie [496].

We consider a system consisting of $N$ particles that can be considered identical, for example atoms or rigid molecules, in a fixed volume $V$. As in the previous section, we omit the momenta of the particles (we assume that they can be integrated out), and represent a state as a point $\mathbf{r}$ in the $3N$-dimensional space that is defined by the $N$ atomic positions $\mathbf{r} = \{\mathbf{r}_i\}$. With $d\mathbf{r} = d\mathbf{r}_1 d\mathbf{r}_2 \ldots d\mathbf{r}_N$ we can write the probability of finding the system in a specific conformation as

$$p_\beta(\mathbf{r}_1, \mathbf{r}_2, \ldots \mathbf{r}_N) \, d\mathbf{r}_1 d\mathbf{r}_2 \ldots d\mathbf{r}_N = \frac{e^{-\beta E(\mathbf{r}_1, \mathbf{r}_2, \ldots \mathbf{r}_N)} d\mathbf{r}_1 d\mathbf{r}_2 \ldots d\mathbf{r}_N}{Z}. \tag{3.10}$$

The *n-particle distribution function*, $\rho_{(n)}$ describes the particle density over a subset of the particles, and can be obtained by integrating over $\mathbf{r}_{n+1} \ldots \mathbf{r}_N$,

$$\rho_{(n)}(\mathbf{r}_1, \mathbf{r}_2, \ldots \mathbf{r}_n) = \frac{N!}{(N-n)!} \frac{\int \ldots \int e^{-\beta E(\mathbf{r}_1, \mathbf{r}_2, \ldots \mathbf{r}_N)} d\mathbf{r}_{n+1} d\mathbf{r}_{n+2} \ldots d\mathbf{r}_N}{Z}, \tag{3.11}$$

where the factor in front accounts for the combinatorics of selecting the $n$ particles. For non-interacting particles, $E \equiv 0$, we have $Z = V^N$, and with $n \ll N$ we obtain

$$\tilde{\rho}_{(n)}(\mathbf{r}_1, \mathbf{r}_2, \ldots \mathbf{r}_n) = \frac{N!}{(N-n)!} \frac{V^{N-n}}{V^N} \approx \frac{N^n}{V^n} = \rho^n, \quad E \equiv 0 \tag{3.12}$$

where $\tilde{\rho}_{(n)}$ refers to the interaction-free distribution and $\rho = N/V$ is the average density of the system. The *n-body correlation function*, $g_{(n)}(\mathbf{r}_1, \mathbf{r}_2, \ldots \mathbf{r}_n)$, is now obtained as the ratio of $\rho_{(n)}$ to $\tilde{\rho}_{(n)}$

$$g_{(n)}(\mathbf{r}_1, \mathbf{r}_2, \ldots \mathbf{r}_n) = \frac{\rho_{(n)}(\mathbf{r}_1, \mathbf{r}_2, \ldots \mathbf{r}_n)}{\tilde{\rho}_{(n)}(\mathbf{r}_1, \mathbf{r}_2, \ldots \mathbf{r}_n)} = \frac{\rho_{(n)}(\mathbf{r}_1, \mathbf{r}_2, \ldots \mathbf{r}_n)}{\rho^n} \quad (3.13)$$

The correlation function describes how the particles in the system are redistributed due to interactions between them. Using Eq. 3.11, the correlation function can be written

$$g_{(n)}(\mathbf{r}_1, \mathbf{r}_2, \ldots \mathbf{r}_n) = \frac{V^n}{N^n} \frac{N!}{(N-n)!} \frac{\int \ldots \int e^{-\beta E(\mathbf{r}_1, \mathbf{r}_2, \ldots \mathbf{r}_N)} d\mathbf{r}_{n+1} d\mathbf{r}_{n+2} \ldots d\mathbf{r}_N}{Z}. \quad (3.14)$$

The two-body correlation function $g_{(2)}(\mathbf{r}_1, \mathbf{r}_2)$ is of particular importance in liquid state physics. In a liquid consisting of spherically symmetric molecules, the two-body correlation function only depends on the pairwise distances $r_{ij} = |\mathbf{r}_i - \mathbf{r}_j|$. We omit the subscript and write the two-body radial distribution function as $g(r)$. For a liquid with density $\rho$, the probability of finding a particle at distance $r$ provided there is a particle at the origin is given by $\rho g(r) dr$. Note that the integral of this probability is

$$\int \rho g(r) 4\pi r^2 dr = N - 1 \approx N. \quad (3.15)$$

At large distances, the positions are uncorrelated, and thus $\lim_{r \to \infty} g(r) = 1$, whereas at small $r$, they are modulated by the inter-particle interactions. In Fig. 3.1 the radial distribution function is shown for a model liquid of hard spheres with diameter $\sigma$ and with intermolecular potential $\Phi(r)$ [377]

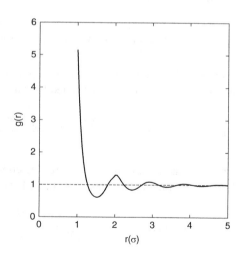

**Fig. 3.1** Radial distribution function for a liquid of hard spheres with diameter $\sigma$ and density $\rho = 0.9$ spheres per unit volume

# 3 On the Physical Relevance and Statistical Interpretation

$$\Phi\left(r\right) = \begin{cases} \infty & r < \sigma \\ 0 & r \geq \sigma \end{cases}. \tag{3.16}$$

It is not possible to calculate the radial distribution function analytically even for this simple model system, but there are approximate solutions that yield results that are very close to results from computer simulations [789]. As seen in Fig. 3.1, the excluded volume of the spheres results in an increased density at close separation. The peak at $r = 2\sigma$ corresponds to the second coordination shell. For distances $r < \sigma$, the pair distribution function is identically zero.

In practice, the radial distribution function can be determined experimentally using X-ray or neutron diffraction [107]. The radial distribution function obtained from diffraction experiments is a superposition of the distributions from different atomic species in the sample, which limits this approach to studies of simple liquids.

Turning back to our $n$-body description of the system, we define a quantity $w_{(n)}\left(\mathbf{r}_1, \mathbf{r}_2, \ldots \mathbf{r}_n\right)$ as

$$e^{-\beta w_{(n)}\left(\mathbf{r}_1, \mathbf{r}_2, \ldots \mathbf{r}_n\right)} \equiv g_{(n)}\left(\mathbf{r}_1, \mathbf{r}_2, \ldots \mathbf{r}_n\right). \tag{3.17}$$

Using Eq. 3.14, the gradient of $w_{(n)}\left(\mathbf{r}_1, \mathbf{r}_2, \ldots \mathbf{r}_n\right)$ with respect to the position of one of the molecules $j \in [1, n]$ can be expressed as

$$\nabla_j w_{(n)}\left(\mathbf{r}_1, \mathbf{r}_2, \ldots \mathbf{r}_n\right) = -\beta \frac{\int e^{-\beta E} \nabla_j E \, \mathrm{d}\mathbf{r}_{n+1} \mathrm{d}\mathbf{r}_{n+2} \ldots \mathrm{d}\mathbf{r}_N}{\int e^{-\beta E} \mathrm{d}\mathbf{r}_{n+1} \mathrm{d}\mathbf{r}_{n+2} \ldots \mathrm{d}\mathbf{r}_N}$$

$$= -\beta \mathbb{E}_{\beta,n}\left[\nabla_j E_N\right]\left(\mathbf{r}_1, \mathbf{r}_2, \ldots \mathbf{r}_n\right), \tag{3.18}$$

where $\mathbb{E}_{\beta,n}$ is the expectation value of a quantity for a fixed set of molecule positions $\mathbf{r}_1, \ldots, \mathbf{r}_n$. In other words, the gradient of $w_{(n)}\left(\mathbf{r}_1, \mathbf{r}_2, \ldots \mathbf{r}_n\right)$ with respect to one particle equals the expectation value of the gradient of the energy over the remaining $N - n$ particles. This corresponds to the force acting on that particle when the remaining particles are distributed according to the canonical distribution. Therefore, $w_{(n)}$ is called the potential for the mean force.

## 3.2.3 Free Energies for General Parameterizations

While Eq. 3.10 is useful in Monte Carlo simulations, it can not be 'inverted'; it is not possible to obtain the detailed energy potential, $E(\mathbf{r})$, by estimating $p(\mathbf{r})$ from observed frequencies, due to the high-dimensional nature of the configuration space $\Omega$. This inversion scheme can only be carried out by studying the probability of the system in some low dimensional subspace, as function of a few descriptors or coarse-grained variables, $\boldsymbol{\lambda}$.

The choice of descriptor could for example be the pairwise distances between atoms, the torsional angles or radius of gyration of a protein molecule (see also

Sect. 3.3.2). It is defined as a function of the positions of the $N$ atoms for a state, e.g.,

$$\lambda = \lambda \left( \mathbf{r}_1, \mathbf{r}_2, ..., \mathbf{r}_N \right), \tag{3.19}$$

and corresponds to a hypersurface in conformational space. In the general case, $\lambda$ may represent a $d$-dimensional function $\lambda = (\lambda_1, \cdots, \lambda_d)$, so the set

$$\Omega(\lambda_0) = \{\mathbf{r} \mid \lambda(\mathbf{r}) = \lambda_0\} \tag{3.20}$$

represents a $3N - d$ dimensional manifold in $\Omega$. The probability distribution as a function of $\lambda$ is the projection of probabilities onto this hypersurface. It can be obtained using the Dirac delta function,

$$p_\beta(\lambda_0) = \int \delta(\lambda_0 - \lambda(\mathbf{r})) \, p_\beta(\mathbf{r}, \mathbf{p}) \, d\mathbf{r} d\mathbf{p} = \int \delta(\lambda_0 - \lambda(\mathbf{r})) \frac{e^{-\beta H(\mathbf{r}, \mathbf{p})}}{Z_{rp}} d\mathbf{r} d\mathbf{p} \tag{3.21}$$

where the integral is over the space of all conformations and momenta. In order to proceed, we need to analyze the metric implied by the Dirac delta function in this equation. Let $\mathbf{q} = (\mathbf{s}, \lambda) = (s_1, s_2, \cdots, s_{3N-d}, \lambda_1, \lambda_2, \cdots, \lambda_d)$ be a re-parameterization of $\Omega(\lambda_0)$ so that for all $\mathbf{s}$ and $\lambda_0$

$$\mathbf{r}(\mathbf{s}, \lambda_0) \in \Omega(\lambda_0). \tag{3.22}$$

The conjugate momenta $\mathbf{p}_q$ of $\mathbf{q}$ are defined as $(p_q)_i = \sum_{j=1}^{3N} g_{ij} \dot{q}_j$, where $\dot{q}_j$ is the time-derivative and $g$ is the *mass metric tensor* [198] given by

$$g_{ij} = \sum_{k=1}^{N} m_k \frac{\partial \mathbf{r}_k(q)}{\partial q_i} \cdot \frac{\partial \mathbf{r}_k(q)}{\partial q_j}, \tag{3.23}$$

The point of introducing conjugate variables is that this construction ensures invariance of phase space volume. In other words

$$p_\beta(\mathbf{r}, \mathbf{p}) d\mathbf{r} d\mathbf{p} = p_\beta(\mathbf{q}, \mathbf{p}_q) d\mathbf{q} d\mathbf{p}_q \tag{3.24}$$

Therefore

$$p_\beta(\mathbf{q}, \mathbf{p}_q) = \frac{e^{-\beta H(\mathbf{q}, \mathbf{p}_q)}}{Z_{rp}} = \frac{e^{-\beta E(\mathbf{q}) - 1/2\beta \sum_{ij} (p_q)_i g_{ij}^{-1} (p_q)_j}}{Z_{rp}} \tag{3.25}$$

The last expression on the right hand side is obtained by expressing the linear momenta $\mathbf{p}_i = m_i \dot{\mathbf{r}}_i$ appearing in the original Hamiltonian, Eq. 3.1, as a function of the time derivative of the generalized coordinate $\dot{q}_j$. From here we obtain

# 3 On the Physical Relevance and Statistical Interpretation

$$p_\beta(\lambda_0) = \int p_\beta(\mathbf{s}, \lambda_0, \mathbf{p}_q) \frac{d\mathbf{s} d\mathbf{p}_q}{h^{3N}} = Z_{rp}^{-1} \left( \frac{\sqrt{2\pi k_b T}}{h} \right)^{3N} \int e^{-\beta E(\mathbf{s}, \lambda_0)} J(\mathbf{s}, \lambda_0) d\mathbf{s}$$

(3.26)

where $J(\mathbf{s}, \lambda_0)$ is a volume factor $J = \sqrt{\det(g)}(\mathbf{s}, \lambda_0)$ coming from the marginalization over the conjugate momenta. As opposed to the case where the configuration space is parameterized using atomic positions, this volume factor will in general depend on the coordinates, $\mathbf{q} = (\mathbf{s}, \lambda_0)$, specifying the configuration. The extra contribution from $J$ is sometimes written in the form $J = \exp(-V)$, where $V$ is called the *Fixman potential* [187, 188, 291, 403]. Taking the Fixman potential into account is important if a reparameterized potential is used for studying protein dynamics for instance. Defining

$$Z(\lambda_0) = e^{-\beta F(\lambda_0)} = \int e^{-\beta E(\mathbf{s}, \lambda_0)} J(\mathbf{s}, \lambda_0) d\mathbf{s}$$

(3.27)

we can express the result, Eq. 3.26, in terms of the conditional probabilities

$$p_\beta(\mathbf{r} | \lambda_0) = \begin{cases} \frac{e^{-\beta E(\mathbf{r})} J(\mathbf{r})}{Z(\lambda_0)} & \text{if } \lambda(\mathbf{r}) = \lambda_0 \\ 0 & \text{otherwise,} \end{cases}$$

(3.28)

where $\mathbf{r} = (\mathbf{r}_1, \cdots, \mathbf{r}_N)$ is a unique function of $(\mathbf{s}, \lambda_0)$. Following Eq. 3.8 the entropy of the conditional distribution $p(\mathbf{r} | \lambda)$ is given by

$$S(\lambda) = -k_b \mathbb{E}_{\beta, \lambda} \left[ \ln p_\beta(\mathbf{r} | \lambda) \right]$$

$$= -k_b \int p_\beta(\mathbf{r}(\mathbf{s}, \lambda) | \lambda) \ln p_\beta(\mathbf{r}(\mathbf{s}, \lambda) | \lambda) \, d\mathbf{s}$$

$$= T^{-1} \left( \mathbb{E}_{\beta, \lambda} [E](\lambda) - F(\lambda) \right),$$

(3.29)

where $\mathbb{E}_{\beta, \lambda} = \mathbb{E}_{p_\beta(\mathbf{r} | \lambda)}$ represents the expectation value over the $3N - d$ dimensional manifold $\Omega(\lambda)$. In analogy to Eq. 3.9 we can write

$$F(\lambda) = \mathbb{E}_{\beta, \lambda} [E](\lambda) - TS(\lambda),$$

(3.30)

i.e., the free energy profile consists of contributions from both the average energy and from the entropy profile, and it can therefore strictly speaking no longer be considered as a potential of mean force.

The above derivation can conveniently be expressed as the following relation between marginalized probability distributions and free energies

$$p_\beta(\lambda) = \frac{Z(\lambda)}{Z} = e^{-\beta(F(\lambda) - F)} = e^{S(\lambda) - \beta \mathbb{E}_{\beta, \lambda} [E] + \beta F}, \quad e^{-\beta F} = \int e^{S(\lambda) - \beta \mathbb{E}_{\beta, \lambda} [E]} d\lambda$$

(3.31)

where $F$ is the free energy of the full system, as defined by Eq. 3.7. When $\lambda$ is chosen to be the subset of atomic positions $\lambda = (\mathbf{r}_1, \mathbf{r}_2, \cdots, \mathbf{r}_n)$ for a liquid system, the Fixman potential will be identically zero and the entropy $S(\lambda)$ is a constant. Hence $F(\lambda)$ *recovers its meaning as the potential of mean force*, consistent with Eq. 3.18. Furthermore, if we set $E \equiv 0$, the corresponding distribution $\tilde{p}(\lambda)$ becomes independent of temperature and will equal the normalized density of states,

$$\tilde{p}(\lambda) = e^{-\beta(\tilde{F}(\lambda)-\tilde{F})} = \frac{e^{S(\lambda)}}{\int e^{S(\lambda)} d\lambda}, \quad E \equiv 0 \tag{3.32}$$

Here, the $\tilde{F}$'s refer to the free energies in the interaction-free ensemble. Again, setting $\lambda = (\mathbf{r}_1, \mathbf{r}_2, \cdots, \mathbf{r}_n)$ in the liquid system this probability is simply given by Eq. 3.12. From Eqs. 3.13 and 3.17 we see that the potential of mean force $w_{(n)}$ can be viewed as the difference between the free energies of the two ensembles in the presence and absence of interactions, respectively.

$$w_{(n)}(\lambda) = (F(\lambda) - F) - (\tilde{F}(\lambda) - \tilde{F}), \quad \lambda = (\mathbf{r}_1, \mathbf{r}_2, \cdots, \mathbf{r}_n)$$

In other words, the interaction-free or *ideal gas* state serves as the reference state for $w_{(n)}$. Since both $F$ and $\tilde{F}$ are constants we can – without loss of generality – redefine $F(\lambda)$ and $\tilde{F}(\lambda)$ so that the above equation simplifies to

$$w_{(n)}(\lambda) = F(\lambda) - \tilde{F}(\lambda), \quad \lambda = (\mathbf{r}_1, \mathbf{r}_2, \cdots, \mathbf{r}_n)$$

By analogy to the definition of the potential of mean force, one can for any choice of $\lambda$ define the reference state for the free energies as the interaction-free ensemble. Consequently, we define

$$\Delta F(\lambda) = F(\lambda) - \tilde{F}(\lambda).$$

Note, that $\tilde{F}(\lambda)$ in general will be $\lambda$-dependent. From Eqs. 3.31 and 3.32 this free energy difference can be expressed as the ratio of 2 probabilities

$$\exp(-\beta \Delta F) = \frac{p_\beta(\lambda)}{\tilde{p}(\lambda)} = \frac{n_{obs}(\lambda)}{n_{exp}(\lambda)}. \tag{3.33}$$

The final expression on the right hand side represents the ratio between the observed statistics in the presence of interactions, $n_{obs}$, to the statistics expected in the absence of interactions, $n_{exp}$. This equation forms the basis for many KBPs that follow Sippl's seminal paper [661]. These KBPs are typically called "potentials of mean force". However, in Eq. 3.33 the reference state has a clear definition and refers to a system with the *same* degrees of freedom (e.g. molecules) taken at $E \equiv 0$ or – equivalently – $\beta = 0$. In contrast, KBPs are typically constructed by comparing statistics across *different* protein systems. We shall return to this and other problems with KBPs in the following sections.

## 3.3 Knowledge Based Potentials for Proteins

The formulation of a potential of mean force for protein molecules is considerably more involved than the case of the homogeneous liquid system due to three distinct complications: a protein molecule is a finite, heterogeneous and polymeric system. Each of these aspects involve modifications of or extensions to the correlation functions of which these potentials are based upon. An additional complication is the fact that the molecules are submerged in a solute, or buffer, the effect of which must be taken into account in order to obtain a useful estimates of the potentials of mean force.

While PMFs in principle can be calculated from a physical force field, the basic philosophy of the knowledge-based approach is to approximate these quantities simply by statistical analysis of known protein structures, an approach sometimes referred to as the '*inverse Boltzmann law*' [662]. The first step in this approach is to decide on the level of details of the structural representation, $\mathbf{r}$, and the type of statistics to be analyzed from the data set, $\boldsymbol{\lambda}(\mathbf{r})$. In the following, we briefly go through these two aspects before discussing how KBPs are constructed in general and the extent to which these are related to thermodynamic quantities.

### 3.3.1 Protein Representation

The choice of representation of the protein structures is intimately linked to the construction of a useful knowledge-based potential. Different representations can be viewed as different ways of restricting the conformational space by reducing the number of degrees of freedom, which in turn simplifies the form of the potential.

In a completely unrestricted model, a conformation is represented by the set of atomic positions and the system has $3N$ degrees of freedom, where $N$ is the number of atoms in the system. There are many approaches for reducing the number of degrees of freedom, a few are presented below, and a more detailed description can be found in the review by Kolinski and Skolnick [385].

A first natural step to reduce conformational space is to replace solvent atoms with an implicit description so that the absence of particles implies the presence of solvent atoms. Solvent effects are included in the potential through the interactions between the protein moieties, or by incorporating energy terms based on solvent exposure [385].

A further simplification arises by using a coarse-grained description, where sets of atoms are grouped into pseudo-atoms [437]. As the number of pairwise interactions scales as $O\left(N_A^2\right)$, where $N_A$ is the number of different kinds of (pseudo-)atoms, coarse-graining can lead to a considerable reduction of the complexity of the potentials with respect to the computational cost of evaluating conformational energies and the requirements on the database for estimating the potential. Hydrogen atoms are therefore often not modeled explicitly, but are implied in energy terms involving hydrogen bonds, e.g. as favorable energies for $O - N$ pairs at a distance of ~2.9 Å.

Many statistical potentials represent each amino acid with a single pseudo-atom located at its $C\alpha$ or side-chain center of mass position [517, 609]. Other potentials use full or partial main chain descriptions but represent the side-chains as pseudo-atoms [186, 259, 793] or ellipsoids [192]. Using the results from coarse-grained modeling, all-atom models can be reconstructed at a later stage [98, 290, 292, 454].

Lattice heteropolymers used in earlier works on protein modeling constitute the limit of this type of approximation scheme, in that the spatial degrees of freedom of each monomer of the polymer chain are also reduced to the vertices of a lattice [305, 384]. However, besides ensuring higher spatial resolution, coarse-graining in off-lattice modeling has been shown to be a more efficient method to reduce the conformational space than their lattice counterparts [559].

For the case of all-atom protein representations, the different atoms must be differentiated further in order to reflect different chemical environments; for example, a carbon atom in a methyl group is different from a carbon atom in an aromatic ring. These methods often involve tabulating the observed frequencies of atom-atom interactions. In that case, the number of different atomic species used is limited by the size of the database used for extracting the parameters. Chen and Shakhnovich developed a potential with 23 different atom types based on chemical equivalence [110], Yang and Zhou used 158 residue-specific atom types for their updated DFIRE energy function, uDFIRE [775] and Lu et al. used 19 rigid-body blocks in their orientation-dependent potential OPUS-PSP [453].

### 3.3.2 Coarse-Grained Variables for Protein Structures

Different choices of descriptors or 'reaction' coordinates, $\lambda$, for the potentials are possible. The appropriate choice depends on the intended application of the potential. In general and as discussed in Sect. 3.2, $\lambda$ will represent some *coarse-grained* variables, meaning that they are deterministic functions of the (fine-grained) structures $\mathbf{r}$, defined by the protein representation discussed in the previous section.

One common type of potential involves contacts between atoms or pseudo-atoms when the distance between them is within a cut-off distance [463, 517, 671]. The energy of a conformation is determined by its set of contacts, the contact map $\lambda = \{(i, j)\}$, where each $(i, j)$ is a pair of contacting amino-acids. This approach is particularly useful for fold recognition, where an amino acid sequence is threaded onto different known folds in order to find the fold that gives the most favorable energy. The contact map description is computationally efficient, and also avoids difficulties with steric clashes that can arise when distance dependent potentials are used. However, the contact map description has been shown to be insufficient for high resolution protein structure prediction [720, 738].

Distance dependent pairwise interactions, $\lambda = \{r_{ij}\}$, where $r$ is the distance between amino-acid $i$ and $j$, are a natural choice of reaction coordinates and are used in many knowledge based potentials [352, 621, 651, 661, 797]. In this case, the statistics required for calculating the potential are usually collected in the form of histograms with discrete distance bins. When calculating these kind of potentials, some care must be taken in order take finite size effects into consideration, otherwise artifacts in the form of long-range repulsions between hydrophilic residues will result, as hydrophilic amino acids are over-represented on the surfaces of folded protein molecules. Another important consideration is to account for the correlations in the pair distributions which arise as a consequence of chemical bonds only [110, 618].

For amino acids that are close in sequence, the possible relative inter-residue distances are mainly determined by the chemical bonds connecting them. A common approach to take these effects into account is to treat local and non-local interactions separately [142, 661, 792]. Another way to achieve a higher accuracy is to use variables that depend on the chemical environment, like solvent accessibility [658], or to construct potentials that depend on orientation in addition to distance [17, 87, 453, 518].

Many common structural motifs in proteins involve clusters of three or more amino acids [360]. These motifs are often functionally important sites, such as metal-binding pockets and active sites. For example, so-called zinc fingers often consist of four amino acids in contact with a zinc ion [395]. In order to model these structures, many-body interactions are required. Even though higher-order interactions are more difficult to model rigorously due to the use of histograms, there are several examples in the literature of many-body knowledge based potentials [142, 178, 491]. In the future, probabilistic models that go beyond histograms will undoubtedly aid the development of adequate higher-order KBPs.

### 3.3.3 Construction of KBPs

While the connection between relative frequencies and free energy differences is apparent from Eq. 3.33, this relation is only valid for two ensembles defined by the *same* degrees of freedom. The database that is used for the construction of a KBP will typically include both structural and sequential variations. Typically, the database will be a set of known native structures from high-resolution data, and may also include a number of decoys depending on the application. Consequently, in order to obtain a general, transferable knowledge based potential (KBP) for protein structures, we need to be able to find a relation between proteins with different amino acid sequences. This is usually carried out by introducing a hypothetical *reference state* which separates energy contributions that are independent of the protein sequence, from the specific, sequence-dependent interactions [661]. Typically, variations in the experimental conditions such as temperature, pH and buffer are averaged out in this approach. As for the construction of the potential of mean force in liquid theory discussed in Sect. 3.2.3, the reference state is *defined* by the statistics one would expect in the absence of the potential.

In order to bring the general formalism in accord with statistical mechanics, we let $S$ denote the database used for the statistical analysis and $\mathbf{s} = (\mathbf{r}, \mathbf{a}) \in S$ be a specific state in $S$ having the structure $\mathbf{r}$ and amino-acid sequence $\mathbf{a}$. Both of these will be determined by the chosen protein representation, as discussed in Sect. 3.3.1. Furthermore, $\lambda(\mathbf{r}) = (\lambda_1, \cdots, \lambda_d)(\mathbf{r})$ will denote the set of chosen structural, coarse-grained variables or 'reaction-coordinates' that form the argument of the desired KBP, $U_s(\lambda, \mathbf{a})$, as discussed in Sect. 3.3.2. In the following we shall use the subscript $s$ on a quantity when it is derived from $S$, to distinguish it from its physical 'counterpart'.

From the choice of $\lambda$ and $S$ a number of 'partition functions' can be constructed

$$\tilde{Z}_s = \sum_{s \in S} 1 \tag{3.34}$$

$$\tilde{Z}_s(\lambda, \mathbf{a}) = \sum_{s \in S} \delta(\lambda, \lambda(\mathbf{s}))\delta(\mathbf{a}, \mathbf{a}(\mathbf{s}))$$

$$Z_s = \sum_{s \in S} \exp\big(-\beta_s E_s(\mathbf{s})\big)$$

$$Z_s(\lambda, \mathbf{a}) = \sum_{s \in S} \delta(\lambda, \lambda(\mathbf{s}))\delta(\mathbf{a}, \mathbf{a}(\mathbf{s})) \exp\big(-\beta_s E_s(\mathbf{s})\big) \tag{3.35}$$

where $\tilde{\ }$ refers to quantities calculated from the reference ensemble and $\delta(\cdot, \cdot)$ is the Kronecker delta. For consistency, we have also introduced the KBP, $E_s(\mathbf{s})$, as function of any specific member $s$ of $S$. In most cases, this function will only be a function of the coarse-grained descriptors $\lambda$, $E_s(\mathbf{s}) = E_s(\lambda(\mathbf{r}), \mathbf{a})$. Note that this 'energy' differs from the physical energy of the particular state $\mathbf{s}$, as we have no *a priori* reason to assume that the structural and sequential variation in $S$ comply with the thermal/canonical distribution, a point we shall return to in Sect. 3.3.6. For the same reason, $\beta_s$ is merely a scaling factor which – however – is often set to the inverse thermodynamic temperature. The corresponding probabilities are

$$\tilde{p}_s(\lambda, \mathbf{a}) = \frac{\tilde{Z}_s(\lambda, \mathbf{a})}{\tilde{Z}_s} = \exp\big(-\beta_s(\tilde{F}_s(\lambda, \mathbf{a}) - \tilde{F}_s)\big)$$

$$p_s(\lambda, \mathbf{a}) = \frac{Z_s(\lambda, \mathbf{a})}{Z_s} = \exp\big(-\beta_s(F_s(\lambda, \mathbf{a}) - F_s)\big). \tag{3.36}$$

Consequently, we find

$$\ln\left(\frac{p_s(\lambda, \mathbf{a})}{\tilde{p}_s(\lambda, \mathbf{a})}\right) = -\beta_s\big((F_s(\lambda, \mathbf{a}) - \tilde{F}_s(\lambda, \mathbf{a})) + (\tilde{F}_s - F_s)\big). \tag{3.37}$$

It is common to neglect the last term, $(\tilde{F}_s - F_s)$, on the right hand side as it is independent of $\mathbf{a}$ and $\lambda$, as discussed in Sect. 3.2.3. In analogy to Eq. 3.33 the KBP,

# 3 On the Physical Relevance and Statistical Interpretation

$U_s$, is defined from the 'free energy' difference, $\Delta F_s(\lambda, \mathbf{a}) = F_s(\lambda, \mathbf{a}) - \tilde{F}_s(\lambda, \mathbf{a})$, as

$$\beta_s U_s(\lambda, \mathbf{a}) = \beta_s \Delta F_s(\lambda, \mathbf{a}) = -\ln\left(\frac{p_s(\lambda, \mathbf{a})}{\tilde{p}_s(\lambda, \mathbf{a})}\right) \approx -\ln\left(\frac{n_{\text{obs}}(\lambda, \mathbf{a})}{n_{\text{exp}}(\lambda, \mathbf{a})}\right). \quad (3.38)$$

Here, we introduce the number of observed occurrences in $S$ and the number of expected occurrences (in the reference state) as function of $(\lambda, \mathbf{a})$. A further approximation that is often made in order to reduce the complexity is to assume independence between different structural variables/reaction coordinates

$$p(\lambda, \mathbf{a}) \approx \prod_\alpha p(\lambda_\alpha, a_\alpha). \quad (3.39)$$

Consequently, the potential for the different $\lambda_\alpha$'s can be inferred separately,

$$\beta_s U_s(\lambda_\alpha, a_\alpha) = \beta_s \Delta F_s(\lambda_\alpha, a_\alpha) = \ln\left(\frac{n_{\text{obs}}(\lambda_\alpha, a_\alpha)}{n_{\text{exp}}(\lambda_\alpha, a_\alpha)}\right). \quad (3.40)$$

Equation 3.39 is an example of an *approximate factorization* or a *mean-field* approximation (see also Chap. 1) to the full probability distribution. In fact, it is only for special choices of $(\lambda_\alpha, a_\alpha)$ that this decomposition scheme does not lead to inconsistencies, namely when the basic degrees of freedom (viz. the individual monomers of a given protein) do not contribute to the statistics of several $\alpha$-components. An example of a valid decomposition would be the construction of a KBP for amino-acid dependent main chain dihedral angles; in that case $a_\alpha$ is the $\alpha$th amino-acid and $\lambda_\alpha$ concerns the main chain dihedral angles for $a_\alpha$. In the context of pairwise potentials, however, the $\alpha$-components would refer to specific pairs of amino-acids, $a_\alpha = (a_i, a_j)$, and $\lambda_\alpha = r_{a_i a_j}$ to their pairwise distance. Here, each monomer will appear in the calculation of several of the $\alpha$-components implying that these strictly speaking never can be treated as statistically independent. The problems of assuming statistical independence for distance dependent KBPs, which we shall return to in the following sections, have been elucidated very clearly by Ben-Naim [32].

In general, the construction of approximate 'free energies', $\beta_s F_{s,\alpha}(\lambda_\alpha, a_\alpha)$ used in the definition of the components of the knowledge based potential, $\beta_s U_{s,\alpha}$, represents a common inference problem, namely how to compute potential functions that consistently reproduce some given observed statistics. From this perspective, the factorized approximation in Eq. 3.39 represents a special case of a wider class of approximation techniques known as region-based approximation methods [786], which also include *Bethe's method* [60] and Kikuchi's *cluster variation method* [372]. In the following we shall focus on various approaches used to derive KBPs in two well-studied cases; the contact potential and the pairwise, distance-dependent potential.

### 3.3.4 The Contact Potential

Using Eqs. 3.38 or 3.40, a knowledge based potential can be obtained from a set of protein structures by extracting the distribution of the descriptor variable $\lambda$ and calculating the expected statistics for the reference set for each sequence. Different choices of descriptors and reference states thus result in different potentials.

Miyasawa and Jernigan (MJ) [517] chose $\lambda$ to be the set of amino-acid contact-pairs. They constructed corresponding contact energies by considering a quasi-chemical equilibrium [260] between residues of species I and J, and solvent S,

$$I : S + J : S \rightleftharpoons I : J + S : S \tag{3.41}$$

where : indicates a contact between two entities. As discussed in [517] the quasi-chemical equilibrium is in this context equivalent to the Bethe free energy approximation method [60]. The KB contact energies, $e_{ij} = \beta_s U_s(\lambda_{ij} = 1, a_i, a_j)$, are extracted from known protein structures through the relation

$$\frac{\bar{n}_{ij}\bar{n}_{ss}}{\bar{n}_{is}\bar{n}_{js}} = e^{-e_{ij}}, \tag{3.42}$$

where $\bar{n}_{kl}$ is the average number of contacts between species $k$ and $l$. In this case, the reference state is isolated and solvated residues, corresponding to an unfolded state. These energies are subsequently used in constructing a second set of contact energies, where the unfolded reference state is replaced with the average residue environment. The quasi-chemical equilibrium under consideration is

$$I : R + J : R \rightleftharpoons I : J + R : R \tag{3.43}$$

where R represents the average residue. Effective interresidue contact energies are extracted through the relation

$$\frac{\bar{n}_{ij}\bar{n}_{rr}}{\bar{n}_{ir}\bar{n}_{jr}} = e^{-(e_{ij}+e_{rr}-e_{ir}-e_{jr})} \tag{3.44}$$

The energy $e_{ij} + e_{rr} - e_{ir} - e_{jr}$ is the energy difference between forming a specific contact between I : J and between the average environments I : R + J : R. In the random mixing approximation, effects due to chain connectivity are neglected so that the expected number of contacts $n_{ij}$ in the reference state is only dependent on the chemical composition of the protein. Although this may seem like a severe approximation, Skolnick et al. compared potentials obtained using the quasichemical approximation with the corresponding potentials using more physical reference states [671]. The reference states used in the comparison were the Gaussian random coil and a reference state based on threading each sequence onto a library of structures with similar compactness as the native conformation. In both cases, the quasi-chemical approximation was shown to be very good. More

3 On the Physical Relevance and Statistical Interpretation 113

recently, Solis and Rackovsky used information-theoretic analysis to demonstrate that the quasichemical reference state results in better-performing contact based potentials in threading-based decoy discrimination [676]. Godzik et al. compared different contact potentials by introducing the excess energy,

$$e_{ij}^{\text{excess}} = e_{ij} - \frac{e_{ii} + e_{jj}}{2}, \tag{3.45}$$

which describes the difference between real proteins and ideal solutions of amino acids. Using this quantity, high correlations between different contact potentials could be demonstrated [233]. Betancourt and Thirumalai re-examined the relations between the MJ potential and the potentials obtained by Skolnick et al., and showed that the potentials were very similar if threonine was chosen as reference solvent in the derivation of the MJ potential [59]. Li, Tang and Wingreen analyzed the MJ interaction matrix using eigenvalue decomposition to show that the major driving force in protein folding are the hydrophobic effect and the force of demixing [439]. Other analyses of the MJ matrix have been carried out in order to partition the amino acids into groups with similar physicochemical characteristics [114, 652, 756].

### 3.3.5 The Pairwise Potential

A more detailed choice of $\lambda$ is to consider pairwise distances. A common approach to construct the corresponding distance-dependent pairwise interactions is to estimate the expected frequency of a specific pair at distance $r$ with the unspecific observed frequency [498, 621, 661, 662],

$$p_{\text{exp}}(r) \approx \frac{\sum_{i,j} n_{\text{obs}}^{i,j}(r)}{\sum_{i,j} \sum_r n_{\text{obs}}^{i,j}(r)}. \tag{3.46}$$

Lu and Skolnick [452] constructed a statistical potential where the reference state was taken as

$$n_{\text{exp}}^{i,j}(r) = \chi_i \chi_j n_{\text{obs}}(r), \tag{3.47}$$

where $\chi_i$ and $\chi_j$ are the mole fractions of species $i$ and $j$, respectively.

The reference state is by definition the conformation where the KBP has its zero point. Zhou and Zhou noted that for reference states where $n_{\text{exp}}^{i,j}(r) \equiv N_{\text{obs}}(r)$, the zero point implicates that the attractive and repulsive interactions of folded proteins average to zero. This observation led to the introduction of the distance-scaled finite ideal-gas reference state (DFIRE)[797]. The DFIRE reference state is based on the uniform distribution of non-interacting points in finite spheres. The number of atomic pairs separated by a distance $r$ is given by the uniform density of pairs times the volume of a spherical shell with radius $r$,

$$n_{\exp}^{i,j}(r) = n_i n_j \left(4\pi r^\alpha \Delta r / V\right), \tag{3.48}$$

where $n_i$ ($n_j$) is the number of entities of species $i$ ($j$), and $V$ is the volume of the system. $\alpha$ is a parameter that reflects that protein molecules have a finite volume and was estimated from a database of folded proteins. For a liquid, $\alpha = 2$, whereas for folded protein molecules it was found to be slightly lower, 1.61, which reflects the existence of cavities. Zhou and Zhou also introduced a cut-off distance $r_{\text{cut}}$ so that the potentials $u(i, j, r) = 0$ for $r > r_{\text{cut}}$. The approach was used to extract two statistical potentials, one all-atom potential and one main chain $+ C\beta$ atoms, and showed good discriminations of decoys, with slightly better results for the former [793, 797].

Chen and Sali used similar reasoning to develop a reference state based on the distribution of distances in spheres [651]. This reference state consist of uniformly distributed atoms in a sphere with the same radius of gyration and density as the native structure. The resulting statistical potential, Discrete Optimized Energy, DOPE, performs very well in decoy discrimination and is used as the energy function in the widely used modeling package MODELLER-8 [619, 651]. Fitzgerald et al. refined the DOPE potential further by reducing the model to a $C\beta$ representation and by separating local from non-local interactions [186]. Rykunov and Fiser also recognized that the sizes of the protein molecules should be taken into account in order to avoid systematic errors in the statistical potentials, and constructed reference ensembles by shuffling the amino acid sequences of protein structures [618].

Most knowledge based potentials are derived following the approach as outlined above. The different potentials are distinguished by the protein representation, the choice of reaction coordinates, the data sets used for extracting the potentials, and in the choice of reference state. The latter is illustrated in Fig. 3.2, where three different potentials are generated using the same data, but with different reference states.

### 3.3.6 Self-consistent Potentials

The database, $S$, used for the construction of the knowledge based potential, $U_s$, will invariably only represent a small fraction of the possible configuration any given sequence can adopt. A requirement which often is not met by these constructions is the self-consistency of the derived potentials. Self-consistency implies that the statistics obtained by sampling from $U_s$ match the statistics observed in $S$.

The question of self-consistency can be addressed by applying efficient sampling algorithms, such as Markov chain Monte Carlo (MCMC) sampling, as discussed in Chap. 2. This method can for instance be employed to sample the distribution of amino acids in some confined volume $V$ according to a given potential, so as to assess the self-consistency of i.e. contact potentials. An alternative self-consistent approach for these potentials has recently been proposed [58], where residue correlation effects are reduced based on a Taylor expansion of the number of observed contacts.

Typically, the use of MCMC-methods for assessing self-consistency has been limited to applications where the potentials are of a physical nature. Here, the statistical ensemble is given by the configuration space $\Omega(\mathbf{a})$ for the given amino-acid sequence, $\mathbf{a}$. As outlined in Chap. 2, the MCMC-method enables one to estimate

$$p_\beta(\boldsymbol{\lambda}|\mathbf{a}) = Z^{-1}(\mathbf{a}) \int_{\Omega(\mathbf{a})} \exp(-\beta E(\mathbf{r},\mathbf{a})) \delta(\boldsymbol{\lambda} - \boldsymbol{\lambda}(\mathbf{r})) d\mathbf{r}, \quad (3.49)$$

where $E$ is a physical energy and $\beta$ is the inverse physical temperature.[2] Since the reference state corresponds to the distribution when $E \equiv 0$ we also have

$$\tilde{p}(\boldsymbol{\lambda}|\mathbf{a}) = \tilde{Z}^{-1}(\mathbf{a}) \int_{\Omega(\mathbf{a})} \delta(\boldsymbol{\lambda} - \boldsymbol{\lambda}(\mathbf{r})) d\mathbf{r} \quad (3.50)$$

Assuming that the potential is only a function of $\boldsymbol{\lambda}$, $E(\mathbf{r},\mathbf{a}) = U(\boldsymbol{\lambda},\mathbf{a})$, Eq. 3.49 becomes

$$p_\beta(\boldsymbol{\lambda}|\mathbf{a}) \propto \exp(-\beta U(\boldsymbol{\lambda},\mathbf{a})) \tilde{p}(\boldsymbol{\lambda}|\mathbf{a}).$$

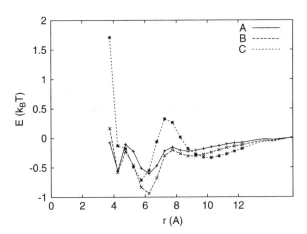

**Fig. 3.2** Distance-dependent potentials between the sidechain centers of mass of valine and phenylalanine, using different reference states. *A*: reference state according to Eq. 3.46, *B*: average frequencies for valine and phenylalanine residues only and *C*: $n_{\text{exp}}(r) \sim r^{-1.61}$, similar to [797]. The resulting scales were translated so that the zero points are at $r = 15.5$ Å in order to facilitate comparison. The data were extracted from structures in a dataset of 4,978 structures at a resolution of 2 Å or better and with sequence similarities of 30% or less. The data set was obtained from the PISCES server [751]

---

[2] Here, we have for simplicity assumed the Fixman potential to be zero.

As a result, and taking Eq. 3.33 into account, self-consistency implies:

$$U(\lambda, \mathbf{a}) \propto -k_B T \ln \left( \frac{p_\beta(\lambda | \mathbf{a})}{\tilde{p}(\lambda | \mathbf{a})} \right) \tag{3.51}$$

However, since the potential is often derived by the wrong assumption of decomposability and/or inaccurate choices of the reference state, Eq. 3.40, self-consistency is not necessarily satisfied. On the other hand, the sampling approach can be extended to a scheme where the KB potential is refined iteratively to eventually ensure self-consistency. This so-called iterative Boltzmann inversion was used by Reith et al. for simulations of polymers [600] and in the context of protein folding in by Thomas and Dill [716] and Májek and Elber [464]. Huang and Zhou used iterative Boltzmann inversion to calculate a statistical potential for protein-ligand interactions [319, 320]. The potential at iteration $i + 1$ is calculated as

$$U_{i+1}(\lambda, \mathbf{a}) = U_i(\lambda, \mathbf{a}) - \beta^{-1} \ln \frac{p^{\text{native}}(\lambda | \mathbf{a})}{p_i(\lambda | \mathbf{a})}. \tag{3.52}$$

Starting with an initial $U_0$, the potential can be refined using Eq. 3.52 until convergence.

Mullinax and Noid have developed a method to extract a statistical potential, termed *generalized potential of mean force*, that is designed to maximize *transferability*, so that samples reproduce structural properties of multiple systems (a so-called extended ensemble) [527]. Initial results for modeling protein structures are promising [528].

Another MCMC-approach to ensure self-consistency is to make use of Boltzmann learning, as proposed by Winther and Krogh [768]. Winther and Krogh used the submanifold of conformational space consisting of native conformations, $\mathcal{N}$, as the structural descriptor, $\lambda$, i.e.

$$\lambda(\mathbf{r}) = \begin{cases} 1 \text{ if } \mathbf{r} \in \mathcal{N} \\ 0 \text{ otherwise} \end{cases} \tag{3.53}$$

The probability for a protein with sequence $\mathbf{a}$ of adopting a native conformation is given by

$$p_\beta(\lambda = 1 \mid \mathbf{a}, \theta) = \frac{\int_{\mathcal{N}} e^{-\beta E_\theta(\mathbf{r}, \mathbf{a})} d\mathbf{r}}{\int e^{-\beta E_\theta(\mathbf{r}, \mathbf{a})} d\mathbf{r}}. \tag{3.54}$$

The integral in the numerator thus runs over all conformations that are considered to be in the native state, while the integral in the denominator is over the entire conformational space. Winther and Krogh proposed to optimize the parameters $\theta$ of the energy function by maximum likelihood,

$$\theta_{\text{ML}} = \text{argmax} \sum_i \ln p_\beta(\lambda_i = 1 \mid \mathbf{a}_i, \theta), \tag{3.55}$$

where the sum runs over all proteins in a training set. If the energy function is differentiable, one can optimize $\theta$ by simple gradient ascent,

$$\theta' = \theta + \eta \nabla_\theta \sum_i \ln p\,(\lambda_i = 1 \mid \mathbf{a}_i, \theta)\,, \tag{3.56}$$

where $\nabla$ is the gradient operator. Using Eq. 3.54, the second term can be written as

$$\eta \beta \sum_i \mathbb{E}_\beta \Big[ \nabla_\theta E_\theta\,(\mathbf{r}_i, \mathbf{a}_i) \Big] - \mathbb{E}_{\beta, \lambda_i = 1} \Big[ \nabla_\theta E_\theta\,(\mathbf{r}_i, \mathbf{a}_i) \Big], \tag{3.57}$$

where $\mathbb{E}_\beta[\cdot]$ is the expectation value over the whole conformational space and $\mathbb{E}_{\beta, \lambda_i = 1}$ is the expectation value over the subspace that makes up the native conformations of protein $i$. The drawbacks with this approach is that it requires a classification of the subspace of native structures $\mathcal{N}_i$, and that it is computationally expensive to obtain expectation values over the entire conformational space. Hinton developed a computationally efficient approximate variant, called *contrastive divergence* [306]. This technique, which is presented in Chap. 5, was used by Podtelezhnikov et al. to model hydrogen bonds [586] and the packing of secondary structure [584].

Irrespective of the choice of statistical descriptors, $\lambda$, the above discussion indicates that the requirement of self-consistency can not be separated from the intended application of the KBPs. Indeed, when KBPs are used to sample plausible protein states $\mathbf{s}$, one must be aware of the details of the sampling scheme involved, a point we shall return to in Sect. 3.4.2.

### 3.3.7 Are KBPs Related to Physical Free Energies?

Up to now we have avoided the discussion as to which extent the statistics of structural features, $\lambda$, of protein states in some given known ensemble, $\mathcal{S}$, directly reveal information of physical free energies. One implicit answer to this question is given by the discussion in the previous section, namely that it relies on how close the extracted KBPs is to the potential that reproduces the observed statistics in a canonical ensemble. The general need of solving this self-consistency requirement by iterative means suggests that KBPs are typically only crude approximations.

In this context, it is somewhat remarkable that there are several cases where the statistics calculated from some set of known native proteins, $\mathcal{S}$, *do* resemble a Boltzmann-type of distribution, i.e.

$$p_s(\lambda) \sim e^{-\beta^* \Delta F^*(\lambda)}. \tag{3.58}$$

Here, $\Delta F^*(\lambda)$ is the typical (w.r.t different sequences) physical free energy for the structural features $\lambda$ and $\beta^*$ represents the inverse 'conformational temperature' for the protein statistics [184]. Before we proceed, it is important to emphasize that $\Delta F^*$ in Eq. 3.58 is obtained from physical principles independent of the statistical features of $S$. The assumption of Boltzmann distributed statistics in $S$ is commonly referred to as the *Boltzmann hypothesis* [656]. For particular choices of $\lambda$ there is compelling empirical evidence of this hypothesis. Bryant and Lawrence analyzed ion-pairs in proteins and found a Boltzmann-like distribution as a function of the electrostatic potential [85], Rashin, Iofin and Honig studied the distribution of cavities and buried waters in globular proteins [597]. Butterfoss and Hermans compared statistics of experimental data with quantum mechanical energies of model compounds and found Boltzmann distributions of sidechain conformations [92].

Finkelstein et al. provided theoretical arguments based on the *random energy model* [145] for the observed Boltzmann statistics [184]. The main lesson from this work may not be the accuracy of the random energy model per se, but rather that the connection between structural features across different folded proteins and physical free energies arises from the restraint that native structures need to have a typical thermodynamical stability. Indeed, such type of restraints automatically produce Boltzmann-like distributions from a general information theoretical perspective [342].

While Boltzmann type of statistics are observed for particular choices of $\lambda$ it does not imply that a particular known set of protein structures, $S$, allows the direct construction of a KBP that would ensure the correct structural stability or 'foldability' of any specific sequence $\mathbf{a}$. This precaution holds true, even in the absence of some erroneous assumptions of decomposability, such as in Eq. 3.40. Indeed, $S$ will invariably cover only a narrow subset of possible structures and sequences, whereas the calculation of physical free energies for any given sequence $\mathbf{a}$ involve a summation over *all* possible structural states, $\Omega(\mathbf{a})$, for this sequence.

This point can be expressed more formally. Specifically, let $\Lambda$ represent the structural restraints embedded in the ensemble $S$. If $S$ represents a data-base of globular native states, $\Lambda$ includes typical coarse-grained features for these, such as compactness and topology. Let $p_s(\Lambda)$ be the distribution of these features. Accordingly, $p_s(\Lambda)$ will have a limited support compared to all possible values of $\Lambda$ a given polypeptide chain $\mathbf{a}$ could adopt. For the simplicity of the argument, we set $p_s(\Lambda) \approx \delta(\Lambda - \Lambda_N)$, for some typical value, $\Lambda_N$ of the native sate. Similarly, we assume that only two values of $\Lambda$ have noticeable probability in the full canonical ensemble, $\Lambda_N$ and $\Lambda_U$, where $\Lambda_U$ represents a typical value for unfolded state. When deriving physical free energies, we consequently have

$$e^{-\beta(F(\lambda,\mathbf{a})-F(\mathbf{a}))} = p_\beta(\lambda|\mathbf{a}) = \int p_\beta(\lambda|\Lambda,\mathbf{a})p_\beta(\Lambda|\mathbf{a})\mathrm{d}\Lambda,$$

$$= p_\beta(\lambda|\Lambda_N,\mathbf{a})p_\beta(\Lambda_N|\mathbf{a}) + p_\beta(\lambda|\Lambda_U,\mathbf{a})p_\beta(\Lambda_U|\mathbf{a}). \quad (3.59)$$

Here,

# 3 On the Physical Relevance and Statistical Interpretation

$$-\ln\left(\frac{p_\beta(\Lambda_N|\mathbf{a})}{p_\beta(\Lambda_U|\mathbf{a})}\right) = \beta(F_N(\mathbf{a}) - F_U(\mathbf{a}))$$

represents the stability free energy of the native state. Using the $\delta$-approximation of $p_s(\Lambda)$ we get for the corresponding construction of the KBP

$$e^{-\beta_s(F_s(\lambda,\mathbf{a})-F_s(\mathbf{a}))} = p_s(\lambda|\mathbf{a}) \approx p_s(\lambda|\Lambda_N,\mathbf{a}) \qquad (3.60)$$

where

$$p_s(\lambda|\mathbf{a}) = \frac{p_s(\lambda,\mathbf{a})}{p_s(\mathbf{a})}$$

represents the statistics obtained from $\mathcal{S}$. The difference between Eqs. 3.59 and 3.60 elucidates the problems of identifying KBPs with physical free energies. Even if one assumes that structural features $\lambda$ of $\mathcal{S}$ (i.e. across different sequences) are *representative for the conditional Boltzmann distribution*, ie. so $p_s(\lambda|\mathbf{a}) \approx p_\beta(\lambda|\Lambda_N,\mathbf{a})$, Eq. 3.59 shows that one still need to include the *stability of the native state* as well as the *statistical features of the unfolded state* in order to fully translate KBPs into physical free energies. We believe that these two points are too often overlooked when physical concepts are brought into play to justify knowledge-based approaches.

To conclude this section, there is no a priori reason to believe that a particular KBP is connected to physical free energies. For some simple choice of $\lambda$ the statistics observed among known protein structures, $\mathcal{S}$, will presumably just reflect the overall restraint of native stability rather than other more specific properties of $\mathcal{S}$. In this case, the observed Boltzmann-like statistics should come as no surprise, as discussed above. However, irrespective of the choice of $\lambda$, the relation between KBPs and free energies can in general only be assessed by using the KBPs in a full canonical calculation. This is required both in order to probe the statistics of non-native states as well as the shortcomings of the typical assumption of decomposability as in Eq. 3.40.

## 3.4 Probabilistic Views on KBPs

Although we have already made extensive use of probabilistic formulations, all KBPs have been considered from an 'energy' point of view, using analogies to statistical mechanics. For many applications there is in fact no obvious gain of expressing the observed to expected probability ratios in terms of energies. Furthermore, conditional or prior information is often more naturally treated from a probabilistic point of view.

### 3.4.1 The Rosetta Explanation

The first purely probabilistic formulation of knowledge based potentials was pioneered by Simons et al. [658, 659], and was used in the groundbreaking protein structure prediction program Rosetta. Although their explanation is purely qualitative and does not consider issues such as self-consistency, it showed for the first time that expressions resembling knowledge based potentials could be obtained from Bayesian reasoning. According to the Bayesian probability calculus, the posterior probability of a structure $\mathbf{r}$ given the sequence $\mathbf{a}$ is:

$$p\left(\mathbf{r} \mid \mathbf{a}\right) = \frac{p\left(\mathbf{a} \mid \mathbf{r}\right) p\left(\mathbf{r}\right)}{p\left(\mathbf{a}\right)} \propto p\left(\mathbf{a} \mid \mathbf{r}\right) p\left(\mathbf{r}\right). \tag{3.61}$$

The factor $p\left(\mathbf{a}\right)$ is constant when considering a specific sequence and can be left out. The prior $p\left(\mathbf{r}\right)$ can incorporate general features of folded protein structures, such as compactness, sterical clashes and solvent exposure. For example, the Rosetta scoring function [658, 659] includes topological information in the form of $\beta$-strand pairing in the prior, which results in a low probability of sampling structures with unpaired $\beta$-strands.

Taking the negative logarithm of both sides of Eq. 3.61,

$$- \ln p\left(\mathbf{r} \mid \mathbf{a}\right) = - \ln p\left(\mathbf{a} \mid \mathbf{r}\right) - \left(- \ln p\left(\mathbf{r}\right)\right) + \text{const}, \tag{3.62}$$

and comparing to Eq. 3.38, we can identify the negative log likelihoods and negative log prior as free energy of the observed and the reference state respectively.

For the likelihood $p\left(\mathbf{a} \mid \mathbf{r}\right)$, an unfounded but convenient assumption of amino acid pair independence leads to the expression:

$$p\left(\mathbf{a} \mid \mathbf{r}\right) = \prod_{i,j} p\left(a_i, a_j \mid r_{ij}\right) \tag{3.63}$$

where the product runs over all amino acid pairs. Bayes' theorem is then applied to the pairwise factors, which results in:

$$p\left(a_i, a_j \mid r_{ij}\right) = p\left(a_i, a_j\right) \frac{p\left(r_{ij} \mid a_i, a_j\right)}{p\left(r_{ij}\right)} \propto \frac{p\left(r_{ij} \mid a_i, a_j\right)}{p\left(r_{ij}\right)}. \tag{3.64}$$

Combining this likelihood with Eq. 3.62 obviously results in the "potentials of mean force" based on pairwise distances as proposed by Sippl [661].

The likelihood actually used in the Rosetta force field $p\left(\mathbf{a} \mid \mathbf{r}\right)$ in turn, is of the form

$$p\left(\mathbf{a} \mid \mathbf{r}\right) \approx \prod_{i} p\left(a_i \mid e_i\right) \prod_{i,j} \frac{p\left(a_i, a_j \mid e_i, e_j, r_{ij}\right)}{p\left(a_i \mid e_i, e_j, r_{ij}\right) p\left(a_j \mid e_i, e_j, r_{ij}\right)}, \tag{3.65}$$

3 On the Physical Relevance and Statistical Interpretation 121

where $r_{ij}$ is the distance between residue $i$ and $j$ and $e_i$ represents the environment of residue $i$ and depends on solvent exposure and secondary structure. The formal justification of this expression is not discussed by Simons et al., but similar expressions involving pairwise and univariate marginal distributions result from the so-called Bethe approximation for joint probability distributions of graphical models [274, 784].

### 3.4.2 The Reference Ratio Method

While the Bayesian formulation by Simons et al. had a great impact on protein structure prediction, the theory rests upon the assumption of pairwise decomposability and only provides a qualitative explanation. Furthermore, the need of applying Bayes' theorem to obtain Eq. 3.64 is quite unclear, since the expression $p(a_i, a_j | r_{ij})$ is already directly available. In the following, we present a Bayesian formulation of knowledge based potentials that does not assume pairwise independence and also provides a quantitative view. This method, called the *reference ratio method*, is also discussed from a purely probabilistic point of view in Chap. 4. Our discussion has a more physical flavor and builds on the concepts introduced earlier in this chapter. As is also discussed in Chap. 4, we show that when a KBP is applied for the purpose of sampling plausible protein conformations, the reference state is uniquely determined by the proposal distribution [276] of the sampling scheme

Assume that some prior distribution, $q(\mathbf{r}|\mathbf{a})$ is given. In protein structure prediction, $q(\mathbf{r}|\mathbf{a})$ is often embodied in a fragment library; in that case, $\mathbf{r}$ is a set of atomic coordinates obtained from assembling a set of polypeptide fragments. Of course, $q(\mathbf{r}|\mathbf{a})$ could also arise from a probabilistic model or a pool of known protein structures or any other conformational sampling method. Furthermore, we consider some probability distribution $p_s(\boldsymbol{\lambda}|\mathbf{a}) = p(\boldsymbol{\lambda}|\mathbf{a})$, given from a set of known protein structures. Here, $\boldsymbol{\lambda}$ again represents some particular set of coarse-grained structural variables, which are deterministic functions of the structure $\boldsymbol{\lambda} = \boldsymbol{\lambda}(\mathbf{r})$. Consequently, we can express the prior distribution for the joint variables $(\boldsymbol{\lambda}, \mathbf{r})$ as

$$q(\mathbf{r}, \boldsymbol{\lambda}|\mathbf{a}) = q(\boldsymbol{\lambda}|\mathbf{r}, \mathbf{a})q(\mathbf{r}|\mathbf{a}) = \delta(\boldsymbol{\lambda} - \boldsymbol{\lambda}(\mathbf{r}))q(\mathbf{r}|\mathbf{a}) \qquad (3.66)$$

Next, we note that $q(\mathbf{r}|\mathbf{a})$ implies a matching marginal probability distribution $\tilde{q}(\boldsymbol{\lambda}|\mathbf{a})^3$

$$\tilde{q}(\boldsymbol{\lambda}|\mathbf{a}) = \int q(\mathbf{r}, \boldsymbol{\lambda}|\mathbf{a})d\mathbf{r} = \int q(\mathbf{r}|\mathbf{a})\delta(\boldsymbol{\lambda} - \boldsymbol{\lambda}(\mathbf{r}))d\mathbf{r}. \qquad (3.67)$$

We consider the case where $\tilde{q}(\boldsymbol{\lambda}|\mathbf{a})$ differs substantially from the known $p(\boldsymbol{\lambda}|\mathbf{a})$; hence, $\tilde{q}(\boldsymbol{\lambda}|\mathbf{a})$ can be considered as incorrect. On the other hand, we also assume

---

[3] The tilde, ~, refers to the fact that $\tilde{q}$ corresponds to the reference state as will be shown.

that the conditional distribution $q(\mathbf{r} \mid \lambda, \mathbf{a})$ is indeed meaningful and informative. The question is now how to combine these two distributions – each of which provide useful information on $\mathbf{r}$ and $\lambda$ – in a meaningful way.

The product rule of probability theory allows us to write:

$$p(\mathbf{r}, \lambda \mid \mathbf{a}) = p(\lambda \mid \mathbf{a}) p(\mathbf{r} \mid \lambda, \mathbf{a}), \qquad (3.68)$$

As only $p(\lambda \mid \mathbf{a})$ is given, we need to make a reasonable choice for $p(\mathbf{r} \mid \lambda, \mathbf{a})$. The choice most consistent with our prior knowledge is to set it equal to $q(\mathbf{r} \mid \lambda, \mathbf{a})$ which leads to:

$$p(\mathbf{r}, \lambda \mid \mathbf{a}) = p(\lambda \mid \mathbf{a}) q(\mathbf{r} \mid \lambda, \mathbf{a}). \qquad (3.69)$$

In the next step, we apply the product formula of probability theory to obtain

$$q(\mathbf{r} \mid \lambda, \mathbf{a}) = \frac{q(\lambda, \mathbf{r} \mid \mathbf{a})}{q(\lambda \mid \mathbf{a})} = \frac{q(\mathbf{r} \mid \mathbf{a}) \delta(\lambda - \lambda(\mathbf{r}))}{\tilde{q}(\lambda \mid \mathbf{a})} \qquad (3.70)$$

and consequently,

$$p(\mathbf{r}, \lambda \mid \mathbf{a}) = p(\lambda \mid \mathbf{a}) \frac{\delta(\lambda - \lambda(\mathbf{r})) q(\mathbf{r} \mid \mathbf{a})}{\tilde{q}(\lambda \mid \mathbf{a})} \qquad (3.71)$$

Finally, we integrate out the, now redundant, coarse-grained variable $\lambda$ from the expression:

$$p(\mathbf{r} \mid \mathbf{a}) = \int p(\mathbf{r}, \lambda \mid \mathbf{a}) d\lambda = \frac{p(\lambda(\mathbf{r}) \mid \mathbf{a})}{\tilde{q}(\lambda(\mathbf{r}) \mid \mathbf{a})} q(\mathbf{r} \mid \mathbf{a}) \qquad (3.72)$$

The distribution $p(\mathbf{r} \mid \mathbf{a})$ clearly has the right marginal distribution $p(\lambda \mid \mathbf{a})$. In fact, it can be shown that Eq. 3.72 minimizes the Kullback-Leibler divergence between $p(\mathbf{r} \mid \mathbf{a})$ and $q(\mathbf{r} \mid \mathbf{a})$ under the constraint of having the correct marginal distribution $p(\lambda(\mathbf{r}))$ (see Chap. 4 for a proof). The influence of the fine-grained distribution $q(\mathbf{r}, \lambda \mid \mathbf{a})$ is apparent in the fact that $p(\mathbf{r} \mid \lambda, \mathbf{a})$ is equal to $q(\mathbf{r} \mid \lambda, \mathbf{a})$. The ratio in this expression corresponds to the usual probabilistic formulation of a knowledge based potential where the distribution $\tilde{q}(\lambda \mid \mathbf{a})$ uniquely defines the reference state. We refer to this explicit construction as the *reference ratio method* [276].

It may be instructive to translate Eq. 3.72 into the 'energy'-language. Suppose, a prior energy is given, according to $-\beta_s E_s(\mathbf{r}, \mathbf{a}) = \ln q(\mathbf{r} \mid \mathbf{a})$. Then Eq. 3.72 states that the new energy, $E'_s$ that correctly reproduces the observed statistics, $p(\lambda \mid \mathbf{a})$, with minimal correction of the original energy function, is given as

$$-\beta_s E'_s(\mathbf{r}, \mathbf{a}) = -\beta_s \big( E_s(\mathbf{r}, \mathbf{a}) + U_s(\lambda(\mathbf{r}), \mathbf{a}) \big),$$

where the KBP, $U_s$, is obtained from the difference of two 'free energies'

# 3 On the Physical Relevance and Statistical Interpretation

$$\beta_s U_s(\boldsymbol{\lambda}, \mathbf{a}) = \beta_s \left( F_s(\boldsymbol{\lambda}, \mathbf{a}) - \tilde{F}_s(\boldsymbol{\lambda}, \mathbf{a}) \right) \tag{3.73}$$

$$\beta_s F_s(\boldsymbol{\lambda}, \mathbf{a}) = -\ln p(\boldsymbol{\lambda}|\mathbf{a}) \tag{3.74}$$

$$\beta_s \tilde{F}_s(\boldsymbol{\lambda}, \mathbf{a}) = -\int \delta(\boldsymbol{\lambda} - \boldsymbol{\lambda}(\mathbf{r})) \exp(-\beta_s E_s(\mathbf{r}, \mathbf{a})) d\mathbf{r}. \tag{3.75}$$

In other words, the reference free energy is the free energy associated with the prior energy function $E_s$. In the absence of prior knowledge, $E_s \equiv 0$, or equivalently if $q(\mathbf{r}|\mathbf{a})$ is constant, the reference state reduces to the interaction-free state discussed hitherto.

For cases when $\boldsymbol{\lambda}$ is of high dimensionality, it may become intractable to determine $\tilde{q}(\boldsymbol{\lambda}|\mathbf{a})$ directly. This problem can be overcome by applying Eq. 3.72 iteratively [276]. In the first iteration ($i = 0$), we simply set $\tilde{q}_{i=0}(\boldsymbol{\lambda}|\mathbf{a})$ equal to the uniform distribution. In iteration $i + 1$, the distribution $p_i(\boldsymbol{\lambda}|\mathbf{a})$ is improved using the samples generated in iteration $i$:

$$p_{i+1}(\mathbf{r}|\mathbf{a}) = \frac{p(\boldsymbol{\lambda}(\mathbf{r})|\mathbf{a})}{\tilde{q}_i(\boldsymbol{\lambda}(\mathbf{r})|\mathbf{a})} p_i(\mathbf{r}|\mathbf{a}) \tag{3.76}$$

where $\tilde{q}_i(\boldsymbol{\lambda}|\mathbf{a})$ is estimated from the samples generated in the $i$-th iteration and $p_0(\mathbf{r}|\mathbf{a}) = q(\mathbf{r} \mid \mathbf{a})$. After each iteration, the reference distribution $\tilde{q}_i(\boldsymbol{\lambda}|\mathbf{a})$ can be progressively estimated more precisely.

In many applications, a KBP is used in conjunction with an MCMC-type of sampling method, where putative conformational changes, $\mathbf{r}'$, is proposed according to some conditional distribution $q(\mathbf{r}'|\mathbf{r})$, where $\mathbf{r}$ is the current state. Again, this proposal function may be derived from assembling fragments from a fragment library, a generative probabilistic model as discussed in Chap. 10, or some other valid sampling algorithm [276]. Setting the potential to zero, the sampling scheme will lead to some stationary distribution, $q(\mathbf{r}|\mathbf{a})$, which now *implicitly* represents the prior structural knowledge of the problem at hand. Again, Eq. 3.72 shows that the proper reference distribution for KBP-construction in this case is obtained simply by calculating the marginal distribution, $\tilde{q}(\boldsymbol{\lambda}|\mathbf{a})$, implied by the proposal distribution alone.[4]

In a canonical sampling, the MCMC-algorithm is constructed to ensure a uniform sampling, $q(\mathbf{r}|\mathbf{a}) = $ const., when the potential is set to zero, $E \equiv 0$. In this case, as discussed in Sect. 3.2.3, $\tilde{q}(\boldsymbol{\lambda}|\mathbf{a})$ will simply be the normalized density of states, $\tilde{q}(\boldsymbol{\lambda}|\mathbf{a}) \propto e^{S(\boldsymbol{\lambda})}$ for the given amino acid sequence and Eq. 3.76 reduces to the iterative Boltzmann inversion, Eq. 3.52. Thus, the present probabilistic formulation demonstrates, that the iterative Boltzmann inversion is necessitated by any choice of reference distribution that differs from the true density of states.

It is important to stress, however, that one does not have to insist that the KBPs should correspond to physical free energies. Indeed, Eq. 3.72 shows that the

---

[4]The specific requirement for this statement to be true in general is that $q(\mathbf{r}|\mathbf{a})$ satisfies the detailed balance equation $q(\mathbf{r}|\mathbf{a})q(\mathbf{r}'|\mathbf{r}) = q(\mathbf{r}'|\mathbf{a})q(\mathbf{r}|\mathbf{r}')$ [276].

KBP is uniquely defined for any well-defined choice of $\lambda$ and prior distribution $q$, which may either be defined explicitly or implicitly through a given sampling procedure. This latter point deserves special attention, since different applications of KBPs, from threading, fold recognition to structure prediction, typically involve very different sampling methodologies. Therefore, we conclude that KBPs can not in general be defined independently from their domain of application.

## 3.5 Summary

Knowledge based potentials (KBP) have proven to be surprisingly useful not only for protein structure prediction but also for quality assessment, fold recognition and threading, protein-ligand interactions, protein design and prediction of binding affinities. The construction of KBPs is often loosely justified by analogy to the potential of mean force in statistical physics. While the two constructions are formally similar, it is often unclear how to define the proper reference state in the knowledge-based approach. Furthermore, these constructs are typically based on some unfounded assumptions of statistical independence, the implications of which are far from trivial. Therefore, KBPs are in general neither real potentials, free energies or potentials of mean force in the ordinary statistical mechanics sense.

If KBPs are intended to be used as physical potentials, they most often need to be iteratively refined to ensure self-consistency within the canonical ensemble. This can for instance be achieved by means of Markov chain Monte Carlo sampling. However, KBPs have many practical applications which do not rely on their specific link to physical energies. The fact that appropriate KBPs can be constructed from any well-defined sampling procedure and any choice of coarse-grained variables, $\lambda$, as shown in Sect. 3.4.2, opens up for a wide range of possible applications based on sound probabilistic reasoning [276].

The steady increase in available structural data will enable even more detailed potentials, including for example detailed many-body effects or the presence of metal ions and cofactors. However, the development of better potentials is not so much limited by the amount of available data, as by better formulations that use the data in a more efficient manner. This is an area where methods from the field of machine learning are extremely promising [274], and it will indeed be interesting to see whether the next generation of KBPs can take advantage of modern statistical methods.

**Acknowledgements** We acknowledge funding by the Danish *Program Commission on Nanoscience, Biotechnology and IT (NABIIT)* (Project: Simulating proteins on a millisecond time-scale) and the *Danish Research Council for Technology and Production Sciences (FTP)* (Project: Data driven protein structure prediction).

# Chapter 4
# Towards a General Probabilistic Model of Protein Structure: The Reference Ratio Method

**Jes Frellsen, Kanti V. Mardia, Mikael Borg, Jesper Ferkinghoff-Borg, and Thomas Hamelryck**

## 4.1 Introduction

The recently introduced *reference ratio method* [276] allows combining distributions over fine-grained variables with distributions over coarse-grained variables in a meaningful way. This problem is a major bottleneck in the prediction, simulation and design of protein structure and dynamics. Hamelryck et al. [276] introduced the reference ratio method in this context, and showed that the method provides a rigorous statistical explanation of the so called *potentials of mean force* (PMFs). These potentials are widely used in protein structure prediction and simulation, but their physical justification is highly disputed [32, 390, 715]. The reference ratio method clarifies, justifies and extends the scope of these potentials.

In Chap. 3 the reference ratio method was discussed in the contexts of PMFs. As the reference ratio method is of general relevance for statistical purposes, we present the method here in a more general statistical setting, using the same notation as in our previous paper on the subject [803]. Subsequently, we discuss two example applications of the method. First, we present a simple educational example, where the method is applied to independent normal distributions. Secondly, we reinterpret an example originating from Hamelryck et al. [276]; in this example, the reference ratio method is used to combine a detailed distribution over the dihedral angles of a protein with a distribution that describes the compactness of the protein using

---

J. Frellsen (✉) · M. Borg · T. Hamelryck
The Bioinformatics Centre, University of Copenhagen, Copenhagen, Denmark
e-mail: frellsen@binf.ku.dk; borg@binf.ku.dk; thamelry@binf.ku.dk

K.V. Mardia
Department of Statistics, University of Leeds, Leeds, LS2 9JT, UK
e-mail: k.v.mardia@leeds.ac.uk

J. Ferkinghoff-Borg
Department of Electrical Engineering, Technical University of Denmark, Lyngby, Denmark
e-mail: jfb@elektro.dtu.dk

T. Hamelryck et al. (eds.), *Bayesian Methods in Structural Bioinformatics*,
Statistics for Biology and Health, DOI 10.1007/978-3-642-27225-7_4,
© Springer-Verlag Berlin Heidelberg 2012

126 J. Frellsen et al.

the radius of gyration. Finally, we outline the relation between the reference ratio method and PMFs, and explain the origin of the name "reference ratio". For clarity, the formal definitions of the probability density functions used in the text, as well as their assumed properties, are summarized in an appendix at the end of the chapter.

## 4.2 The Reference Ratio Method

We start by introducing the reference ratio method in a general statistical setting, following the notation of our previous paper [803]. Consider the two random variables $\mathbf{X}$ and $Y$. We assume that

(i) the probability density function (pdf) $g(\mathbf{x})$ of $\mathbf{X}$ is specified, and
(ii) the pdf $f_1(y)$ of $Y = m(\mathbf{X})$ is given, where $m(\cdot)$ is a known many-to-one function.

In the work of Hamelryck et al. [276], $\mathbf{X}$ is denoted the *fine-grained variable* and $Y$ the *coarse-grained variable* due to their functional relation. Both variables can take values in multidimensional spaces, but to indicate that the range of $\mathbf{X}$ typically is of higher dimension than the range of $Y$, we will use vector notation for $\mathbf{X}$ and scalar notation for $Y$. Specially for $g(\mathbf{x})$ we will also assume that

(iii) the pdf $g_1(y)$ of the coarse-grained variable $Y$ can be obtained from $g(\mathbf{x})$, while
(iv) the conditional pdf $g_2(\mathbf{x}|y)$ of $\mathbf{X}$ given $Y$ is unknown and not easily obtained from $g(\mathbf{x})$.

Now, we want to construct a new density $\hat{f}(\mathbf{x})$ such that

(v) the pdf of the coarse-grained variable $Y$ for $\hat{f}(\cdot)$ is equal to $f_1(y)$ and
(vi) the conditional pdf of $\mathbf{X}$ given $Y = y$ for $\hat{f}(\cdot)$ is equal to $g_2(\mathbf{x}|y)$.

In other words $\hat{f}(\mathbf{x})$ should have the properties that

$$\hat{f_1}(y) = f_1(y) \quad \text{and} \quad \hat{f_2}(\mathbf{x}|y) = g_2(\mathbf{x}|y) , \qquad (4.1)$$

where $\hat{f_1}(y)$ and $\hat{f_2}(\mathbf{x}|y)$ respectively denotes the distribution of the coarse-grained variable $Y$ and the conditional distribution of $\mathbf{X}$ given $Y$ for $\hat{f}(\cdot)$.

It would be straightforward to construct $\hat{f}(\mathbf{x})$ if the conditional pdf $g_2(\mathbf{x}|y)$ was known. In particular, generation of samples would be efficient, since we could sample $\bar{y}$ according to $f_1(\cdot)$ and subsequently sample $\tilde{\mathbf{x}}$ according to $g_2(\cdot|\bar{y})$, if efficient sampling procedures were available for the two distributions. However, as previously stated $g_2(\mathbf{x}|y)$ is assumed unknown. An approximate solution for sampling could be to approximate the density $g_2(\mathbf{x}|y)$ by drawing a large amount of samples according to $g(\mathbf{x})$ and retain those with the required value of $Y$. Obviously, this approach would be intractable for a large sample space. The solution to the

# 4 Towards a General Probabilistic Model of Protein Structure

problem was given by Hamelryck et al. [276] and is summarized in the following theorem.

**Theorem 4.1.** *The conditions (v) and (vi) are satisfied for the pdf given by*

$$\hat{f}(\mathbf{x}) = \frac{f_1(y)}{g_1(y)} g(\mathbf{x}),$$ (4.2)

*where* $y = m(\mathbf{x})$.

*Proof.* First consider an arbitrary pdf $h(\mathbf{x})$ of $\mathbf{X}$. Since $Y$ is a function of $\mathbf{X}$, we can express the density of $\mathbf{X}$ in terms of the pdf $h_1(y)$ of the coarse-grained variable $Y$ and the conditional pdf $h_2(\mathbf{x}|y)$ of $\mathbf{X}$ given $Y$ by

$$h(\mathbf{x}) = h_1(m(\mathbf{x})) h_2(\mathbf{x}|m(\mathbf{x})).$$ (4.3)

This means that the pdf $\hat{f}(\mathbf{x})$ of $\mathbf{X}$ can be written as

$$\hat{f}(\mathbf{x}) = \hat{f}_1(m(\mathbf{x})) \hat{f}_2(\mathbf{x}|m(\mathbf{x})),$$ (4.4)

where $\hat{f}_1(y)$ and $\hat{f}_2(\mathbf{x}|y)$ denotes the pdf of $Y$ and the conditional pdf of $\mathbf{X}$ given $Y$ implied by $\hat{f}(\mathbf{x})$. By inserting the desired pdfs from Eq. 4.1 in the expression above we obtain

$$\hat{f}(\mathbf{x}) = f_1(m(\mathbf{x})) g_2(\mathbf{x}|m(\mathbf{x})) = \frac{f_1(m(\mathbf{x}))}{g_1(m(\mathbf{x}))} g(\mathbf{x}),$$

where we used Eq. 4.3 to expand the term $g_2(m(\mathbf{x})|y)$. By construction $\hat{f}(\mathbf{x})$ satisfies the conditions (v) and (vi). □

## 4.2.1 Kullback-Leibler Optimality

Another way to look at the reference ratio method, is to consider it as a technique for modifying the pdf $g(\mathbf{x})$ such that it attains the pdf $f_1(y)$ of the coarse-grained variable. In this view, the reference ratio distribution represents the minimal modification of $g(\mathbf{x})$ in terms of the Kullback-Leibler divergence. We will show this in the theorem below.

**Theorem 4.2.** *Consider the set of all pdfs of* $\mathbf{X}$ *that imply the pdf* $f_1(y)$ *of* $Y$, $D = \{h(\mathbf{x}) \,|\, \forall \tilde{y} : \int_{\mathbf{x} \in \{\tilde{\mathbf{x}} \,|\, m(\tilde{\mathbf{x}}) = \tilde{y}\}} h(\mathbf{x}) \, d\mathbf{x} = f_1(\tilde{y})\}$. *The density* $\hat{f}(\mathbf{x})$ *constructed by the reference ratio method is the pdf in* $D$ *with the minimal Kullback-Leibler divergence from* $g$, *that is* $\hat{f} = argmin_{h \in D} \, \mathrm{KL}[g \,\|\, h]$.

*Proof.* We want to find $\hat{h} \in D$ that minimizes the Kullback-Leibler divergence between $g$ and $\hat{h}$. Using the definition of Kullback-Leibler divergence and Eq. 4.3 we have

$$\hat{h} = \underset{h \in D}{\arg\min} \, \text{KL}[g \parallel h]$$

$$= \underset{h \in D}{\arg\min} \int_{\mathbf{x}} g(\mathbf{x}) \log \frac{g(\mathbf{x})}{h(\mathbf{x})} \, d\mathbf{x}$$

$$= \underset{h \in D}{\arg\min} \int_{y} \int_{\mathbf{x} \in \{\tilde{\mathbf{x}} \mid m(\tilde{\mathbf{x}}) = y\}} g_1(y) g_2(\mathbf{x}|y) \left[ \log \frac{g_1(y)}{h_1(y)} + \log \frac{g_2(\mathbf{x}|y)}{h_2(\mathbf{x}|y)} \right] d\mathbf{x} \, dy$$

$$= \underset{h \in D}{\arg\min} \underbrace{\int_{y} g_1(y) \log \frac{g_1(y)}{f_1(y)} \, dy}_{A} + \underbrace{\int_{y} g_1(y) \int_{\mathbf{x} \in \{\tilde{\mathbf{x}} \mid m(\tilde{\mathbf{x}}) = y\}} g_2(\mathbf{x}|y) \log \frac{g_2(\mathbf{x}|y)}{h_2(\mathbf{x}|y)} \, d\mathbf{x} \, dy}_{B}$$

where we have used $h_1(y) = f_1(y)$ in the term $A$. The first term $A$ is does not depend on $h$. It follows from Jensen's inequality [344] that the integral $B$ is non-negative, and consequently it obtains the minimal value of zero only when $h_2 = g_2$. This means that the whole expression is minimized when the conditional pdf of $\mathbf{X}$ given $Y$ for $h$ is equal to $g_2$. As this conditional density is indeed equal to $g_2$ for the reference ratio density, this shows that the reference ratio density minimizes the Kullback-Leibler divergence to $g$. $\qquad\square$

In the following sections we will present two applications of the reference ratio method.

## 4.3 Example with Independent Normals

The purpose of our first example is purely educational. It is a simple toy example based on independent normal distributions, which simplifies the functional form of the pdfs involved. Let $\mathbf{X} = (X_1, X_2)$, where $X_1$ and $X_2$ are independent normals with

$$X_1 \sim \mathcal{N}(\mu, 1) \quad \text{and} \quad X_2 \sim \mathcal{N}(0, 1).$$

Accordingly, the pdf of $\mathbf{X}$ is given by

$$f(\mathbf{x}) = c \, e^{-\frac{1}{2}(x_1 - \mu)^2 - \frac{1}{2}x_2^2},$$

where $\mathbf{x} = (x_1, x_2)$ and $c$ is the normalizing constant. In this example we assume that we not only know the density, $f_1(y)$, of the coarse-grained $Y$, but the full density, $f(\cdot)$, for $\mathbf{X}$. For the distribution $g(\mathbf{x})$ let $X_1$ and $X_2$ be independently distributed as

$$X_1 \sim \mathcal{N}(0, 1) \quad \text{and} \quad X_2 \sim \mathcal{N}(0, 1).$$

Note the different means of the distributions of $X_1$. Consequently the pdf of $\mathbf{X}$ is given by

$$g(\mathbf{x}) = d \ e^{-\frac{1}{2}x_1^2 - \frac{1}{2}x_2^2} \ ,$$

where $d$ is the normalizing constant. Suppose that $Y = m(\mathbf{X}) = X_1$. This means that the pdf of $Y$ for $f(\cdot)$ is

$$f_1(y) = c' \ e^{-\frac{1}{2}(x_1-\mu)^2} \ ,$$

and for $g(\cdot)$ the density of $Y$ is

$$g_1(y) = d' \ e^{-\frac{1}{2}x_1^2} \ ,$$

where $c'$ and $d'$ are the appropriate normalizing constants. Note that for both $f(\cdot)$ and $g(\cdot)$ the conditional density of $\mathbf{X}$ given $Y$ is the same and equal to the pdf of the normal distribution $\mathcal{N}(0, 1)$.

By applying the ratio method from Eq. 4.2, we obtain the expression

$$\hat{f}(\mathbf{x}) = \frac{c' \ e^{-\frac{1}{2}(x_1-\mu)^2} \ d \ e^{-\frac{1}{2}x_1^2-\frac{1}{2}x_2^2}}{d' \ e^{-\frac{1}{2}x_1^2}} = c \ e^{-\frac{1}{2}(x_1-\mu)^2-\frac{1}{2}x_2^2} \ . \tag{4.5}$$

In this example we observed that $\hat{f}(\cdot) = f(\cdot)$, which is expected since the conditional distribution of $\mathbf{X}$ given $Y$ is the same for both $f(\cdot)$ and $g(\cdot)$. Accordingly, it is now trivial to check that the distribution of $Y$ for $\hat{f}(\cdot)$ is equal to $f_1(\cdot)$ and that the conditional distribution of $\mathbf{X}$ given $Y$ is $g_2(\mathbf{x}|y)$, as stated in (v) and (vi).

In most relevant applications of the reference ratio method, the conditional density $f_2(\mathbf{x}|y)$ is unknown. In the next section we will consider such an example.

## 4.4 Sampling Compact Protein Structures

A more realistic application of the reference ratio method is given by Hamelryck et al. [276]. In this example the method is used to sample compact protein structures. The fine-grained variable in this example will be the dihedral angles in the protein main chain. In Chap. 10 we discussed TORUSDBN [68], which is a probabilistic model of the dihedral angles in the protein main chain. TORUSDBN captures the structure of proteins on a local length scale. However, it does not capture global properties, such as the compactness. The compactness of a protein can be roughly described by its radius of gyration, which is defined as the root mean square distance between the atoms in the protein and the geometric center of the atoms. In this example the reference ratio method is used to combine TORUSDBN with a normal distribution over radius of gyration. The setup is as follows (see also Fig. 4.1):

(a) Let $\mathbf{X} = ((\phi_i, \psi_i))_{i=1,\dots,n}$ be a sequence of dihedral angle pairs in a protein with a known sequence of $n$ amino acids.

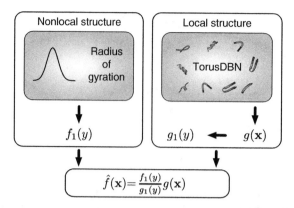

**Fig. 4.1** An application of the reference ratio method. The purpose of this application is to sample protein structures with a given distribution over the radius of gyration and a plausible local structure. The desired radius of gyration is given by the pdf $f_1(y)$ (*left box*). In this example $f_1(y)$ is the normal distribution $\mathcal{N}(22\,\text{Å}, 4\,\text{Å}^2)$, but typically this distribution would be derived from known structures in the protein data bank (PDB) [45]. TORUSDBN is a distribution, $g(\mathbf{x})$, over the sequence of dihedral angles, $\mathbf{X}$, in the protein main chain (*right box*). TORUSDBN describes protein structure on a local length scale. The distribution over the radius of gyration imposed by TORUSDBN is $g_1(y)$. The desired distribution over the radius, $f_1(y)$, and TORUSDBN, $g(\mathbf{x})$, can be combined in a meaningful way using the reference ratio method (*formula at the bottom*). The reference ratio distribution, $\hat{f}(\mathbf{x})$, has the desired distribution for the radius of gyration (The figure is adapted from Fig. 1 in Hamelryck et al. [276])

(b) The pdf, $g(\mathbf{x})$, of the fined grained variable $\mathbf{X}$ is given by TORUSDBN.
(c) Let $Y = m(\mathbf{X})$ be the radius of gyration of the protein, and assume that the pdf $f_1(y)$ of $Y$ is the normal distribution $\mathcal{N}(22\,\text{Å}, 4\,\text{Å}^2)$.
(d) The density $g_1(y)$ of the coarse-grained variable is obtained by generalized ensemble sampling [180] from $g(\mathbf{x})$ [276], which can be done since TORUSDBN is a generative model.

The reference ratio method is now applied to construct the density $\hat{f}(\cdot)$, based on the normal distribution over the radius of gyration, $f_1(y)$, the TORUSDBN distribution, $g(\mathbf{x})$, and the distribution over the radius of gyration for TORUSDBN, $g_1(y)$. It is important to stress that typical samples generated from TORUSDBN, $g(\mathbf{x})$, are unfolded and non-compact, while typical samples from $\hat{f}(\mathbf{x})$ will be more compact as the radius of gyration is controlled by the specified normal distribution. Accordingly, samples from the reference ratio distribution, $\hat{f}(\mathbf{x})$, are expected to look more like folded structures than samples from $g(\mathbf{x})$.

Hamelryck et al. [276] test this setup on the protein ubiquitin, which consists of 76 amino acids. Figure 4.2 shows the distribution over $Y$ obtained by sampling from $g(\mathbf{x})$ and $\hat{f}(\mathbf{x})$, respectively. The figure also shows the normal density $f_1(y)$. We observe that samples from $g(\mathbf{x})$ have an average radius of gyration around 27 Å,

# 4 Towards a General Probabilistic Model of Protein Structure

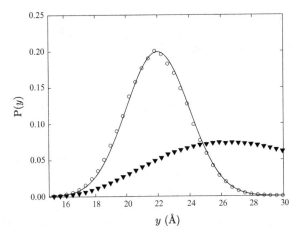

**Fig. 4.2** The reference ratio method applied to sampling protein structures with a specified distribution over the radius of gyration ($Y$). The distribution over the radius of gyration $Y$ for samples from TORUSDBN, $g(\mathbf{x})$, is shown as *triangles*, while the distribution for samples from the ratio distribution, $\hat{f}(\mathbf{x})$, is shown as *circles*. The pdf, $f_1(y)$, for the desired normal distribution over $Y$, $\mathcal{N}(22\,\text{Å}, 4\,\text{Å}^2)$, is shown as a *solid line*. The samples are produced using the amino acid sequence of ubiquitin (The figure is adapted from Fig. 3 in Hamelryck et al. [276])

while samples from $\hat{f}(\mathbf{x})$ indeed have a distribution very near $f_1(y)$. As expected, samples from $\hat{f}(\mathbf{x})$ are compact, unlike samples from $g(\mathbf{x})$.

A key question here is how to sample from $\hat{f}(\mathbf{x})$ efficiently? From a generative point of view, we could use Eq. 4.4 directly and generate a sample, $\tilde{\mathbf{x}}$, using the two steps:

1. Sample $\tilde{y}$ according to $f_1(y)$ and
2. Sample $\tilde{\mathbf{x}}$ according to $g_2(\mathbf{x}|\tilde{y})$.

However, the problem lies in step 2, as there is no efficient way to sample from $g_2(\mathbf{x}|y)$; TORUSDBN only allows efficient sampling from $g(\mathbf{x})$. One could consider using rejection sampling or the Approximate Bayesian computation (ABC) method [29,486,591] for step 2, but either method would be very inefficient. Hamelryck et al. [276] have given a highly efficient method, which does not, in principle, involve any approximations. The idea is to use the Metropolis-Hastings algorithm with $g(\mathbf{x})$ as proposal distribution and $\hat{f}(\mathbf{x})$ as target distribution. In this case, the probability of accepting a proposed value $\mathbf{x}'$ given a previous value $\mathbf{x}$ becomes

$$\alpha(\mathbf{x}'|\mathbf{x}) = \min\left(1, \frac{f_1(y')g(\mathbf{x}')/g_1(y')}{f_1(y)g(\mathbf{x})/g_1(y)}\frac{g(\mathbf{x})}{g(\mathbf{x}')}\right) = \min\left(1, \frac{f_1(y')}{f_1(y)}\frac{g_1(y)}{g_1(y')}\right), \quad (4.6)$$

where $y = m(\mathbf{x})$ and $y' = m(\mathbf{x}')$. In practice, the proposal distribution in the MCMC algorithm would only change a randomly chosen consecutive subsequence of $\mathbf{X}$ using TORUSDBN (see Chap. 10 or supporting information in Boomsma et al. [68] for details), as this leads to a higher acceptance rate. It can be shown that the acceptance probability in this case also is given by Eq. 4.6.

## 4.5 The Reference Ratio Method Explains PMFs

Methods for predicting the structure of proteins rely on an energy function or probability distribution that describes the space of possible conformations. One approach to constructing such energies or distributions is to estimate them from a set of experimental determined protein structures. In this case they are called *knowledge based potentials*. Chap. 3 discusses knowledge based potentials in-depth. In this section we outline the relation between the reference ratio method and knowledge based potentials.

A subclass of the knowledge based potentials are based on probability distributions over pairwise distances in proteins. These are called *potentials of mean force* (PMFs) and are loosely based on an analogy with the statistical physics of liquids [32, 390]. The potential of mean force, $W(\mathbf{r})$, associated with a set of pairwise distances $\mathbf{r}$ is given by an expression of the form

$$W(\mathbf{r}) \propto -\log \frac{f_1(\mathbf{r})}{g_1(\mathbf{r})} \, ,$$

where $f_1(\mathbf{r})$ is a pdf estimated from a database of known protein structures, and $g_1(\mathbf{r})$ is the pdf for a so-called *reference state*. The reference state is typically defined based on physical considerations. The pdf $f_1(\mathbf{r})$ is constructed by assuming that the individual pairwise distances are conditionally independent, which constitutes a crude approximation. In practice, the potential of mean force is combined with an additional energy function, that is concerned with the local structure of proteins. This additional energy term is typically brought in via sampling from a *fragment library* [658] – a set of short fragments derived from experimental protein structures – or any other sampling method that generates protein-like conformations. From a statistical point of view, this means that the samples are generated according to the pdf

$$\hat{f}(\mathbf{x}) \propto \frac{f_1(\mathbf{r})}{g_1(\mathbf{r})} g(\mathbf{x}) \, , \tag{4.7}$$

where $\mathbf{x}$ are the dihedral angles in the protein, $\mathbf{r}$ are the pairwise distances implied by $\mathbf{x}$, and $g(\mathbf{x})$ is the pdf of the dihedral angles embodied in the sampling method.

In this formulation, it can be seen that PMFs are justified by the reference ratio method; their functional form arises from the combination of the sampling method (which concerns the fine-grained variable) with the pairwise distance information (which concerns the coarse-grained variable). This interpretation of PMFs also provides some surprising new insights. First, $g_1(\mathbf{r})$ is uniquely defined by $g(\mathbf{x})$, and does not require any external physical considerations. Second, if the three involved probability distributions are properly defined, the PMF approach is entirely rigorous and statistically well justified. Third, the PMF approach generalizes beyond pairwise distances to arbitrary coarse-grained variables. Fourth, PMFs should not be seen as physical potentials, but rather as statistical constructs.

Obviously, the name "reference ratio" refers to the fact that we now understand the nature of the reference state, and why the ratio arises in the first place. In conclusion, the reference ratio method settles a dispute over the validity of PMFs that has been going on for more than twenty years, and opens the way to efficient and well-justified probabilistic models of protein structure.

## 4.6 Conclusions

The reference ratio method is important for the development of a tractable and accurate probabilistic model of protein structure and sequence. Probabilistic models such as those described in Chap. 10 are tractable and computationally efficient, but only capture protein structure on a local length scale. It is important to point out that the two properties are closely linked: these models are computationally attractive *because* they only capture dependencies on a fairly local length scale. The reference ratio method provides a convenient, mathematically rigorous and computationally tractable way to "salvage" such models by including information on nonlocal aspects of protein structure.

A surprising aspect of the reference ratio method is that it finally provides the mathematical underpinnings for the knowledge based potentials that have been widely used – and hotly debated – for more than twenty years. In addition, the method opens the way to "potentials" that go beyond pairwise distances. We illustrated this in this chapter by using the radius of gyration; in ref. [276] the reference ratio method is also applied to hydrogen bonding. The latter application also illustrates how the method can be applied in an interactive way.

Finally, we would like to end by pointing out that the reference ratio method could find wide applications in statistical modelling. The method reconciles the advantages of relatively simple models that capture "local" dependencies with the demand for capturing "nonlocal" dependencies as well. For example, the reference ratio method makes it possible to correctly combine two priors that respectively bring in information on fine- *and* coarse-grained variables. The method might thus be applied to image recognition, document classification or statistical shape analysis in general.

**Acknowledgements** The authors acknowledge funding by the Danish Research Council for Technology and Production Sciences (FTP, project: Protein structure ensembles from mathematical models, 09-066546).

# Appendix

In this appendix, we given an overview of all the pdfs mentioned in the chapter, their formal definitions and their assumed properties. We also briefly outline the ratio method at the end.

We start by noting that $\mathbf{X}$ and $Y$ are random variables, with $Y = m(\mathbf{X})$, where $m(\cdot)$ is a many-to-one function. The random variables $\mathbf{X}$ and $Y$ are called *fine-grained* and *coarse-grained*, respectively. Normally, the dimensionality of $Y$ is smaller than that of $\mathbf{X}$. To emphasize this, we will use a bold vector notation for $\mathbf{X}$ but not for $Y$, though $Y$ is often a vector as well.

The pdfs are:

- $f_1(y)$: the pdf over the coarse-grained variable $Y = m(\mathbf{X})$. This pdf is assumed to be "true" for practical purposes. We assume that this pdf can be evaluated.
- $g(\mathbf{x})$: the approximate pdf over $X$. This pdf is assumed to be approximative, in the sense that its associated *conditional* pdf $g_2(\mathbf{x}|y)$ is "true" for practical purposes. However, its associated pdf $g_1(y)$ of the coarse-grained variable $Y$ differs from the desired pdf $f_1(y)$. One way to view $g(\mathbf{x})$ is as approximately correct on a local scale, but incorrect on a global scale. The pdfs $g_1(y)$ and $g_2(\mathbf{x}|y)$ are defined below. We assume that $g(\mathbf{x})$ can be simulated.
- $g_1(y)$: the pdf for $Y$ implied by $g(\mathbf{x})$. We assume that this pdf can be correctly obtained from $g(\mathbf{x})$, and can be evaluated. However, $g_1(y)$ is not "true" in the sense that it significantly differs from the desired pdf $f_1(y)$. The pdf $g_1(y)$ is given by

$$g_1(y) = \int_{\mathbf{x} \in \{\mathbf{x}' \, | \, m(\mathbf{x}') = y\}} g(\mathbf{x}) \, d\mathbf{x}.$$

- $g_2(\mathbf{x}|y)$: the conditional pdf of $\mathbf{X}$ given $Y$, as implied by $g(\mathbf{x})$. This pdf is also assumed to be "true" for practical purposes. However, this pdf cannot be easily evaluated or simulated. Formally, the distribution is given by

$$g_2(\mathbf{x} \mid y) = \begin{cases} 0 & \text{if } y \neq m(\mathbf{x}) \\ \frac{1}{Z(y)} g(\mathbf{x}) & \text{if } y = m(\mathbf{x}) \end{cases}$$

where the normalization factor is given by

$$Z(y) = \int_{\mathbf{x} \in \{\mathbf{x}' \, | \, m(\mathbf{x}') = y\}} g(\mathbf{x}) \, d\mathbf{x}.$$

- $f(\mathbf{x})$: this pdf is unknown and needs to be constructed. This pdf is given by

$$f(\mathbf{x}) = f_1(y) g_2(\mathbf{x}|y).$$

The pdf $f(\mathbf{x})$ results from combining the correct distribution over $Y$ with the correct conditional over $\mathbf{X}$. As explained above, the problem is that $g_2(\mathbf{x}|y)$ cannot be easily evaluated or simulated. The reference ratio method re-formulates $f(\mathbf{x})$ as

$$f(\mathbf{x}) = \frac{f_1(y)}{g_1(y)} g(\mathbf{x}),$$

where $y = m(\mathbf{x})$. In this way, $f(\mathbf{x})$ can be evaluated and simulated, because $g(\mathbf{x})$ allows simulation and the pdfs in the ratio $\frac{f_1(y)}{g_1(y)}$ can be evaluated.

# Chapter 5
# Inferring Knowledge Based Potentials Using Contrastive Divergence

**Alexei A. Podtelezhnikov and David L. Wild**

## 5.1 Introduction

Interactions between amino acids define how proteins fold and function. The search for adequate potentials that can distinguish the native fold from misfolded states still presents a formidable challenge. Since direct measurements of these interactions are impossible, known native structures themselves have become the best experimental evidence. Traditionally, empirical 'knowledge-based' *statistical potentials* were proposed to describe such interactions from an observed ensemble of known structures. This approach typically relies on some unfounded assumptions of statistical independence as well as on the notion of a reference state defining the expected statistics in the absence of interactions. We describe an alternative approach, which uses a novel statistical machine learning methodology, called *contrastive divergence*, to learn the parameters of statistical potentials from data, thus inferring force constants, geometrical cut-offs and other structural parameters from known structures. Contrastive divergence is intertwined with an efficient Metropolis Monte Carlo procedure for sampling protein main chain conformations. Applications of this approach have included a study of protein main chain hydrogen bonding, which yields results which are in quantitative agreement with experimental characteristics of hydrogen bonds. From a consideration of the requirements for efficient and accurate reconstruction of secondary structural elements in the context of protein structure prediction, we demonstrate the applicability of the framework

---

A.A. Podtelezhnikov (✉)
Michigan Technological University, Houghton, MI, USA
e-mail: apodtele@gmail.com

D.L. Wild
University of Warwick, Coventry, CV4 7AL, UK
e-mail: d.l.wild@warwick.ac.uk

T. Hamelryck et al. (eds.), *Bayesian Methods in Structural Bioinformatics*,
Statistics for Biology and Health, DOI 10.1007/978-3-642-27225-7_5,
© Springer-Verlag Berlin Heidelberg 2012

to the problem of reconstructing the overall protein fold for a number of commonly studied small proteins, based on only predicted secondary structure and contact map.

## 5.2 Background

Computer simulations continue to shed light on the phenomena of protein folding and function [424, 492]. Protein modeling and structure prediction from sequence faces two major challenges. The first is the difficulty of efficient sampling in the enormous conformational space, which is especially critical for molecular dynamics and Markov Chain Monte Carlo simulations (MCMC) [385, 631]. The techniques of threading [75, 352] and homology modeling [64, 620] avoid sampling altogether by only considering known structures as templates. The second challenge is the development of the energy function describing molecular interactions for the problem at hand. All major paradigms of structural modeling rely on either physical or statistical potentials to guide the simulations or assess the quality of prediction.

In protein modeling, physical potentials between atoms or residues cover covalent bonding, hydrogen bonding, electrostatic interactions, van der Waals repulsion and attraction, hydrophobic interactions, etc. The exact formulation of these potentials is usually taken from *first principles* or experimental evidence. Although some progress has been achieved in experimental studies of protein energetics, the current interpretation of protein interactions is still not sufficiently detailed [20]. Whilst ab initio computer simulations are in qualitative agreement with the theoretical view of the folding landscape [650], atomistic force fields based on first principles lack the precision to quantitatively describe experimental observations [20]. In the absence of adequate theoretical potentials, the field of protein structure prediction has shifted to empirical potentials [423]. In particular, knowledge-based potentials and scoring functions derived from the statistics of native structures have gained popularity and recognition [88] as discussed in detail in Chap. 3. Empirical potentials, or their parameters, can also be optimized to discriminate native conformations from decoy alternatives [670]. Although such knowledge-based potentials are responsible for recent advances in protein structure prediction [608], they still require significant improvement [670].

In this chapter, we will focus on a methodology called *contrastive divergence* (CD) learning. After originating in the machine learning and artificial intelligence community just a few years ago [306], contrastive divergence learning has mostly been used in the fields such as computer vision and microchip design that are very far from biological applications [190, 614]. It belongs to the family of maximum likelihood (ML) methods. Surprisingly, although ML methods are very commonplace in computational genetics [321], protein secondary structure prediction [113, 633], and protein crystal structure refinement [532], they are used very rarely for the purposes of protein folding and tertiary structure prediction [378, 768]. The performance of contrastive divergence learning makes it an attractive option for biological computer simulations.

# 5 Inferring Knowledge Based Potentials Using Contrastive Divergence

Knowledge based potentials are typically derived from simple statistical analysis of the observed protein structures. CD learning is, on the other hand, a very powerful technique to optimize the interaction parameters of an arbitrarily chosen protein model. CD learning achieves thermodynamic stability of the model in agreement with a set of observed protein structures. In this chapter, we describe a protein model which features an all atom representation of rigid peptide bonds elastically connected at $C\alpha$ atoms [583, 585]. Amino acid side-chains are reduced to single $C\beta$ atoms. CD learning relies heavily on the performance of a Monte Carlo (MC) sampler. The reduced number of degrees of freedom in our model helps speed up the simulations [385]. In addition, the conformations of just a few amino acids are perturbed locally on each step, leaving the rest of the chain intact, which increases the acceptance probability of the attempted moves and the efficiency of the Metropolis MC procedure [172]. We model the polypeptide chain in Cartesian space using efficient *crankshaft moves* [61].

The remainder of this chapter is structured as follows. Because of significant overlap between the assumptions behind the CD learning approach and other methodologies, we start from a brief overview of the previous chapter regarding statistical potentials and native structure discriminants. We then describe the CD learning approach and how it can be applied to a physically motivated Boltzmann distribution. A critical comparison of the CD learning procedure with traditional knowledge-based potentials is an important focus of this chapter. In the second half of the chapter, we introduce our protein model and MC sampler and provide an overview of the application of CD learning to protein interactions. In particular, we investigate hydrogen bonding and the side-chain interactions that stabilize secondary structure elements. We test our approach by reconstructing the tertiary structure of two proteins: protein G and SH3 domain. We conclude by discussing outstanding questions in protein energetics that can potentially be addressed by contrastive divergence learning.

## 5.3 Statistical Potentials

We assume that the overall protein conformation $r = \{r_1, \cdots, r_N\} \in \Omega$ is governed by the Boltzmann distribution. Common to the standard knowledge-based potential approach and the approach of contrastive divergence is the parameterization of energies according to a subset of structural features, $\lambda(r) = \{\lambda_1(r), \lambda_2(r), \dots\}$, such as dihedral angles $\varphi$ and $\psi$, separations between $C\alpha$ atoms, burial of a particular residue in the protein core, direct contact between particular residues, or the presence of a hydrogen bond. Correspondingly, we define $\Omega(\lambda) = \{r | \lambda(r) = \lambda\}$ to be the set of conformations adopting a particular structural feature $\lambda$.

Knowledge based potentials (KBP) are typically derived by assuming that the statistics observed in native states comply with the Boltzmann distribution (the *Boltzmann hypothesis*) and that the selected structural features are statistically independent. In that case, an expression for the (free)-energies

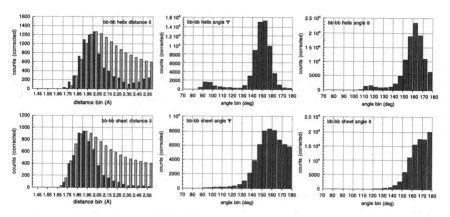

**Fig. 5.1** Distribution of O-H distances and angles that characterize hydrogen bonding in globular proteins. Distributions observed in helices and sheets are significantly different (Reproduced from [392] with permission)

$U(\lambda_i)$ is obtained as

$$U(\lambda_i) = -k_B T \ln \frac{n_{\text{obs}}(\lambda_i)}{n_{\text{exp}}(\lambda_i)}, \qquad (5.1)$$

where $n_{\text{obs}}(\lambda_i)$ is the number of observations in crystal structures taken from the Protein Data Bank (PDB) [45], $n_{\text{exp}}(\lambda_i)$ is the expected (reference) number of observations, and $k_B T$ is the Boltzmann constant times the temperature. This framework was first introduced by Pohl to analyze dihedral angle potentials using a uniform reference [587]. It was later proposed to study residue and atomic pairing frequencies [516, 706], with more detailed distance-dependent pairing potentials introduced much later [661]. Subsequently, statistical potentials were applied to residue contacts in protein complexes [519]. Figure 5.1 [392] illustrates the application of statistical potentials to distance- and angle-dependent hydrogen bonding.

Shortle [656] argues that the Boltzmann hypothesis represents a genuine evolutionary equilibrium, which has maintained the stability of each protein within a narrow set of values for these parameters. Bastolla et al. [24, 25] also proposed the idea that protein sequences have evolved via a process of neutral evolution about an optimal mean hydrophobicity profile. According to Jaynes [342], the maximum entropy principle (MaxEnt) offers a way to choose a probability distribution for which constraints imply that only the averages of certain functions (such as the hydrophobicity profile) are known. In the present application these may be considered to be evolutionary constraints on the protein sequence. Application of the MaxEnt principle would then lead to a Boltzmann-like distribution, where the partition function and "inverse temperature" are set to satisfy these constraints [342, 457].

There are several objectionable assumptions underlying the standard knowledge-based potentials, and their physical interpretation has been the subject of much criticism and debate [32,276,715]. First of all, the *ansatz* of statistical independence is generally very poor. Secondly, the reference state $n_{\exp}(\lambda_i)$ lacks a rigorous definition. However, recent research has vindicated these potentials as approximations of well-justified statistical quantities, clarified the nature of the reference state and extended their scope beyond pairwise interactions [276]. We refer to Chaps. 3 and 4 for a thorough discussion of the statistical and physical interpretation and validity of these potentials.

## 5.4 Native Structure Discriminants

Unlike polypeptides with a random sequence, proteins have a unique native state with energy much lower than any alternative conformation [11]. Furthermore, protein sequences have evolved so that they quickly find their native state as if following a funnel-like energy landscape [86]. According to the model proposed by Rose and co-workers, on the folding pathways, protein segments rapidly twist into compact $\alpha$-helices, or accumulate into $\beta$-sheets [21,22]. The entropy losses are compensated by favorable short-range interactions when the secondary structure elements are formed by amino acids meticulously joined in the sequence.

Figure 5.2 illustrates the folding funnel and a few selected meta-stable energy levels. This is considered a fundamental property of proteins; any model system attempting to mimic folding should mimic this property.

A simple model of a protein with sequence $a$ assumes that its energy is completely determined by the pairwise contacts between residues. In this model, the protein conformation is reduced to a binary contact map (a proximity matrix) between its $N$ residues, $\lambda = \{C_{ij}\}$. The total energy is thus given be the following equation

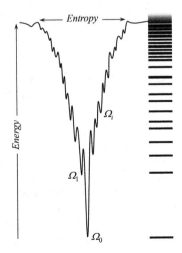

**Fig. 5.2** Folding funnel. This is a cartoon view of the protein energy landscape showing that the native conformation $\Omega_0$ has very low energy that is reached through gradual loss of entropy, sometimes encountering "glassy" meta-stable states $\Omega_i$, in the course of folding

$$E(\lambda,\theta) = \sum_{i=1}^{N}\sum_{j=1}^{i-1} E_{ij}$$

$$E_{ij} = \kappa_{a_i,a_j} C_{ij} \tag{5.2}$$

where the interaction parameters $\kappa_{a_i,a_j}$ are defined by the types of residues $a_i$ and $a_j$, giving the set of 210 parameters for 20 amino acid types. It is feasible that different adjustable parameter vectors, $\theta = \{\kappa_1,\dots,\kappa_{210}\}$, may render model proteins with different folding properties. The goal is to find the vector that reproduces the fundamental characteristics of protein folding and gives the native state the lowest energy. The optimal parameter values, in this case, may reflect the relative strength of pairwise residue interactions. In general,

$$\theta = \arg\min_{\theta} \frac{E(\lambda_0,\theta) - D(\lambda,\theta)}{\delta E(\lambda,\theta)} \tag{5.3}$$

where $E(\lambda_0,\theta)$ is the energy of the native state $\lambda_0$, $D(\lambda,\theta)$ is the energy of alternative (decoy) conformations, and $\delta E(\lambda,\theta)$ is a scaling factor. This approach has been proposed in a number of studies since the early 1990s, with implementation details largely driven by limited computer power. Instead of properly sampling alternative conformations in continuous space, researchers either generated decoys by threading the sequence onto the fixed main chain of other proteins [238, 463] or simulated hypothetical proteins on a cubic lattice [289, 512]. The set of alternative conformations was then reused throughout the optimization protocol.

In one implementation of this optimization, $D(\lambda,\theta)$ was estimated as the mean energy of all decoy conformations and the scaling factor $\delta E(\lambda,\theta)$ was equated to the standard deviation of the decoy energies [238, 289, 512]. In this approach Eq. 5.3 minimizes the negative Z-score of the native energy against the decoy energies. The solution draws heavily on the linear dependence of Eq. 5.2 with respect to the parameters. The numerator and the denominator in Eq. 5.3 can then be expressed as a dot product $A \cdot \theta$ and a quadratic form $\sqrt{\theta \cdot \mathbf{B}\theta}$. The vector $A$ and the matrix $\mathbf{B}$ can be evaluated for any native structure and the corresponding set of decoys. This optimization is reminiscent of feed-forward neural network approximations and explicitly gives $\theta = \mathbf{B}^{-1}A$ [238]. Gradient descent schemes have also been deployed for this optimization [289].

Another implementation sought to guarantee that the native energy is below *any* decoy energy [463, 522, 736]. As above, since the energy in Eq. 5.2 is a linear function of $\kappa$, this is equivalent to a number of simultaneous dot-product inequalities, $A_K \cdot \theta > 0$, where $A_K$ is defined by the difference between the contact maps of the $K$th decoy and the native state. Each of these inequalities dissects the 210-dimensional parameter space with a hyperplane, resulting in only a small region of allowed parameters. It was recognized early that only a small number of inequalities define the region boundaries. This means that the native energy has to be compared to a few low-energy decoys that define $D(\lambda,\theta)$ in Eq. 5.3,

5   Inferring Knowledge Based Potentials Using Contrastive Divergence          141

effectively maximizing the gap between the native state and the low-energy decoys. The denominator $\delta E(\lambda, \theta)$ in this approach is kept constant and disregarded.

The parameter optimization in this approach was done either by linear programming [463,522], support vector machine learning [317], or perceptron neural network learning algorithms [736]. The latter is an iterative scheme of cyclically presenting each $A_K$ and updating the parameters if they appear contradict the inequality, $A_K \cdot \theta \not> 0$.

$$\theta^{(i+1)} := \frac{\theta^{(i)} + \eta A_K}{|\theta^{(i)} + \eta A_K|} \tag{5.4}$$

where $\eta$ is a small positive learning rate. The procedure essentially moves the vector $\theta$ towards the hyperplane along its normal $A_K$ and normalizes $\theta$ afterwards. The procedure is supposed to converge when the inequalities for all decoys are satisfied. Note that, in 210-dimensional space, the solution does not necessarily exist when the number of inequalities is 210 or greater. Indeed, Vendruscolo and Domany [736] found that this procedure did not always converge, especially if the training set contained several hundreds of decoys. The pairwise contact potentials are, therefore, demonstrably insufficient (rather than unsuitable) to accurately discriminate the native state. The later inclusion of solvent accessibility surface and secondary structure information in a neural network framework greatly improved the accuracy of discrimination [750].

In the discussion so far we have given Eq. 5.3 and outlined the optimization procedures in applications to a single native state. For a dataset of a few hundreds of proteins the solution is obtained by either cycling through different proteins in iterative schemes or averaging the final results for individual proteins. Parameter optimization was also performed for more complex models of protein interactions, where the total energy was not necessarily a linear function of parameters, or contained a distance dependence instead of using a binary contact definition. In this case, gradient ascent methodologies were used to find the optimal energy discriminant [506,611].

Finally, we note that in describing native structure discriminant methods we never mentioned the thermal energy, $k_B T$. Fundamentally, discriminant analysis does not address the question of thermodynamic stability of proteins. The parameters that are optimal as native structure discriminants may only reflect the relative strength of interactions.

## 5.5   Contrastive Divergence

In general, the energy of a polypeptide $E(\lambda, \theta)$ is defined by its conformation $\lambda$ and arbitrary interaction parameters $\theta$. The interaction parameters may be as diverse as force constants, distance cut-offs, dielectric permittivity, atomic partial charges, etc. This energy, in turn, defines the probability of a particular conformation $\lambda$ via the Boltzmann distribution:

$$p(\lambda|\theta) = \frac{1}{Z(\theta)} \exp\left[-E(\lambda, \theta)\right]$$

$$Z(\theta) = \int d\lambda \exp\left[-E(\lambda, \theta)\right] \tag{5.5}$$

where $Z(\theta)$ is the partition function. Here, the energy is expressed in units of $k_B T$. Assuming that $\lambda_0$ is an *observed* conformation with energy near the minimum, the inverse problem of estimating the values of the parameters, $\theta$, can be solved by maximum likelihood (ML) optimization using the gradient ascent method [223]:

$$\theta^{(i+1)} := \theta^{(i)} + \eta \frac{\partial}{\partial\theta} \ln p(\theta|\lambda_0) = \theta^{(i)} + \eta \left[-\frac{\partial \ln Z(\theta)}{\partial\theta} - \frac{\partial E(\lambda_0, \theta)}{\partial\theta}\right] \tag{5.6}$$

where we used Bayesian equality $p(\theta|\lambda_0)p(\lambda_0) = p(\lambda_0|\theta)p(\theta)$ and differentiated Eq. 5.5, disregarding the prior $p(\theta)$. A positive learning rate $\eta$ needs to be small enough for the algorithm to converge. In general, the first term in the square brackets is equal to the expectation value for the energy gradient with respect to parameters, $\theta$,

$$-\frac{\partial \ln Z(\theta)}{\partial\theta} = -\frac{1}{Z(\theta)} \frac{\partial Z(\theta)}{\partial\theta}$$

$$= \frac{1}{Z(\theta)} \int d\lambda \frac{\partial E(\lambda, \theta)}{\partial\theta} \exp\left[-E(\lambda, \theta)\right]$$

$$= \left\langle \frac{\partial E(\lambda, \theta)}{\partial\theta} \right\rangle_\infty. \tag{5.7}$$

Here, $\langle\cdot\rangle_\infty = \mathbb{E}_p(\cdot)$ is the expectation of a quantity with respect to the equilibrium distribution $p$, corresponding to an infinitely long sampling time. After substituting Eq. 5.7 into Eq. 5.6 we obtain the generalized Boltzmann machine learning rule [307]:

$$\theta^{(i+1)} := \theta^{(i)} + \eta \frac{\partial}{\partial\theta} \ln p(\theta|\lambda_0) = \theta^{(i)} + \eta \left[\left\langle \frac{\partial E(\lambda, \theta)}{\partial\theta} \right\rangle_\infty - \frac{\partial E(\lambda_0, \theta)}{\partial\theta}\right] \tag{5.8}$$

Equation 5.8 can be generalized for the case of multiple observed protein structures, $\lambda_0$. From information theory, ML optimization by gradient ascent follows the gradient and minimizes the Kullback-Leibler divergence,

$$\mathrm{KL}[p(\lambda_0) \parallel p(\lambda_0, \theta)] = \sum_{\lambda_0} p(\lambda_0) \ln \frac{p(\lambda_0)}{p(\lambda_0, \theta)} \tag{5.9}$$

which reflects the difference between model distribution $p(\lambda_0, \theta)$ and the distribution of observations $p(\lambda_0)$. Differentiating this equation produces essentially the same result,

# 5 Inferring Knowledge Based Potentials Using Contrastive Divergence

$$\theta^{(i+1)} := \theta^{(i)} + \eta \frac{\partial}{\partial \theta} \text{KL}[p(\lambda_0) \parallel p(\lambda_0, \theta)]$$

$$= \theta^{(i)} + \eta \left[ \left\langle \frac{\partial E(\lambda, \theta)}{\partial \theta} \right\rangle_\infty - \left\langle \frac{\partial E(\lambda, \theta)}{\partial \theta} \right\rangle_0 \right] \tag{5.10}$$

where the subscript 0 signifies averaging over the dataset of originally observed conformations. This expression is as attractive as it is difficult to use in practice, because evaluating the expectation value of the energy gradient may require tremendously long equilibrium simulations. There was, however, a recent attempt to undertake such a feat [768].

Instead of extensively sampling conformations to determine the expectation value, Hinton [306] proposed an approximate ML algorithm called *contrastive divergence* (CD) learning. The intuition behind CD learning is that it is not necessary to run the MCMC simulations to equilibrium. Instead, just after a few steps, we should notice that the conformations start to diverge from the initial distribution. Iterative updates to the model interaction parameters will eventually reduce the tendency of the chain to leave the initial distribution. CD learning follows the gradient of a difference between two KL divergences,

$$\text{CD}_K = \text{KL}[p(\lambda_0) \parallel p(\lambda_0, \theta)] - \text{KL}[p(\lambda_K) \parallel p(\lambda_0, \theta)] \tag{5.11}$$

To obtain $\lambda_K$ in this expression, the original conformations are perturbed in the field of model potentials $E(\lambda, \theta)$ using a $K$-step MCMC procedure. The CD learning rule can be obtained by differentiating Eq. 5.11:

$$\theta^{(i+1)} := \theta^{(i)} + \eta \frac{\partial}{\partial \theta} \text{CD}_K = \theta^{(i)} + \eta \left[ \left\langle \frac{\partial E(\lambda, \theta)}{\partial \theta} \right\rangle_K - \left\langle \frac{\partial E(\lambda, \theta)}{\partial \theta} \right\rangle_0 \right] \tag{5.12}$$

In CD learning, therefore, the expectation value of the energy gradient is approximated as the energy gradient after a very small number of MCMC steps. In principle the number of steps can be as low as $K = 1$. A larger number of steps lead to a more accurate convergence towards the ML estimate of model parameters [100].

Let us emphasize that, in this formulation, contrastive divergence is a very general methodology for the iterative optimization of interaction parameters. The methodology requires a dataset of known equilibrium conformations $\lambda_0$ and a Metropolis Monte Carlo procedure to produce perturbed conformations $\lambda_K$. In the contrastive divergence approach, unlike the traditional approach of statistical potentials, no assumptions are made regarding the a priori distribution of conformations in the absence of interactions. The possible applications of this methodology reach far beyond biological molecules into the realms of nanotechnology and material science.

## 5.6 Protein Model

The Boltzmann distribution defines the probability $p(\lambda, a)$ that a protein sequence $a$ adopts a conformation $\lambda$. This probability can be factorized into the product of the sequence likelihood for a given conformation and the prior distribution of conformations, $p(\lambda, a) = p(a|\lambda)p(\lambda)$. This can be rewritten in energetic terms

$$E(\lambda, a) = -\ln p(a|\lambda) + E(\lambda) \qquad (5.13)$$

where sequence-*dependent* and sequence-*independent* contributions to the energy are separated. We assume that the sequence-independent term, $E(\lambda)$, is defined by short-range interactions between the polypeptide main chain and $C\beta$ atoms. At room temperature, van der Waals repulsions and covalent bonding between atoms are extremely rigid interactions that contribute to this energy. Another large contribution comes from hydrogen bonding, but the magnitude of this interaction is vaguely understood. The sequence-dependent part of the potential (the negative log-likelihood) can be approximated by pair-wise interactions between side-chains that make the largest contribution to this term.

We modeled van der Waals repulsions with hard-sphere potentials with prohibitively large energetic cost of overlaps between atoms. We used values of hard-sphere atomic radii close to a lower limit of the range found in the literature [314, 558, 595, 771]: $r(C\alpha) = r(C\beta) = 1.57\,\text{Å}$, $r(C) = 1.42\,\text{Å}$, $r(O) = 1.29\,\text{Å}$, $r(N) = 1.29\,\text{Å}$. We modeled the polypeptide as a chain of absolutely rigid peptide groups elastically connected at $\alpha$-carbons, with the valence angles constrained to $111.5° \pm 2.8°$. The positions of all peptide bond atoms including hydrogen were specified by the orientations of the peptide bonds. We fixed the peptide bond lengths and angles at standard values [80, 173, 174]. The distance between $C\alpha$ atoms separated by trans peptide bonds was fixed at $3.8\,\text{Å}$. The $C\beta$ positions were stipulated by the tetrahedral geometry of the $C\alpha$ atoms and corresponded to L-amino acids. Most of the conformational variability of polypeptides comes from relatively free rotation around $N - C\alpha$ and $C\alpha - C$ bonds characterized respectively by dihedral angles $\varphi$ and $\psi$. These rotations are least restricted in glycine that lacks $C\beta$. The dihedral angles $\varphi$ in proline were elastically constrained to $-60° \pm 7°$ by covalent bonding [308]. We introduced a harmonic potential $E_i^B$ to impose these and other elastic constraints. The atomic radii and peptide bond geometry were constant parameters in our model because they are well established experimentally and are not so interesting from the machine learning perspective. A more detailed description of the model was given in our previous work [585].

Hydrogen bonding is a major polar interaction between the $NH$ and $CO$ groups of the polypeptide main chain. Based on surveys of the Protein Data Bank (PDB) [45], important reviews of hydrogen bonding in globular proteins have formulated the basics of the current understanding of hydrogen bond geometry and networking [18, 493, 626, 690]. We considered the hydrogen bond formed when three distance and angular conditions were satisfied: $r(O, H) < \delta$, $\angle OHN > \Theta$, and $\angle COH > \Psi$,

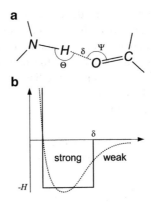

**Fig. 5.3** (**a**) Hydrogen bond geometry. The distance and two angular parameters of hydrogen bonds are shown. (**b**) Schematic one-dimensional approximation of hydrogen bond energy with a square-well potential. This approximation sharply discriminates between strong and weak hydrogen bonds. Weak bonds do not contribute to the total energy and are dropped from consideration in this work. The hydrogen bond strength $H$ corresponds to an average strength of hydrogen bonds (Reproduced from [586] with permission)

where $r(O, H)$ is the distance between oxygen and hydrogen, and symbol $\angle$ denotes the angle between the three atoms (see Fig. 5.3a). The lower bound on the separation between the atoms ($r(O, H) > 1.8$ Å) was implicitly set by the hard-sphere collision between oxygen and nitrogen. We used the same hydrogen bond potential regardless of the secondary structure adopted by the peptide main chain. The energy of the hydrogen bond (Fig. 5.3b) was described by a square-well potential,

$$E_{ij}^{HB} = -n_h H \qquad (5.14)$$

where $H$ is the strength of each hydrogen bond, and $n_h$ is the number of hydrogen bonds between the amino acids $i$ and $j$. Determining the strength of the hydrogen bonds, $H$, as well as the three cutoff parameters, $\delta$, $\Theta$, $\Psi$, is the task of the CD-learning procedure.

The sequence-*dependent* part of the potential (the negative log-likelihood) was approximated in our model by pair-wise interactions between side-chains. Our main focus was on the resulting effect of these interactions and how they stabilize secondary structural elements. We did not consider the detailed physical nature of these forces or how they depend on the amino acid types. We introduced these interactions between the polypeptide side chains as an effective Gō-type potential [230] dependent on the distance between $C\beta$ atoms,

$$E_{ij}^{SC} = \kappa C_{ij} r_{ij}^2 \qquad (5.15)$$

where $r_{ij}$ is a distance between non-adjacent $C\beta$ atoms, $|i - j| > 1$; and $\kappa$ is a force constant. In [584] we introduced a "regularized contact map", $C_{ij}$. In this binary matrix, two types of contacts were defined in the context of protein secondary structure. First, only lateral contacts in the parallel and anti-parallel $\beta$-sheets were indicated by 1s. Second, the contacts between amino-acids $i$ and $i + 3$ in $\alpha$-helices were also represented by 1s. The contacts of the first and second type typically have the closest $C\beta - C\beta$ distance among non-adjacent contacts in native proteins. The force constants depended on the secondary structure type, introducing positive $\kappa_\alpha$ and $\kappa_\beta$. Non-adjacent contacts in secondary structural elements were, therefore, stabilized by attracting potentials.

We also modeled interactions between sequential residues. This interaction was defined by the mutual orientation of adjacent residues that are involved in secondary structure elements,

$$E_{i,i+1}^{SC} = \eta \cos \gamma_{i,i+1} \tag{5.16}$$

where $\gamma_{i,i+1}$ is the dihedral angle $C\beta - C\alpha - C\alpha - C\beta$ between the adjacent residues. The purpose of this interaction is to bias the conformation towards the naturally occurring orientations of residues in secondary structural elements. In $\alpha$-helices, adjacent residues adopt a conformation with positive $\cos \gamma$. In $\beta$-sheets, $\cos \gamma$ is negative. We, therefore, used two values of the force constant: negative $\eta_\alpha$ and positive $\eta_\beta$.

To summarize, the total energy of a polypeptide chain with conformation $\lambda$ was calculated as follows

$$E(R, \lambda) = \sum_{i=1}^{N} E_i^B + \sum_{i=1}^{N} \sum_{j=1}^{i} (E_{ij}^{vdW} + E_{ij}^{HB} + E_{ij}^{SC}) \tag{5.17}$$

where we consider harmonic valence elasticity, $E_i^B$, hard-sphere van der Waals repulsions, $E_{ij}^{vdW}$, and square-well hydrogen bonding, $E_{ij}^{HB}$. The valence elasticity, van der Waals repulsions, and hydrogen bonding that contribute to this potential have a clear physical meaning and are analogous to traditional ab initio approaches. The side-chain interactions, $E_{ij}^{SC}$ in this model were introduced as a long-range quadratic Gō-type potential based on the contact map and secondary structure assignment. This pseudo-potential had two purposes: it was needed to stabilize the secondary structure elements and to provide a biasing force that allows reconstruction of the main chain conformation in the course of Metropolis Monte Carlo simulations [585, 586].

## 5.7   Monte Carlo Procedure

Because peptide bonds are rigid and flat, polypeptides are often modeled in the space of $\varphi$-$\psi$ angles, which reduces the number of degrees of freedom and speeds up MC simulations [385]. As an alternative, we proposed a sampler that utilized local crankshaft rotations of rigid peptide bonds in Cartesian space. In our model,

# 5 Inferring Knowledge Based Potentials Using Contrastive Divergence

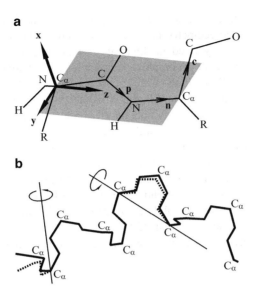

**Fig. 5.4** (a) Polypeptide model. The orientations of perfectly planar and rigid peptide bonds are given by the orthonormal triplets (**x**, **y**, **z**), with **z** pointing along the $C\alpha - C\alpha$ direction. Other peptide bond atoms lie in the plane $yz$. The position of the side-chain atoms R is specified by the vectors **n** and **c**. (b) Local Metropolis moves. Two types of moves are used in this work: a crankshaft rotation around the line connecting two $C\alpha$ atoms in the middle of the chain, and a random rotation at the termini around a random axis passing through the $C\alpha$ atom (Reproduced from [585] with permission)

the primary descriptors of the polypeptide chain conformation were the orientations of the peptide bonds in the laboratory frame (Fig. 5.4a). For a chain of $N$ amino acids the orientations of the peptide bonds were specified by the orthonormal triplets, $(\mathbf{x}_i, \mathbf{y}_i, \mathbf{z}_i), i = 0 \ldots N$. In this representation most of the uninteresting covalent geometry was frozen by fixing the positions of peptide bond atoms in these local coordinate frames. In principle, any mutual orientation of peptide bonds was allowed. In practice, they were governed by Boltzmann distribution with energy given by Eq. 5.17.

To obtain the canonical ensemble of polypeptide conformations, we developed a novel MCMC procedure. New chain conformations were proposed by rotating a few adjacent peptide bonds and applying regular Metropolis-Hastings acceptance criteria [293, 501]. We used crankshaft rotations in the middle of the chain and pivotal rotations at the termini to preserve chain connectivity. Each move was local and the conformation of the rest of the chain was not altered. Local moves are extremely important in achieving an efficient sampling of dense polypeptide conformations [172]. To satisfy detailed balance any rotation and its inverse are picked with equal probability. Figure 5.4b illustrates the rotations that we used in our procedure. We refer the reader to Chap. 2 for a general discussion on polymer sampling using MCMC-algorithms and to [585] for greater details of the sampler employed in the current work.

This Metropolis procedure in Cartesian space would be impossible without making the $C\alpha$ valence geometry flexible in our model. This is an extra degree of freedom that is usually fixed in dihedral space simulations. Dihedral space simulations, however, require complex computations of dihedral angle moves (so called "re-bridging" or "loop-closure") so that the structure only perturbed locally [231, 760]. In addition, sampling in dihedral space requires the calculation of a Jacobian correction to satisfy the microscopic reversibility principle and achieve unbiased sampling of dihedral angles [153, 309, 310]. The crankshaft rotations in Cartesian space are, on the other hand, trivially and equiprobably reversible. Crankshaft rotations have been used in polymer simulations for decades [61, 739] and rigorously proven to achieve unbiased sampling [582]. Therefore, the Metropolis acceptance criterion did not require the Jacobian correction. In our previous work, this was further validated by demonstrating unbiased sampling of $\varphi$-$\psi$ angles using crankshaft rotations [585]. We believe that the simplicity of our Metropolis MC procedure is well worth adding of an extra degree of freedom to our model.

## 5.8 Learning and Testing Protein Interaction Parameters

Secondary structure elements appear early in protein folding [21,22]. They are stabilized by both sequence-dependent side-chain interactions and sequence-independent interactions dominated by hydrogen bonds between main chain atoms [565, 566]. Careful balance between the two contributions is crucial for secondary structure element stability at room temperature. In the context of our protein model this requires careful optimization of hydrogen bonding parameters and interactions between side-chains as mimicked by Gō-type interactions between $C\beta$ atoms (see Sect. 5.6). Overall, eight model parameters were simultaneously optimized using contrastive divergence: four parameters characterizing the hydrogen bonding and four parameters characterizing side-chain interactions.

The strength of hydrogen bond is a subject of ongoing discussions in the literature (see recent review [189]). Pauling et al. [566] suggested that the strength of the hydrogen bond is about 8 kcal/mol. Some experimental evidence suggests that the strength is about 1.5 kcal/mol [533,630]. Others suggest that hydrogen bonding has a negligible or even a destabilizing effect [19]. At present the consensus is that the strength of the hydrogen bond is in the range of 1–2 kcal/mol [189]. Figure 5.5 shows the parameter learning curves produced by the iterative CD learning procedure. We found that the hydrogen bond strength converges to $H/k_B T = 1.85$, or 1.1 kcal/mol, in excellent agreement.

In the literature, geometric criteria for hydrogen bonding have been designed to capture as many reasonable hydrogen bonds as possible, which in general produced rather loose criteria [18, 189, 690]. For the first time, to our knowledge, we were able to simultaneously optimize hydrogen bond geometry and strength using the CD learning procedure. We found the $H \cdots O$ distance, $\delta < 2.14$ Å, and minimum

**Fig. 5.5** Contrastive divergence optimization of the model parameters. The *top* panel shows iterative convergence of four parameters of side-chain interactions: $\kappa_\alpha$ in *red*, $\kappa_\beta$ in *black*, $\eta_\alpha$ in *green*, and $\eta_\beta$ in *blue*. The *bottom* panel shows the convergence of hydrogen bond parameters: $H$ in *blue*, $\delta$ in *black*, $\Theta$ in *green*, and $\Psi$ in *red* (Reproduced from [584] with permission)

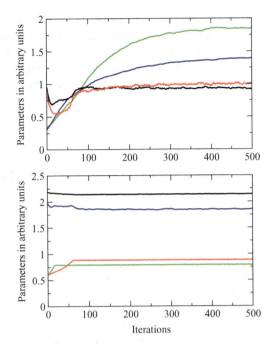

allowed angles $\angle COH$ and $\angle OHN$, $\Theta > 140°$ and $\Psi > 150°$ respectively. This means that all four atoms seem to be approximately co-linear when they form a *strong* hydrogen bond. The established view until now was that only $N - H \cdots O$ are co-linear. We determined that the same hydrogen bonding potential works reasonably well in both $\alpha$-helices and $\beta$-sheets, validating the assumption that the intra-peptide hydrogen bonding is independent of secondary structure.

We also found the value of the force constant for the attracting potential between amino acids in secondary structure elements to be equal to $\kappa_\alpha/k_BT = 0.10\,\text{Å}^{-2}$ and $\kappa_\beta/k_BT = 0.09\,\text{Å}^{-2}$. The two values are very close to each other indicating that these interactions are indeed similar in both helices and sheets. This is an expected result because the separation between the $C\beta$ atoms in both $\alpha$-helices and $\beta$-sheets is about 5.4 Å. The effective force constants for the interactions between adjacent residues were determined to be $\eta_\alpha/k_BT = -0.6$ and $\eta_\beta/k_BT = 4.5$. In agreement with our expectations, these force constants have opposite signs, stabilizing alternative mutual orientations of adjacent residues in $\alpha$-helices and $\beta$-strands.

Spontaneous folding of $\alpha$-helices and $\beta$-hairpins structures is difficult under the influence of hydrogen bonds alone and greatly facilitated by side-chain interactions [585]. To improve the stability of $\alpha$-helices and $\beta$-sheets in these and further simulations, we moderately adjusted some force constants ($\kappa_\alpha/k_BT = 0.12\,\text{Å}^{-2}$, $\kappa_\beta/k_BT = 0.11\,\text{Å}^{-2}$, and hydrogen bonding parameters ($H/k_BT = 2.5$, $\delta = 2.19\,\text{Å}$, $\Theta = 130°$ and $\Psi = 140°$). The difference between these values and those determined in the contrastive divergence procedure is less than 30%. Unfortunately, without this modest modification, the formation of persistent and hydrogen-bonded $\alpha$-helices and $\beta$-sheets became unlikely in our simulations, with lateral $C\beta - C\beta$ distances noticeably exceeding the native 5.4 Å [584]. Unavoidable systematic errors in the dataset may explain the underestimation of force parameters in a contrastive divergence procedure that assumes a correct representation of thermal fluctuations in the dataset. Another justification for the small adjustment is the necessity to compensate for other interactions that were not considered in our model.

## 5.9 Reconstructing Protein Tertiary Structure

Another test of CD utility involved the reconstruction of secondary structure elements and an overall 3D fold based on the preliminary prediction of secondary structure and a contact map. Recognizing the fundamental importance of secondary structure, along with long-range contacts in $\beta$-sheets, we extended the hierarchical folding theory of Baldwin and Rose [21, 22] in this work. We hypothesized that lateral contacts between $\beta$-strands and between the turns of an $\alpha$-helix are most important for stabilizing secondary structure elements. The reconstruction of tertiary structure usually utilizes residue-independent empirical potentials in the course of a simulated annealing or energy minimization protocol [555, 678, 737]. In contrast, our Metropolis Monte Carlo procedure does not rely on annealing. It is, therefore, important to use the strength of the contacts in $\alpha$-helices and $\beta$-sheets and their stability at room temperature before employing the contact potentials in a reconstruction.

We attempted to reconstruct the conformation of protein G (PDB code 1PGB) by utilizing predicted secondary structure and contact map. The prediction was provided by the segmental semi-Markov modeling procedure of [113] and required some manual interpretation of the prediction results to resolve ambiguities, and thus be usable in the reconstruction of 3D structure. Figure 5.6a illustrates the three-step interpretation of the prediction results. First, the predicted helical region in the middle of the protein specified helical contacts and corresponding Gō-type potentials in our protein model. Second, the pairing between central $\beta$-strand residues was specified based on the position of a local maximum on the predicted contact map. Third, the corresponding "regularized" contacts were diagonally extended in a parallel or anti-parallel direction to the boundaries of the reliable prediction, where the predicted probability drops to the background level.

For protein G, the predicted $\alpha$-helix was slightly shorter than the native one. This ambiguity in predicted positions corresponded to a plausible 2- or 4-residue shift between the $\beta$-strands . The orientation of the contacts that appeared close to the main diagonal necessarily corresponded to anti-parallel $\beta$-hairpins. The predicted contact between N- and C-terminus could be both parallel (as in the native protein structure) or anti-parallel. We only show simulations that correspond to the parallel orientation as represented in Fig. 5.6. The evolution of the total energy and the

5 Inferring Knowledge Based Potentials Using Contrastive Divergence  151

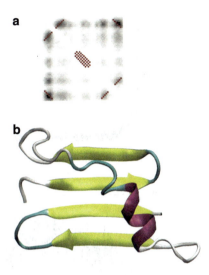

**Fig. 5.6** Reconstruction of the protein G fold by specifying predicted interactions. *Panel A* shows the regularized contact map with predicted interactions in $\alpha$-helix and $\beta$-sheets with the predicted contact map in the background. The *grey* levels in the predicted contact map represent the predicted probability of a particular contact. The regularized diagonal contacts pass through the local maxima on the predicted contact map and extend until the predicted contact probability levels off. The best structure corresponding to the maximum fraction of predicted contacts at relatively low energy is shown in the *panel B* (Reproduced from [584] with permission)

fraction of predicted contacts specified in the regularized contact map during the simulations can be found in Fig. 5.7.

Figure 5.6b shows the structure that corresponds to the maximum fraction of predicted contacts at relatively low total energy. The fold of the structure corresponds to the native fold of the protein G, although the RMSD with the native structure is 10 Å. Because of the underprediction of the length of both the $\alpha$-helix and $\beta$-sheets, larger portions on the chain were left as coils in comparison to the native structure. This can partly explain why the $\alpha$-helix does not pack against the $\beta$-sheets in our simulated structures. Both the anti-parallel and parallel $\beta$-sheets between the termini of the chain were able to form in our simulations. It is, therefore, impossible to rule out the anti-parallel conformation based on the contact map prediction alone. Our results indicate that it is crucial to obtain a good quality predicted contact map and secondary structure to faithfully reconstruct the 3D fold of a protein.

In another example, the general applicability of the modeling procedure described above was further demonstrated by modeling Src tyrosine kinase SH3 domain (SH3, PDB code 1SRL) [788]. This protein is often used as folding model in simulation systems [529, 650]. Native 56-residue SH3 has a 5-stranded $\beta$-barrel structure. The simulation results for the SH3 domain are shown in Fig. 5.8. The quality of the secondary structure prediction was comparable to that of protein G: $\beta$-strand locations in the sequence were correctly predicted, whereas their

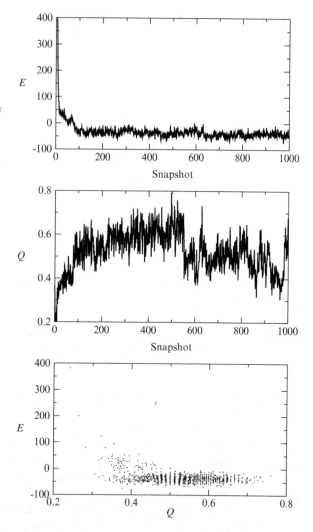

**Fig. 5.7** Reconstruction of the protein G fold by specifying predicted interactions (Case 1). The graphs demonstrate the evolution of the microscopic energy, $E$, and fraction of the native contacts, $Q$, during the simulation run in the *top* and *middle* panel respectively. The *bottom* panel demonstrated the relationship between $Q$ and $E$ (Reproduced from [584] with permission)

length was slightly underpredicted. The contact map prediction (shown in panel A) presented a challenge for our further modeling because all possible combinations between the $\beta$-strands were predicted with comparable probability. The correctly predicted $\beta$-hairpin contacts allowed unambiguous interpretation in terms of residue contacts and Gō-type potentials in our protein model (see protein G modeling above). On the contrary, the other predicted contacts were mostly false positives. For further simulations, we used three $\beta$-hairpin contacts and one longer-range contact between N- and C-terminus that were reliably predicted and corresponded to the native structure.

The structures shown in the Fig. 5.8b were selected based on the maximum fraction of predicted contacts at relatively low total energy. In the case of the

5  Inferring Knowledge Based Potentials Using Contrastive Divergence        153

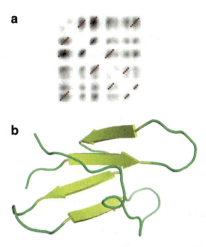

**Fig. 5.8** Reconstruction of Src tyrosine kinase SH3 domain. *Panel A* shows the regularized contact map with the predicted contact map in the background. The regularized diagonal contacts pass through the local maxima on the predicted contact map and extend until the predicted contact probability levels off. The selected regularized contacts correspond to the native fold of SH3. False positive predictions are not included in the reconstruction. The best structure corresponding to the maximum fraction of predicted contacts at relatively low energy is shown in the *panel B* (Reproduced from [584] with permission)

SH3 domain, four out of five strands correctly packed in the $\beta$-sheet with small misalignments of up to two residues. The C-terminus was correctly packed against the rest of the structure but the hydrogen bonds did not completely form to complete the barrel. The RMSD of the structure shown with the native fold was equal to 5.8 Å. In all these cases, we can speculate that the resulting structures are stable folding intermediates, with the native structure being adopted as a result of the formation of the final contacts and small adjustments in alignment between the $\beta$-strands.

## 5.10   Conclusions

Interactions that stabilize proteins have been an ongoing challenge for the scientific community for decades [20]. The microscopic size of protein molecules makes it impossible to measure these interactions directly. Several major contributors to protein stability have, however, been identified. Non-specific interactions that are only weakly determined by amino acid type have received the most attention. In particular, main chain hydrogen bonding [189] and hydrophobic core interactions [106] have been extensively studied both theoretically and experimentally. Although non-specific interactions may explain why proteins adopt a compact shape, they are not sufficiently detailed to explain how proteins adopt a particular fold. Only specific interactions defined by the amino acid type define the secondary structure and the tertiary fold for each particular sequence.

Non-specific interactions are ubiquitous and usually well represented in datasets of even a moderate size. Studies of specific interactions between particular types of amino acids may require significantly larger datasets of known structures to be well represented and to allow contrastive divergence to properly converge. Optimization of potentials has to be done on a large set of 500–1,000 known structures. The data set has to be representative of the structures and sequences observed in the native proteins. CulledPDB [752] and ASTRAL [78] contain such diverse structures with low sequence identity.

The results from contrastive divergence optimization of interaction parameters may provide valuable insights into protein energetics. They will help select the model of protein interactions that best corresponds to the native protein structures. For example, the value of dielectric permittivity, $\varepsilon$, inside a protein is hotly debated in the scientific community with experimental and theoretical estimates ranging from 2 to 20 [637]. CD learning can also optimize the Kauzmann hydrophobicity coefficient, $k_h$, between the hydrophobic energy and the buried surface area, $\Delta S$,

$$E_i^{HP} = k_h \Delta S \tag{5.18}$$

Current estimates for this parameter range from 9 to $20\,\mathrm{kJ \cdot mol^{-1} \cdot nm^{-2}}$ [106]. Machine learning of these parameters with contrastive divergence should provide important information about the magnitude of the corresponding interactions in native proteins.

As with statistical potentials, CD learning relies on the Boltzmann hypothesis that the observed protein structures and their details correspond to the canonical ensemble. The success of either methodology depends on the magnitude of systematic errors in the crystal structures. Based on our preliminary studies, we have no reason to believe that this presents significant obstacles for the contrastive divergence approach. The discrepancy between the stability of the model and actual protein may result from systematic errors in the original data set of crystal structures.

Contrastive divergence learning could also be applied to study protein-protein and protein-ligand interactions, where statistical potentials have been successfully used as scoring functions for docking [519, 675]. In these applications, the docking fundamentals are similar to protein folding approaches, and include system representation with reduced dimensionality, global conformational space search, and evaluation of conformations using a scoring function [273]. As discussed above, the scoring functions based on statistical potentials currently typically rely on assumptions regarding the reference state [511, 519, 798], which are often dubious [715]. This problem and its recent solution [276] are discussed extensively in Chaps. 3 and 4. The scoring function could also be optimized using a CD learning procedure that does not depend on any reference state, but requires a small number of MC steps in the chosen protein interface representation. A suitable dataset of crystallized protein complex structures are available in the Protein Data Bank, with non-redundant subsets discussed in the literature (see [604] and references therein).

It should be noted that the CD learning is agnostic to the choice of interaction model. The procedure only attempts to optimize the interaction parameters, and the functional form and details of the interaction model must be carefully thought through. In the worst case scenario, the parameters of a bad model may not even converge to optimal values. The quality of the optimization results may help make a decision regarding the quality of the model under consideration. If, however, the model is chosen carefully, CD learning is capable of deducing the interaction details with remarkable efficiency.

**Acknowledgements** This work was supported by a grant from the National Institutes of Health (1 P01 GM63208). DLW acknowledges support from an EU Marie-Curie IRG Fellowship (46444).

# Part III
# Directional Statistics for Biomolecular Structure

# Chapter 6
# Statistics of Bivariate von Mises Distributions

**Kanti V. Mardia and Jes Frellsen**

## 6.1 Introduction

Circular data arises in many areas of science, including astronomy, biology, physics, earth science and meteorology. In molecular biology, circular data emerges particularly in the study of macromolecules. One of the classical examples in this field is the Ramachandran map [595], which describes dihedral angles in the protein main chain.

When dealing with circular data, conventional statistical methods are often inadequate. A classical way of illustrating this is to consider the arithmetic mean of the two angles $1°$ and $359°$, which is $180°$. The fact that this is not a sensible mean clearly shows that linear methods are not applicable to circular data; the periodic nature of the data has to be taken into consideration.

The Ramachandran map is a plot of two circular stochastic variables, namely the two consecutive dihedral angles $\phi$ and $\psi$ in the protein main chain, where each angle lies in the interval $(-\pi, \pi]$. The conventional Ramachandran map is depicted as a square in the plane. However, due to the circular nature of the data, one should envision that the opposite edges of this square are "glued" together. In other words, the square has opposite sides identified, corresponding to a torus. Accordingly, the natural way of parameterizing the pairs of conformational angles $(\phi, \psi)$ in the Ramachandran map is as points on the torus (Fig. 6.1). In order to describe distributions of the conformational angles $(\phi, \psi)$, a bivariate distribution on the torus is required.

---

K.V. Mardia (✉)
Department of Statistics, University of Leeds, Leeds, United Kingdom
e-mail: k.v.mardia@leeds.ac.uk

J. Frellsen
The Bioinformatics Centre, University of Copenhagen, Copenhagen, Denmark
e-mail: frellsen@binf.ku.dk

T. Hamelryck et al. (eds.), *Bayesian Methods in Structural Bioinformatics*,
Statistics for Biology and Health, DOI 10.1007/978-3-642-27225-7_6,
© Springer-Verlag Berlin Heidelberg 2012

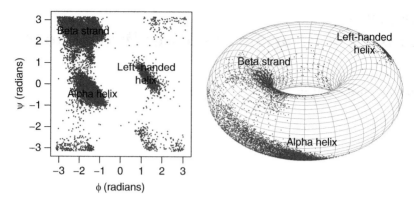

**Fig. 6.1** Two Ramachandran plots for the 100 protein structures in the *Top100 database* [771]. The *left plot* is a conventional scatter plot in the plane. The *right plot* is a scatter plot of the same data on the torus, where $\phi$ describes the rotation about the axis through the center of the hole in the torus and $\psi$ describes the rotation about the axis through the center of the torus tube. Areas for $\beta$-strands, right-handed $\alpha$-helices and left-handed $\alpha$-helices are indicated in both plots

In this chapter, we will discuss how to describe distributions on the torus using directional statistics. We will focus on bivariate von Mises distributions, which in several studies have been successfully applied in modelling the Ramachandran map [68, 478]. We will start by presenting some basic concepts in directional statistics in Sect. 6.2, including an short description of the univariate von Mises distribution. Following this, we will introduce the full bivariate von Mises distribution and a number of submodels in Sect. 6.3. Just as the standard bivariate normal distribution, these submodels have five parameters: two means, two concentrations, and a parameter controlling "correlation". In Sect. 6.4, we will describe some of the key properties of these distributions, and in Sects. 6.5 and 6.6 we will discuss inference and simulation. In Sect. 6.7, we will introduce a multivariate extension of the bivariate von Mises distribution, and finally in Sect. 6.8 we describe conjugate priors for the von Mises distributions.

## 6.2 Basic Concepts in Directional Statistics

One of the fundamental concepts in directional statistics is an appropriate definition of the mean of circular data. Given a set of angles $\theta_1, \ldots, \theta_n$ and corresponding set of vectors $\mathbf{x}_1, \ldots, \mathbf{x}_n$ pointing to the unit circle, the mean direction $\bar{\theta}$ is defined as the direction of the center of mass $(\mathbf{x}_1 + \ldots + \mathbf{x}_n)/n$ of this set of vectors. The Cartesian coordinates $(\bar{C}, \bar{S})$ of the center of mass can be calculated from the angles as

$$\bar{C} = \frac{1}{n} \sum_{i=1}^{n} \cos \theta_i, \qquad \bar{S} = \frac{1}{n} \sum_{i=1}^{n} \sin \theta_i . \qquad (6.1)$$

6 Statistics of Bivariate von Mises Distributions 161

Accordingly, the mean direction $\bar{\theta}$ is given by the solution to the equations

$$\bar{C} = \bar{R} \cos \bar{\theta}, \qquad \bar{S} = \bar{R} \sin \bar{\theta} , \qquad (6.2)$$

where $\bar{R} = (\bar{C} + \bar{S})^{\frac{1}{2}}$ is the *mean resultant length*. When the mean resultant length is positive, $\bar{R} > 0$, the mean direction given by the expression

$$\bar{\theta} = \begin{cases} \arctan(\bar{S}/\bar{C}) & \text{if } \bar{C} \geq 0 \\ \arctan(\bar{S}/\bar{C}) + \pi & \text{if } \bar{C} < 0 \end{cases} , \qquad (6.3)$$

while for a zero mean resultant length, $\bar{R} = 0$, the mean direction is undefined. If we return to the example from the introduction and consider the mean of the two angles $1°$ and $359°$ using this definition, we obtain the more intuitive value of $0°$.

### 6.2.1 Univariate von Mises Distribution

The univariate von Mises distribution [743] is the most well known circular distribution from directional statistics. It can be considered the circular analogue of the univariate normal distribution. Akin to the normal distribution, the von Mises distribution has two parameters, a mean and a concentration parameter; the latter can be considered as an anti-variance. The density of the von Mises distribution $M(\mu, \kappa)$ is given by (see for example [474])

$$f(\theta) = \{2\pi I_0(\kappa)\}^{-1} \exp\{\kappa \cos(\theta - \mu)\} , \qquad (6.4)$$

where $I_0(\cdot)$ denotes the modified Bessel function of the first kind and order 0, the parameter $-\pi < \mu \leq \pi$ is the mean direction, and $\kappa \geq 0$ is the concentration parameter. In fact, the von Mises distribution can be approximated by the normal density for high concentration, which corresponds to a small variance in the normal distribution [474].

For a set of angular observations $\theta_1, \ldots, \theta_n$, it can be shown that the maximum likelihood estimate $\hat{\mu}$ of the mean parameter in the von Mises distribution is given by the mean direction $\bar{\theta}$ of the observations [474], as described in Eq. 6.3. Furthermore, the maximum likelihood estimate $\hat{\kappa}$ of the concentration parameter is given by the solution to

$$\frac{I_1(\hat{\kappa})}{I_0(\hat{\kappa})} = \bar{R} , \qquad (6.5)$$

where $\bar{R}$ is the mean resultant length of the observations and $I_1(\cdot)$ denotes the modified Bessel function of the first kind and order 1. However, there does not exist a general closed for solution to this equation. Accordingly, the maximum likelihood estimate of the concentration parameter is normally obtained by a numerical or an

**Algorithm 1** A sampling algorithm for the univariate von Mises distribution. This algorithm was originally developed by Best and Fisher [56].

---

**Require:** Mean value $\mu \in [0; 2\pi)$ and concentration parameter $\kappa > 0$
1  Set $a = 1 + \sqrt{1 + 4\kappa^2}$
2  Set $b = \frac{a - \sqrt{2a}}{2\kappa}$
3  Set $r = \frac{1 + b^2}{2b}$
4  **repeat**
5      Sample $u_1, u_2 \sim \text{Uniform}(0, 1)$
6      Set $z = \cos(u_1 \pi)$
7      Set $f = \frac{1 + rz}{r + z}$
8      Set $c = (r - f)\kappa$
9  **until** $c(2 - c) - u_2 > 0$ **or** $\log(c/u_2) + 1 - c \geq 0$
10  Sample $u_3 \sim \text{Uniform}(0, 1)$
11  Set $\theta = \mu + \text{sign}(u_3 - 1/2) \arccos(f)$
12  **return** $\theta$

---

approximative solution to Eq. 6.5. Several different approximations are given in the literature, here we give the approximation described by Lee [428]

$$
\hat{\kappa} \approx \begin{cases} \bar{R}\left(\frac{2 - \bar{R}^2}{1 - \bar{R}^2}\right) & \text{if } \bar{R} \leq 2/3 \\ \frac{\bar{R} + 1}{4\bar{R}(1 - \bar{R})} & \text{if } \bar{R} > 2/3 \end{cases} . \tag{6.6}
$$

There is no closed form expression for the distribution function of the von Mises distribution [56]. This complicates simulation of samples from the distribution. However, an efficient acceptance/rejection algorithm has been given by Best and Fisher [56, 474], see Algorithm 1.

## 6.3   Bivariate von Mises Distributions

A general bivariate extension of the univariate von Mises distribution was introduced by Mardia in 1975 [468]. This model, which we will call the *full bivariate von Mises* distribution, has probability density function proportional to

$$
f(\phi, \psi) \propto \exp \{\kappa_1 \cos(\phi - \mu) + \kappa_2 \cos(\psi - \nu) +
$$
$$
(\cos(\phi - \mu), \sin(\phi - \mu)) \, \mathbf{A} \, (\cos(\psi - \nu), \sin(\psi - \nu))^T \} , \tag{6.7}
$$

where $\mathbf{A}$ is a $2 \times 2$ matrix. The distribution has eight parameters: $\mu, \nu$ can be described as the mean values, $\kappa_1, \kappa_2$ the concentrations and $\mathbf{A}$ allows for dependencies between the two angles. Mardia has derived the normalizing constant for this full bivariate von Mises distribution in a compact way [472].

# 6 Statistics of Bivariate von Mises Distributions

With eight parameters this model appears to be overparameterized compared to the analogous bivariate normal distribution in the plane with only five parameters (two mean parameters, two parameters for variance and a parameter which determines the correlation). In fact, the parameters are known to be redundant for high concentrations, which leads to difficulties in fully interpreting the meaning of the parameters [478]. Several submodels have been proposed in order to address this issue.

The basis for these submodels is a subclass of the full bivariate von Mises model originally proposed by Mardia in 1975 [467] and further studied by Rivest in 1988 [605]. This subclass is called the *6-parameter model* and it is constructed by fixing the off-diagonal elements of $\mathbf{A}$ to zero in Eq. 6.7. The probability density function for this model is proportional to

$$f(\phi, \psi) \propto \exp\left\{\kappa_1 \cos(\phi - \mu) + \kappa_2 \cos(\psi - \nu)+\right.$$
$$\left. \alpha \cos(\phi - \mu)\cos(\psi - \nu) + \beta \sin(\phi - \mu)\sin(\psi - \nu)\right\}, \quad (6.8)$$

where $\alpha$ and $\beta$ correspond to the diagonal elements of $\mathbf{A}$, that is $\alpha = \mathbf{A}_{11}$ and $\beta = \mathbf{A}_{22}$.

In the following, we will describe four submodels of the 6-parameter model: the sine model, the cosine model with positive interaction, the cosine model with negative interaction and the hybrid model. Each of these submodels are constructed by removing one degree of freedom from the 6-parameter model, resulting in models with five parameters analogous to the bivariate normal distribution.

In the following sections we will also give the marginal and conditional densities for the sine model and the cosine model with positive interaction. For these two models, the conditional distributions are von Mises while the marginal distributions are generally not. In fact, Mardia [468] has proved that there cannot be any exponential family of bivariate distributions on the torus with marginals and conditionals that are *all* von Mises.

## 6.3.1 Sine Model

In 2002 Singh et al. [660] presented a special case of the 6-parameter model, where $\alpha = 0$ and $\beta = \lambda$ in Eq. 6.8. We call this the *sine model*, and the probability density function is given by

$$f_s(\phi, \psi) = C_s \exp\{\kappa_1 \cos(\phi - \mu) + \kappa_2 \cos(\psi - \nu) + \lambda \sin(\phi - \mu)\sin(\psi - \nu)\}, \quad (6.9)$$

for $\kappa_1, \kappa_2 \geq 0$. The normalizing constant for the sine mode is given by

$$C_s^{-1} = 4\pi^2 \sum_{m=0}^{\infty} \binom{2m}{m} \left(\frac{\lambda^2}{4\kappa_1\kappa_2}\right)^m I_m(\kappa_1) I_m(\kappa_2), \quad (6.10)$$

where $I_m(\cdot)$ is the modified Bessel function of the first kind and order $m$. The marginal and conditional distributions of the sine model were also given by Singh et al. [660].

The marginal density of $\phi$ is given by the expression

$$f_s(\phi) = C_s 2\pi I_0\{\kappa_{2\lambda}(\phi)\} \exp\{\kappa_1 \cos(\phi - \mu)\}, \tag{6.11}$$

where $\kappa_{2\lambda}(\phi)^2 = \kappa_2^2 + \lambda^2 \sin^2(\phi - \mu)$. Note that the marginal density is symmetric about $\phi = \mu$ but not von Mises, except for the trivial case of $\lambda = 0$. The marginal probability density of $\psi$ is given by an analogous expression.

It can be shown from Eqs. 6.9 and 6.11 that the conditional probability of $\Psi$ given $\Phi = \phi$ is the von Mises distribution $M(\nu + \phi_\mu, \kappa_{2\lambda}(\phi))$, where $\tan\phi_\mu = (\lambda/\kappa_2)\sin(\phi - \mu)$ [478]. Similarly, the conditional probability of $\Phi$ given $\Psi = \psi$ is also a von Mises distribution with analogous parameters.

### 6.3.2 Cosine Models

The *cosine model with positive interaction* was introduced and studied by Mardia et al. in 2007 [478], while the naming convention was given by Kent et al. [369]. This model is obtained by setting $\alpha = \beta = -\kappa_3$ in the 6-parameter model given in Eq. 6.8 and has the probability density function

$$f_{c+}(\phi, \psi) = C_c \exp\{\kappa_1 \cos(\phi-\mu)+\kappa_2 \cos(\psi-\nu)-\kappa_3 \cos(\phi-\mu-\psi+\nu)\}, \tag{6.12}$$

where $\kappa_1, \kappa_2 \geq 0$. The normalizing constant is given by

$$C_c^{-1} = (2\pi)^2 \left\{ I_0(\kappa_1)I_0(\kappa_2)I_0(\kappa_3) + 2\sum_{p=1}^{\infty} I_p(\kappa_1)I_p(\kappa_2)I_p(\kappa_3) \right\}. \tag{6.13}$$

For the cosine model with positive interaction, the marginal probability density of $\psi$ is given by

$$f_{c+}(\psi) = C_c 2\pi I_0(\kappa_{13}(\psi)) \exp\{\kappa_2 \cos(\psi - \nu)\}, \tag{6.14}$$

where $\kappa_{13}(\psi)^2 = \kappa_1^2 + \kappa_3^2 - 2\kappa_1\kappa_3 \cos(\psi - \nu)$ [478]. The marginal distribution of $\psi$ is symmetric about $\nu$ and for small values of $\kappa_3$ it is approximately a von Mises distribution. For $\kappa_1 = \kappa_3 = 0$ the marginal distribution is von Mises with mean angle $\nu$ and concentration parameter $\kappa_2$, and trivially the marginal is uniform for $\kappa_1 = \kappa_2 = \kappa_3 = 0$. The marginal density of $\phi$ is given by an expression analogous to Eq. 6.14.

It can also be shown that the conditional distribution of $\phi$ given $\Psi = \psi$ is a von Mises distribution $M(\psi_\nu, \kappa_{13}(\psi))$, where $\tan\psi_\nu = -\kappa_3 \sin(\psi - \nu)/(\kappa_1 - \kappa_3$

6 Statistics of Bivariate von Mises Distributions

$\cos(\psi - \nu))$ [478]. The the conditional distribution of $\psi$ given $\Phi = \phi$ is von Mises with analogous parameters.

An alternative cosine model was also given by Mardia et al. in 2007 [478]. This model is called the *cosine model with negative interaction*. It is obtained by setting $\alpha = -\kappa_3'$ and $\beta = \kappa_3'$ in the 6-parameter model and has the probability density function

$$f_{c-}(\phi, \psi) = C_c \exp\{\kappa_1 \cos(\phi - \mu) + \kappa_2 \cos(\psi - \nu) - \kappa_3' \cos(\phi - \mu + \psi - \nu)\},$$

where $\kappa_1, \kappa_2 \geq 0$ and the normalizing constant is the same as for the model with positive interaction given in Eq. 6.13. Note that the cosine model with negative interactions can be obtained by applying the transforming $(\phi, \psi) \mapsto (\phi, -\psi)$ in the model with positive interactions, which corresponds to a rotation of the density function in Eq. 6.14. So far the cosine model with negative interaction has only been discussed briefly in the literature. This will also be reflected in this chapter, where we will primarily be concerned with the model with positive interaction.

### 6.3.3 Hybrid Model

In 2008 Kent et al. [369] suggested a new model which is a hybrid between the sine and cosine models. The authors gave the following motivation for the model. Unimodal cosine and sine models have elliptical equiprobability contours around the mode. Generally, this elliptical pattern becomes distorted away from the mode. However, for the cosine model with positive interaction this pattern becomes the least distorted under positive correlation, that is $\kappa_3 < 0$, while for the cosine model with negative interaction the pattern is least distorted under negative correlation, that is $\kappa_3' < 0$ (see Fig. 6.2). Thus, to attain the least distortion in the contours of constant probability, it would be ideal to use the cosine model with positive correlation for positively correlated $\sin \phi$ and $\sin \psi$ and the cosine model with negative interaction for negatively correlated $\sin \phi$ and $\sin \psi$.

To address this issue, Kent et al. [369] suggested a *hybrid model* that provides a smooth transition between the two cosine models via the sine model. The probability density function for this hybrid model is given by

$$f(\phi, \psi) \propto \exp \{\kappa_1 \cos \phi + \kappa_2 \cos \psi$$
$$+ \epsilon[(\cosh \gamma - 1) \cos \phi \cos \psi + \sinh \gamma \sin \phi \sin \psi]\},$$

where $\epsilon$ is a tuning parameter which Kent et al. suggest setting to 1 for simplicity. If $\epsilon$ was a free parameter, the hybrid model would just be a reparameterization of the 6-parameter model [369].

For large $\gamma > 0$ the hybrid model is approximately a cosine model with positive interaction where $\kappa_3 \approx -\epsilon \exp(\gamma)/2$, while for large $-\gamma > 0$ the hybrid model is approximately a cosine model with negative interaction where $\kappa_3' \approx -\epsilon \exp(-\gamma)/2$. For $\gamma \approx 0$ the hybrid model is approximately a sine model with $\lambda \approx \epsilon \gamma$. In other

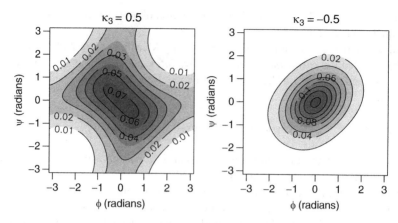

**Fig. 6.2** An illustration of the distortion in the contours of constant probability density in the density function for the cosine model with positive interaction. The densities in the two plots have the same values for the parameters $\mu, \nu = 0$ and $\kappa_1, \kappa_2 = 1$, but different values for $\kappa_3$. The *left plot* shows the density with $\kappa_3 = 0.5$ and the *right plot* show the density with $\kappa_3 = -0.5$. In general, the elliptical pattern is distorted least for the positive interaction cosine model when $\kappa_3 < 0$

words, for small correlations the hybrid model is approximately a sine model, and in line with the motivation above, the hybrid model is approximately a cosine model of suitable choice for large correlations [369].

The hybrid model is an promising construction. However, we will not discuss this model further in the remaining of the chapter, since at the time of writing the model has not been fully developed.

## 6.4 Properties of Bivariate von Mises Models

In this section we will give some of the properties of the bivariate von Mises models. We will discuss the conditions under which the models become bimodal, the approximative normal behavior of the models and finally give some interim conclusions on how to choose a model. As we will see in Sects. 6.5 and 6.6, these properties become important in parameters estimation and sampling for the models.

### 6.4.1 Bimodality Conditions

Here, we will state some of the key results on bimodality for the sine and cosine models. The proofs are given by Mardia et al. [478], except the proof for Theorem 6.3 which is given by Singh et al. [660].

# 6 Statistics of Bivariate von Mises Distributions

The following two theorems describe the conditions under which the sine model and the cosine model with positive interaction are bimodal.

**Theorem 6.1.** *The joint density function of the sine model in Eq. 6.9 is unimodal if $\kappa_1\kappa_2 > \lambda^2$ and is bimodal if $\kappa_1\kappa_2 < \lambda^2$ when $\kappa_1 > 0$, $\kappa_2 > 0$ and $-\infty < \lambda < \infty$.*

**Theorem 6.2.** *The joint density function of the positive interaction cosine model in Eq. 6.12 is unimodal if $\kappa_3 < \kappa_1\kappa_2/(\kappa_1+\kappa_2)$ and is bimodal if $\kappa_3 > \kappa_1\kappa_2/(\kappa_1+\kappa_2)$ when $\kappa_1 > \kappa_3 > 0$ and $\kappa_2 > \kappa_3 > 0$.*

Now we will consider the conditions under which the marginal distributions for these two models are bimodal. It turns out that these conditions in general are different from those of the joint densities. This may not be directly apparent, but there exist sets of parameters for these models, where the marginal density is unimodal although the bivariate density is bimodal. An example of this is illustrated in Fig. 6.3. The following two theorems state the conditions under which the marginal distributions are bimodal for the sine model and the positive interaction cosine model.

**Theorem 6.3.** *For the sine model given in Eq. 6.9 with $\lambda \neq 0$, the marginal distribution of $\Phi$ is symmetric around $\phi = \mu$ and unimodal (respectively bimodal) with mode at $\mu$ (respectively with the modes at $\mu - \phi^*$ and $\mu + \phi^*$) if and only if*

$$A(\kappa_2) \leq \kappa_1\kappa_2/\lambda^2$$

*(respectively $A(\kappa_2) > \kappa_1\kappa_2/\lambda^2$),*

*where $\phi^*$ is given by the solution to $\cos(\phi - \mu)A(\kappa_{2\lambda}(\phi))/\kappa_{2\lambda}(\phi) = \kappa_1/\lambda^2$, and $A(\kappa) = I_1(\kappa)/I_0(\kappa)$.*

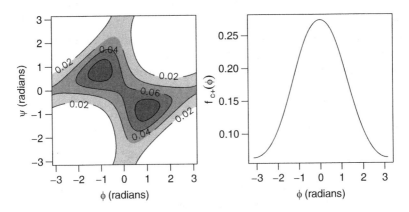

**Fig. 6.3** An example where the bivariate density of a von Mises distribution is bimodal while the marginal density is unimodal. The *left plot* shows the contours of the bimodal density for the positive interaction cosine model with parameters $\kappa_1 = 1.5$, $\kappa_2 = 1.7$, $\kappa_3 = 1.3$ and $\mu, \nu = 0$. The *right plot* shows the unimodal density for the corresponding marginal distribution of $\phi$

**Theorem 6.4.** *For the positive interaction cosine model given in Eq. 6.12 with $\kappa_3 \neq 0$, the marginal distribution of $\Phi$ is symmetric around $\phi = \mu$ and unimodal (respectively bimodal) with mode at $\mu$ (respectively with the modes at $\mu - \phi^*$ and $\mu + \phi^*$) if and only if*

$$A(|\kappa_1 - \kappa_3|) \leq |\kappa_1 - \kappa_3|\kappa_2/(\kappa_1\kappa_3)$$

$$(respectively \ A(|\kappa_1 - \kappa_3|) > |\kappa_1 - \kappa_3|\kappa_2/(\kappa_1\kappa_3)) \,,$$

*where $\phi^*$ is given by the solution to $\kappa_1\kappa_3 A(\kappa_{13}(\phi))/\kappa_{13}(\phi) - \kappa_2 = 0$, and $A(\kappa) = I_1(\kappa)/I_0(\kappa)$.*

## 6.4.2 Normal Behavior under High Concentration

The sine and cosine models all behave as bivariate normal distributions under high concentrations, and we have approximately

$$(\Phi - \mu, \Psi - \nu) \sim \mathcal{N}_2(0, \Sigma) \,,$$

where by high concentration is understood that the applicable parameters $\kappa_1$, $\kappa_2$, $\lambda$, $\kappa_3$, $\kappa_3'$ become large but remaining in constant proportion to each other [369]. The corresponding inverse covariance matrices for the sine model ($\Sigma_s^{-1}$), the cosine model with positive interaction ($\Sigma_{c+}^{-1}$) and the cosine model with negative interaction ($\Sigma_{c-}^{-1}$) are given by

$$\Sigma_s^{-1} = \begin{bmatrix} \kappa_1 & -\lambda \\ -\lambda & \kappa_2 \end{bmatrix}, \quad \Sigma_{c+}^{-1} = \begin{bmatrix} \kappa_1 - \kappa_3 & \kappa_3 \\ \kappa_3 & \kappa_2 - \kappa_3 \end{bmatrix}, \quad \Sigma_{c-}^{-1} = \begin{bmatrix} \kappa_1 - \kappa_3' & -\kappa_3' \\ -\kappa_3' & \kappa_2 - \kappa_3' \end{bmatrix}.$$

$$(6.15)$$

These matrices must be positive definite in order to ensure the existence of the covariance matrix. For the three models it can be shown [369] that the restriction to positive definite matrices is equivalent to the following constraints

sine model: $\quad \kappa_1 > 0, \quad \kappa_2 > 0, \quad \lambda^2 < \kappa_1\kappa_2 \,,$

positive cosine model: $\quad \kappa_1 - \kappa_3 > 0, \quad \kappa_2 - \kappa_3 > 0, \quad \kappa_3^2 < (\kappa_1 - \kappa_3)(\kappa_2 - \kappa_3) \,,$

negative cosine model: $\quad \kappa_1 - \kappa_3' > 0, \quad \kappa_2 - \kappa_3' > 0, \quad \kappa_3'^2 < (\kappa_1 - \kappa_3')(\kappa_2 - \kappa_3') \,.$

Note, that for the sine model it is possible to choose the parameters in such a way that the matrix above, $\Sigma_s^{-1}$, can match any given positive definite inverse covariance. However, by definition the cosine models have the additional restriction $\kappa_1, \kappa_2 \geq 0$, which limits the set of positive definite inverse covariance matrices that can be matched by the cosine models.

### 6.4.3 Choosing a Model

At the time of writing, a complete comparison between the different submodels is not available in the literature. However, Kent et al. [369] gives some preliminary conclusions about the submodels: to a large extent the sine and cosine models are quite similar. Each of these models provides a satisfactory analog to the bivariate normal distribution, and under high concentrations they all approximately behave as bivariate normal distributions. In terms of practical applications the sine model can be easier to work with, since it can match any given positive definite inverse covariance matrix. However, in some cases the cosine or hybrid models may provide a better fit.

As we will see in Sect. 6.5, the methods used for statistical inference in all the models are basically the same and similar in efficiency. However, it is proposed by Kent et al. [369] that the marginal distribution is closer to von Mises for the cosine models than for the sine model. If this is the case, estimation using maximum pseudolikelihood will be more efficient for the cosine model than for the sine model (see sect. 6.5.3).

## 6.5 Aspects of Estimation

In this section we will consider estimation of parameters in some of the bivariate von Mises models we have introduced in Sect. 6.3. Parameter estimation is not straightforward for these models, since closed expressions cannot be derived for the maximum likelihood estimators. However, three general strategies are applicable to these models:

- Maximum likelihood estimation using numerical methods
- Method of moments
- Maximum pseudolikelihood estimation

In the following sections, we will describe each of these strategies and we will give examples from the literature of how they have been applied to the sine and cosine models.

### 6.5.1 Maximum Likelihood Estimation Using Numerical Methods

Closed expressions for the maximum likelihood estimators cannot be derived for any of the models we have considered. However, the value of the normalizing constant can be calculated using numerical integration. This means that maximum likelihood estimates can be obtained using numerical optimization (e.g. the Nelder-Mead downhill simplex algorithm [540]).

In order to use a numerical optimization algorithm, a good initial guess of the parameters is needed. Furthermore, at each optimization step in these algorithms, the data likelihoods (including the normalizing constant) has to be calculated. The advantage of the numerical maximum likelihood approach is that it is quite accurate, while the clear drawback is that it is computationally more demanding than the alternative methods.

### 6.5.1.1 Maximum Likelihood Estimation for the Cosine Model

Here, we will give an example of maximum likelihood estimation for the cosine model with positive interaction using numerical optimization and integration. The approach is similar for the other models.

The density for the cosine model with positive interaction is given in Eq. 6.12. For a set of observations $\{(\phi_i, \psi_i)\}_{i=1}^n$ this gives us the log-likelihood function

$$LL_{c+}(\kappa_1, \kappa_2, \kappa_3 | \{(\phi_i, \psi_i)\}_{i=1}^n) =$$

$$n \log C_c + \sum_{i=1}^n \{\kappa_1 \cos(\phi_i - \mu) + \kappa_2 \cos(\psi_i - \nu) - \kappa_3 \cos(\phi_i - \mu - \psi_i + \nu)\}.$$

Using Eq. 6.14, the normalizing constant can be expressed as the single integral

$$C_c^{-1} = \int_0^{2\pi} 2\pi I_0(\kappa_{13}(\psi)) \exp\{\kappa_2 \cos(\psi)\} \, \mathrm{d}\psi . \tag{6.16}$$

This expression is somewhat simpler to calculate than the original expression for the normalization in Eq. 6.13 or the double integral of the bivariate density function from Eq. 6.12.

The log-likelihood function $LL_{c+}$ can be optimized using a numerical optimization method. It is important for such method to have good initial values. These values can be obtained by assuming that the marginal distributions are von Mises, that is $\phi \sim M(\mu, \kappa_1)$ and $\psi \sim M(\nu, \kappa_2)$. The initial values are then set to the maximum likelihood estimate under this assumption. This means that the initial values for the estimates $\hat{\mu}$ and $\hat{\nu}$ are given respectively by the circular means of $\{\phi_i\}_{i=1}^n$ and $\{\psi_i\}_{i=1}^n$, as described in Sect. 6.2.1. Similarly, the initial values for the estimates $\hat{\kappa}_1$ and $\hat{\kappa}_2$ can be obtained using the approximation to the maximum likelihood estimate of the concentration parameter for the von Mises distribution given in Eq. 6.6. For the estimate $\hat{\kappa}_3$ the mean of the initial values of $\hat{\kappa}_1$ and $\hat{\kappa}_2$ can be used a starting value.

The approach described here was used in a study by Mardia et al. to fit the dihedral angles in the Ramachandran plot for a small set of proteins [478].

## 6.5.2 Method of Moments

While the maximum likelihood estimates using numerical optimization are quite accurate, the method is highly computational demanding. A faster strategy might therefore be preferable in some cases. The method of moments offers such an alternative. The basic idea behind the method of moments is to equate the sample moments with the corresponding distribution moments. Estimators for the distribution's parameters are then constructed by solving this set of equations with respect to the parameters.

Let us consider how to obtain moment estimates for the sine and the cosine models. Assume that we have a set of observation $\{(\phi_i, \psi_i)\}_{i=1}^n$. Using the first moment, the means $\mu$ and $\nu$ can simply be estimated by the marginal circular means $\bar{\phi}$ and $\bar{\psi}$, as given in Eq. 6.3.

For highly concentrated data, the models behave approximately as normal distributions (see Sect. 6.4.2). The moment estimate for the covariance matrix in the normal distribution is given by the second moment of the population, that is

$$\hat{\Sigma} = \bar{\mathbf{S}} = \begin{bmatrix} \bar{S}_1 & \bar{S}_{12} \\ \bar{S}_{12} & \bar{S}_2 \end{bmatrix}, \tag{6.17}$$

where

$$\bar{S}_1 = \frac{1}{n} \sum_{i=1}^n \sin^2(\phi_i - \bar{\mu})$$

$$\bar{S}_2 = \frac{1}{n} \sum_{i=1}^n \sin^2(\psi_i - \bar{\nu})$$

$$\bar{S}_{12} = \frac{1}{n} \sum_{i=1}^n \sin(\phi_i - \bar{\mu}) \sin(\psi_i - \bar{\nu}) .$$

By taking the inverse of $\hat{\Sigma}$ and equating it to the expression for the inverse covariance matrix from Eq. 6.15, we can obtain estimates of the concentrations parameters $\kappa_1$, $\kappa_2$ and $\lambda/\kappa_3/\kappa_3'$ (depending on choice of model). For the sine model the moment estimates become

$$\kappa_1 = \frac{\bar{S}_2}{\bar{S}_1 \bar{S}_2 - \bar{S}_{12}^2}, \quad \kappa_2 = \frac{\bar{S}_1}{\bar{S}_1 \bar{S}_2 - \bar{S}_{12}^2}, \quad \lambda = \frac{\bar{S}_{12}}{\bar{S}_1 \bar{S}_2 - \bar{S}_{12}^2},$$

and for the cosine model with positive interaction the estimates become

$$\kappa_1 = \frac{\bar{S}_2 - \bar{S}_{12}}{\bar{S}_1 \bar{S}_2 - \bar{S}_{12}^2}, \quad \kappa_2 = \frac{\bar{S}_1 - \bar{S}_{12}}{\bar{S}_1 \bar{S}_2 - \bar{S}_{12}^2}, \quad \kappa_3 = \frac{-\bar{S}_{12}}{\bar{S}_1 \bar{S}_2 - \bar{S}_{12}^2} .$$

A similar result can be obtained for the cosine model with negative interaction.

The moment estimates are not as accurate as the maximum likelihood estimates obtained by numeric optimization, but they are considerably faster to calculate and have shown to be sufficiently accurate in some areas of application [68]. Alternatively, the moment estimates can be used as initial values in the numerical maximum likelihood approach. Moment estimates was used for fitting the TORUSDBN model described in Chap. 10.

### 6.5.3 Maximum Pseudo-likelihood Estimation

Another less computationally demanding alternative to the maximum likelihood approach is the maximum pseudolikelihood method. The concept of pseudolikelihood (also know as composite likelihoods) dates back to Besag in 1975 [55]. Recently Mardia et al. have studied the use of maximum pseudolikelihood for exponential families [480]. Mardia et al. showed that for certain exponential families that are said to be "closed" under marginalization, the maximum pseudolikelihood estimate equals the estimate obtained by maximizing the regular likelihood. For other distributions, the maximum pseudolikelihood estimate can be considered an approximation to the regular maximum likelihood estimate.

Consider a bivariate distribution of two random variables $(\Phi, \Psi)$ with joint probability density function $f(\phi, \psi | q)$, where $q$ is the vector of parameters. Based on a set of observations $\{(\phi_i, \psi_i)\}_{i=1}^{n}$, we define the pseudolikelihood as

$$\mathrm{PL} = \prod_{i=1}^{n} f(\phi_i | \psi_i, q) f(\psi_i | \phi_i, q) \,,$$

where $f(\cdot | \cdot, q)$ are the marginal probability densities. This is also known as *full conditional composite likelihood*, which in the bivariate case also equals the *pairwise conditional composite likelihood* [480]. The maximum pseudolikelihood method proceeds by maximizing this pseudolikelihood with respect to the parameter vector, which yields the estimate of $q$.

It turns out that the sine and cosine models are non-closed exponential family models [480]. However, they can be regarded as approximately closed, and the estimates obtained by maximizing the pseudolikelihood are an approximation to the maximum likelihood estimate. For the sine model the maximum pseudolikelihood estimator has been shown to have high efficiency in most cases [480].

#### 6.5.3.1 Maximum Pseudo-likelihood Estimator for the Sine Model

For the sine model the conditional distributions are von Mises distributed, as described in Sect. 6.3.1. With the parameter vector $q_s = (\mu, \nu, \kappa_1, \kappa_2, \lambda)$, the

6 Statistics of Bivariate von Mises Distributions

pseudolikelihood expression becomes

$$
PL_s = \prod_{i=1}^{n} f(\phi_i|\psi_i, q_s) f(\psi_i|\phi_i, q_s)
$$

$$
= \prod_{i=1}^{n} \frac{\exp\{\kappa_{1\lambda}(\psi_i)\cos(\phi_i - \mu - \psi_{i\nu})\}}{2\pi\, I_0(\kappa_{1\lambda}(\psi_i))} \frac{\exp\{\kappa_{2\lambda}(\phi_i)\cos(\psi_i - \nu - \phi_{i\mu})\}}{2\pi\, I_0(\kappa_{2\lambda}(\phi_i))},
$$

where

$$
\tan\psi_{i\nu} = (\lambda/\kappa_1)\sin(\psi_i - \nu) \qquad \kappa_{1\lambda}(\psi_i)^2 = \kappa_1^2 + \lambda^2\sin^2(\psi_i - \nu)
$$

$$
\tan\phi_{i\mu} = (\lambda/\kappa_2)\sin(\phi_i - \mu) \qquad \kappa_{2\lambda}(\phi_i)^2 = \kappa_2^2 + \lambda^2\sin^2(\phi_i - \mu).
$$

The expression for the pseudolikelihood can be shown to be equal to the somewhat simpler expression [476]

$$
PL_s = \prod_{i=1}^{n} \left[ \frac{\exp\{k_1\cos(\phi_i - \mu) + \lambda\sin(\phi_i - \mu)\sin(\psi_i - \nu)\}}{2\pi\, I_0(\kappa_{1\lambda}(\psi_i))} \cdot \right.
$$

$$
\left. \frac{\exp\{k_2\cos(\psi_i - \nu) + \lambda\sin(\phi_i - \mu)\sin(\psi_i - \nu)\}}{2\pi\, I_0(\kappa_{2\lambda}(\phi_i))} \right]. \tag{6.18}
$$

Finding the maximum pseudolikelihood estimate then becomes a matter of maximizing the expression in Eq. 6.18 with respect to the parameters $q_s$. This is normally done using numerical optimizations methods, as is the case for the regular maximum likelihood method described in Sect. 6.5.1.1. However, the pseudolikelihood expression does note include any complicated normalizing constants that require numerical integration. Hence, the maximum pseudolikelihood method is less computational demanding than the maximum likelihood method. It should be noted that this computational advantage is even more pronounced for the multivariate von Mises distribution described in Sect. 6.7.

## 6.6 Simulation from Bivariate von Mises Models

In this section we will discuss how to sample from the sine model and the cosine model with positive interaction. Generally, there are two approaches: Gibbs sampling and rejection sampling. Both approaches work well in practice, but the latter is the most efficient [478]. They both make use of the fact that there exist an efficient algorithm for simulating from the univariate von Mises distribution, as described in Sect. 6.2.1 and Algorithm 1.

## 6.6.1 Gibbs Sampling

A straightforward approach to generating samples from the sine model and the cosine models is Gibbs sampling. In Gibbs sampling, a sequence of samples is obtained by alternating simulating values from the two conditional distributions $f(\phi|\psi)$ and $f(\psi|\phi)$. That is, we start with an initial random value for $\psi_0$ and then sample $\phi_0 \sim f(\phi|\Psi = \psi_0)$. In the next step we sample $\psi_1 \sim f(\psi|\Phi = \phi_0)$, continue by sampling $\phi_1 \sim f(\phi|\Psi = \psi_1)$ and so forth. This is straightforward, since the conditional distributions are all von Mises for these models.

The main problem with this sampling approach is that a suitable burn-in period is needed to make the samples independent of the starting point, $\psi_0$, and thinning is needed to ensure that there are no sequential dependencies between the samples [478]. This makes the Gibbs sampling approach inefficient.

## 6.6.2 Simulation Using Rejection Sampling

A more efficient approach to sampling from the sine model and the cosine models is to first use rejection sampling to simulate a value $\psi'$ from the marginal density $f(\psi)$, and then simulate a value $\phi'$ from the conditional distribution $f(\phi|\Psi = \psi')$ [68, 478]. The description here is mainly based on the one of Mardia et al. [478].

In order to draw a sample $\psi'$ from the marginal density $f(\psi)$ using rejection sampling, we first need to determine whether the marginal density is unimodal or bimodal. This is done by checking the conditions of respectively Theorems 6.3 or 6.4. Depending on the modality we use either a single von Mises or a mixture of two von Mises as proposal distribution:

- If the marginal distribution is unimodal, we propose a candidate value $\psi'$ by sampling from the univariate von Mises distribution $M(\nu, \kappa^*)$.
- If the marginal distribution is bimodal, we propose a candidate value $\psi'$ by sampling from an equal mixture of the two von Mises densities $M(\mu - \psi^*, \kappa^*)$ and $M(\mu + \psi^*, \kappa^*)$, where $\psi^*$ is given by either Theorems 6.3 or 6.4.

In the expressions above, $\kappa^*$ should be chosen so that the distance between the proposal density and the marginal density $f(\psi)$ is minimized. The next step is to sample a value $u$ from the uniform distribution on $[0, 1]$ and test the condition $u < f(\psi')/(L \cdot g(\psi'))$, where $g$ is the probability density function of the proposal distribution and $L > 1$ is an upper bound on $f(\psi)/g(\psi)$. If the condition is true then $\psi'$ is accepted as a realization of $f(\psi)$, otherwise $\psi'$ is rejected and the candidate sampling step is repeated.

The proportion of proposed samples that are accepted is given by $1/L$, which is also called the efficiency. $L$ should therefore be minimized under the conditions $L > 1$ and $L > f(\psi)/g(\psi)$ for all $\psi \in (-\pi, \pi]$. In practice this can be done by finding $\max(f(\psi)/g(\psi))$ over $\psi \in (-\pi, \pi]$ using numerical optimization.

6 Statistics of Bivariate von Mises Distributions

175

For many sets of parameters the marginal densities are close to von Mises and the rejection sampling algorithm has high efficiency. However, for large values of $\kappa_1 \approx \kappa_2 \approx \lambda, \kappa_3, \kappa_3'$ the algorithm becomes quite inefficient. As an example, for the positive interaction cosine model with $(\kappa_1, \kappa_2, \kappa_3) = (100, 100, 90)$ the density is very bimodal and the efficiency is around 69% [478].

Once a sample $\psi'$ has been obtained from the marginal distribution $f(\psi)$ using rejection sampling, $\phi'$ can be sampled from the conditional distribution $f(\phi|\Psi = \psi')$, which is von Mises for both the sine and cosine model with positive interaction. Using this procedure we can generate a sequence of samples $\{(\phi_i, \psi_i)\}_{i=1}^n$ from either the sine model or cosine models. These sample are independent and contrary to the Gibbs sampler no burn-in period or thinning is required. This sampling approach has been employed for models of local protein structure [68, 478], and in particular it was used for TORUSDBN described in Chap. 10.

## 6.7 The Multivariate von Mises Distribution

In this section, we will look beyond the bivariate angular distribution and consider a general multivariate angular distribution. In 2008, Mardia et al. [479] presented a multivariate extension of the bivariate sine model from Eq. 6.9. This distribution is called the multivariate von Mises distribution and is denoted $\Theta \sim M_p(\mu, \kappa, \Lambda)$ in the $p$-variate case. The probability density function for $\Theta = (\Theta_1, \Theta_2, \ldots, \Theta_p)^T$ is given by

$$f_p(\theta) = \{T(\kappa, \Lambda)\}^{-1} \exp\{\kappa^T c(\theta, \mu) + s(\theta, \mu)^T \Lambda s(\theta, \mu)/2\}, \qquad (6.19)$$

where $-\pi < \theta_i \leq \pi, -\pi < \mu_i \leq \pi, \kappa_i \geq 0, -\infty < \lambda_{ij} < \infty, \kappa = (\kappa_1, \ldots, \kappa_p)$,

$$c(\theta, \mu)^T = (\cos(\theta_1 - \mu_1), \ldots, \cos(\theta_p - \mu_p)),$$

$$s(\theta, \mu)^T = (\sin(\theta_1 - \mu_1), \ldots, \sin(\theta_p - \mu_p)),$$

and $(\Lambda)_{ij} = \lambda_{ij} = \lambda_{ji}, \lambda_{ii} = 0$. The normalizing constant $\{T(\kappa, \Lambda)\}^{-1}$ ensures that the expression in Eq. 6.19 defines a probability density function, however the constant is only known in explicit form for the two cases $p = 1$ and $p = 2$. For $p = 1$ this model is a univariate von Mises model with the normalizing constant given in Eq. 6.4, and for $p = 2$ it is a bivariate sine model with normalizing constant given by Eq. 6.9.

Under large concentrations in the circular variables, the model behaves like a multivariate normal distribution [479]. Without any loss of generality, we assume $\mu = 0$ and have approximately

$$\Theta = (\Theta_1, \Theta_2, \ldots, \Theta_p)^T \sim N_p(0, \Sigma),$$

with

$$(\Sigma^{-1})_{ij} = \begin{cases} \kappa_i & \text{for } i = j \\ -\lambda_{ij} & \text{for } i \neq j \end{cases},$$

where $\mathcal{N}_p(\mathbf{0}, \Sigma)$ denotes a $p$-variate normal distribution with mean $\mathbf{0}$ and covariance matrix $\Sigma$.

Inference in the multivariate von Mises distribution can be done using methods of moments or the maximum pseudolikelihood method. The approaches are similar to those described for the bivariate sine model in Sects. 6.5.2 and 6.5.3. Sampling can be done using Gibbs sampling. We refer to Mardia et al. for further details [479].

As an example of usages in structural bioinformatics, Mardia et al. [479] considered a dataset of gamma turns in protein structures divided into triplets of amino acids. A trivariate von Mises distribution was fitted to the triplets of $\phi$ and $\psi$ separately, and a reasonable data fit was reported.

## 6.8 Conjugate Priors for von Mises Distributions

There has been renewed interest in directional Bayesian analysis since the paper of Mardia and El-Atoum [473], as indicated below. In this section we will consider conjugate priors for the various von Mises distributions, in particular we focus on priors for the mean vector. For a general introduction to conjugate priors we refer to Chap. 1.

### 6.8.1 Univariate Case

Consider the von Mises distribution with probability density function given in Eq. 6.4. Recall that in this expression $\mu$ is the mean direction and $\kappa$ is the concentration (precision) parameter. It has been shown by Mardia and El-Atoum [473] that for a given $\kappa$, the von Mises distribution, $M(\mu^*, \kappa^*)$, is a conjugate prior for $\mu$. In other words, the von Mises distribution is self-conjugate for fixed $\kappa$.

Guttorp and Lockhart [265] have given the joint conjugate prior for $\mu$ and $\kappa$, and Mardia [470] has considered a slight variant. However, the distribution for $\kappa$ is not straightforward. Various suggestions have appeared in the literature; for example, take the prior for $\kappa$ independently as a chi-square distribution, use the non-informative prior and so forth.

### 6.8.2 Bivariate Case

Mardia [472] has shown that a conjugate prior of the full bivariate von Mises distribution for the mean vector $(\mu, \nu)$ given $\kappa_1, \kappa_2$ and $\mathbf{A}$ is the full bivariate von

6 Statistics of Bivariate von Mises Distributions 177

Mises itself, as defined in Eq. 6.7. Furthermore, Mardia has also obtained a compact form of the normalizing constant for this general case in order to write down the full conjugate prior and the posterior. We will now show how this results applies to the bivariate sine model from Sect. 6.3.1. Let $(\phi_i, \psi_i)_{i=1}^n$ be distributed according to a sine model with known concentration $(\kappa_1, \kappa_2)$ and dependence parameter $\lambda$. Denote the center of mass for $(\phi_i)_{i=1}^n$ by $(C_1, S_1)$, and the center of mass for $(\psi_i)_{i=1}^n$ by $(C_2, S_2)$, as defined in Eq. 6.1. Now, let the prior distribution of $(\mu, \nu)$ be given by a sine model with mean $(\mu_0, \nu_0)$, concentration $(\kappa_{01}, \kappa_{02})$ and dependence $\lambda_0$. It can be shown [472] that the posterior density for $(\mu, \nu)$ is given by the full bivariate von Mises distribution with mean $(\mu_0^*, \nu_0^*)$, concentration $(\kappa_1^*, \kappa_2^*)$ and the matrix $\mathbf{A}^*$ as defined below.

$$\kappa_1^* \cos \mu_0^* = \kappa_{01} \cos \mu_0 + \kappa_1 C_1, \qquad \kappa_1^* \sin \mu_0^* = \kappa_{01} \sin \mu_0 + \kappa_1 S_1,$$
$$\kappa_2^* \cos \nu_0^* = \kappa_{02} \cos \nu_0 + \kappa_2 C_2, \qquad \kappa_2^* \sin \nu_0^* = \kappa_{02} \sin \nu_0 + \kappa_2 S_2,$$

$$\mathbf{A}^* = \begin{bmatrix} \lambda_0 \sin \mu_0 \sin \nu_0 + \lambda \sum_{i=1}^n \sin \phi_i \sin \psi_i & -\lambda_0 \sin \mu_0 \cos \nu_0 - \lambda \sum_{i=1}^n \sin \phi_i \cos \psi_i \\ -\lambda_0 \cos \mu_0 \sin \nu_0 - \lambda \sum_{i=1}^n \cos \phi_i \sin \psi_i & \lambda_0 \cos \mu_0 \cos \nu_0 + \lambda \sum_{i=1}^n \cos \phi_i \cos \psi_i \end{bmatrix}$$

This result was also obtained independently by Lennox et al. [435] (see correction). A key point is that the posterior density for $(\mu, \nu)$ is not a sine density. However, for the cosine model, it can be shown that the posterior distribution of the mean is another cosine submodel of the full bivariate von Mises, but with six parameters only; details are given by Mardia [472].

Lennox et al. [434] have provided a template-based approach to protein structure prediction using a semiparametric Bayesian model which uses the Dirichlet process mixture model. This work relies on priors for the bivariate sine model, which has been studied in details by Lennox et al. [435] and Mardia [471, 472].

## 6.8.3 The Multivariate von Mises Distribution

Finally, we will consider priors for the multivariate von Mises distribution from Sect. 6.7. When the multivariate von Mises distribution is used as prior for the mean vector $\mu$ and the likelihood is a multivariate von Mises with know $\kappa$ and $\Lambda$, then the posterior density of $\mu$ has been shown [472] to belong to an extension of the expression given by Mardia and Patrangenaru [475] in their Eq. 3.1. For $\kappa, \Lambda$ we can use an independent prior distribution as Wishart for $\Gamma$, where $(\Gamma)_{ii} = \kappa_i$ and $(\Gamma)_{ij} = -\lambda_{ij}$, following the proposal for $p = 2$ by Lennox et al. [435]. Full details are given by Mardia [472].

## 6.9 Conclusion

The various von Mises distributions are of prime interest in structural bioinformatics as they are extremely suited to formulate probabilistic models that involve dihedral angles, arguably the most important degree of freedom in the parameterization of biomolecules. Together with the Kent distribution discussed in Chap. 7, which concerns unit vectors and thus covers another potentially important and common degree of freedom, these distributions form a powerful tool for the structural biologist in search for rigorous statistical solutions to challenging problems.

# Chapter 7
# Statistical Modelling and Simulation Using the Fisher-Bingham Distribution

John T. Kent

## 7.1 Introduction

One simplified view of a protein main chain is that it consists of a sequence of amino acids, represented by points centered at their $C\alpha$ atoms, separated by bonds of constant length. This representation is called the protein's $C\alpha$ trace. The "shape" of the protein is determined by the angles between two, and the dihedral angles between three, successive bonds. An equivalent view is that the shape is determined by a sequence of unit vectors or points on the sphere. Since the orientation of each bond represents a direction in three-dimensional space, we see that it is important to have statistical tools to understand and fit data on the sphere.

In this paper we review the use of one particular distribution, FB5. Methods of estimation and simulation are discussed. The representation of a protein main chain with $n$ amino acids can then be represented in terms of $n - 1$ spherical directions. The FB5 distribution was used to formulate a probabilistic model of the $C\alpha$ trace of proteins [275]. This model, together with similar models based on graphical models and directional statistics, is discussed in Chap. 10. Here we focus on the properties of the FB5 distribution itself.

## 7.2 Spherical Coordinates

A point on the unit sphere in $\mathbb{R}^3$ can be represented as a vector $\boldsymbol{x} = (x_1, x_2, x_3)^T$, with $x_1^2 + x_2^2 + x_3^2 = 1$. Boldface is used to indicate vectors and the transpose $^T$

---

J.T. Kent (✉)
Department of Statistics, University of Leeds, Leeds LS2 9JT, UK
e-mail: j.t.kent@leeds.ac.uk

T. Hamelryck et al. (eds.), *Bayesian Methods in Structural Bioinformatics*,
Statistics for Biology and Health, DOI 10.1007/978-3-642-27225-7_7,
© Springer-Verlag Berlin Heidelberg 2012

is included because we generally think of vectors as column vectors. Thus the unit sphere can be written as $\Omega_3 = \{x \in \mathbb{R}^3 : x^T x = x_1^2 + x_2^2 + x_3^2 = 1\}$.

It is useful to use polar coordinates to represent $x$,

$$x_1 = \sin\theta\cos\phi, \quad x_2 = \sin\theta\sin\phi, \quad x_3 = \cos\theta.$$

Here $\theta \in [0, \pi]$ and $\phi \in [0, 2\pi)$ define the colatitude and longitude, respectively. For a point on the earth, with the $x_3$-axis pointing towards the north pole, $\theta$ ranges from 0 to $\pi$ as $x$ ranges between the north pole and the south pole. If $\theta = 0$ or $\pi$, the longitude is undefined; otherwise for fixed $\theta$, $0 < \theta < \pi$, the set of points $\{x : x_3 = \cos\theta\}$ defines a small circle, and the longitude $\phi$ identifies the position of $x$ along this small circle.

The uniform measure on the sphere can be written as

$$\omega(dx) = \sin\theta \, d\theta \, d\phi \tag{7.1}$$

The left-hand side is more convenient for mathematical descriptions; the right-hand side is useful for explicit calculations. Note the $\sin\theta$ factor on the right-hand side occurs because a small circle has a smaller radius for $\theta$ near the poles, 0 or $\pi$, than for $\theta$ near the equator, $\theta = \pi/2$. The surface area of the sphere can be found as

$$\int_0^{2\pi} \left\{ \int_0^{\pi} \sin\theta \, d\theta \right\} \, d\phi = 4\pi.$$

In addition to polar coordinates, two other coordinate systems are of interest here. The first is given by the Euclidean coordinates $x_1$ and $x_2$, also known as the orthogonal projection onto tangent coordinates at the north pole. Since $x_3 = \pm\{1 - (x_1^2 + x_2^2)\}^{1/2}$, these coordinates suffice to parameterize either the northern or southern hemisphere. The second coordinate system is given by the equal area projection, with coordinates

$$u_1 = 2\sin(\theta/2)\cos\phi, \quad u_1 = 2\sin(\theta/2)\sin\phi, \tag{7.2}$$

so that the sphere, excluding the south pole, is mapped into the disk $\{(u_1, u_2) : u_1^2 + u_2^2 < 4\}$. The uniform measure on the sphere takes the forms

$$\omega(dx) = \{1 - (x_1^2 + x_2^2)\}^{-1/2} \, dx_1 \, dx_2 = du_1 du_2 \tag{7.3}$$

in these coordinate systems. The lack of a multiplying coefficient in $du_1 du_2$ justifies the name "equal area". The equal area projection is also known as the Lambert azimuthal projection and is sometimes plotted on special graph paper known as a Schmidt net.

## 7.3 The FB5 Distribution

Let us start by recalling the key properties of the bivariate normal distribution for a random vector $x = (x_1, x_2)^T$. This distribution has the density

$$f(x) = |2\pi\Sigma|^{-1/2} \exp\left\{-\frac{1}{2}(x-\mu)^T \Sigma^{-1}(x-\mu)\right\}, \quad x \in \mathbb{R}^2. \quad (7.4)$$

There are five parameters: the means $\mu = (\mu_1, \mu_2)^T$ of $x_1$ and $x_2$, the variances $\sigma_{11}, \sigma_{22}$ of $x_1$ and $x_2$, and the covariance $\sigma_{12} = \sigma_{21}$ between $x_1$ and $x_2$, where the $2 \times 2$ covariance matrix has elements

$$\Sigma = \begin{bmatrix} \sigma_{11} & \sigma_{12} \\ \sigma_{21} & \sigma_{22} \end{bmatrix}.$$

This distribution is the most important distribution for bivariate data. One way to think about this distribution is to plot contours of constant probability. These are given by ellipses centered at the mean vector of the distribution, and with size and orientation governed by $\Sigma$. A typical ellipse is given in Fig. 7.1a, together with the major and minor axes.

The surface of a sphere is a (curved) two-dimensional surface. Hence it is natural to look for an analogue of the bivariate normal distribution. Thus we wish to find a distribution which can have any modal direction, with ellipse-like contours of constant probability about this modal direction.

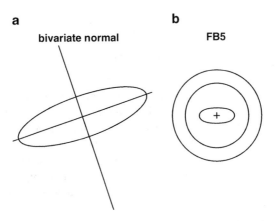

**Fig. 7.1** (a) An ellipse representing a contour of constant probability for the bivariate normal distribution, together with the major and minor axes. (b) A oval-shaped contour of constant probability for the FB5 density (Eq. 7.5). The plot is given in equal area coordinates; the point at the center represents the north pole, the outer *circle* represents the south pole and the intermediate *circle* represents the equator. The distribution is centered at the north pole and the major and minor axes have been rotated about the north pole to coincide with the horizontal and vertical axes

The most promising candidate is the five-parameter Fisher-Bingham, or FB5, distribution, also known as the Kent distribution ([365]; [185, p. 92]; [474, p. 176]). The name Fisher-Bingham is motivated by the fact that the exponent of the density of the Fisher distribution involves a linear function of $x$ (see $x_3$ in Eq. 7.5 below) whereas the Bingham distribution involves a quadratic function of $x$ (see $x_1^2 - x_2^2$ in Eq. 7.5).

The FB5 distribution contains five parameters which play roles analogous to the parameters of the bivariate normal distribution. The density in its simplest form takes the form

$$f(x) = C(\kappa, \beta)^{-1} \exp\left\{\kappa x_3 + \beta(x_1^2 - x_2^2)\right\} \tag{7.5}$$

with respect to the uniform measure on the sphere, and where we usually suppose that $0 \le 2\beta < \kappa$. In this case the formula defines a unimodal density on the sphere. The mode is at the north pole $x_3 = 1$ and the anti-mode is at the south pole $x_3 = -1$. The contours of constant probability are ellipse-like with the major axis pointing along the $x_1$-axis and the minor axis pointing along the $x_2$-axis. The normalizing constant $C(\kappa, \beta)^{-1}$ depends on two concentration parameters, $\kappa$ and $\beta$. As $\kappa > 0$ increases, the distribution becomes more concentrated about its mode at the north pole. If $\beta = 0$ the distribution is rotationally symmetric about the north pole; as $\beta$ increases (with $2\beta < \kappa$), the distribution becomes more elliptical. A typical contour of constant probability is given in Fig. 7.1b.

A more general version of the density allows the mean direction and the major and minor axes to be given by any three orthonormal vectors. Recall that a $3 \times 3$ matrix $\Gamma$ is called orthogonal if its columns are orthonormal, that is if

$$\gamma_{(j)}^T \gamma_{(j)} = 1, \quad \gamma_{(j)}^T \gamma_{(k)} = 0, \quad j, k = 1, 2, 3, \quad j \ne k,$$

where

$$\gamma_{(j)} = \begin{bmatrix} \gamma_{1j} \\ \gamma_{2j} \\ \gamma_{3j} \end{bmatrix}$$

denotes the $j$th column of $\Gamma$. In matrix form the orthogonality condition can be written $\Gamma^T \Gamma = I_3$, the identity matrix in three dimensions, and so the inverse of $\Gamma$ is its transpose $\Gamma^{-1} = \Gamma^T$. Hence the determinant satisfies $|\Gamma| = \pm 1$. If an orthogonal matrix satisfies $|\Gamma| = +1$, it is called a rotation matrix. For our purposes, the sign of the final column $\gamma_{(3)}$ is important However, for the other two columns, only the axis information is relevant; i.e. $\gamma_{(1)}$ and $-\gamma_{(1)}$ contain the same information; similarly for $\gamma_{(2)}$ and $-\gamma_{(2)}$. Hence, without loss of generality, we may restrict $\Gamma$ to be a rotation matrix.

The more general FB5 density takes the form

$$f(x) = C(\kappa, \beta)^{-1} \exp\left\{\kappa \gamma_{(3)}^T x + \beta[(\gamma_{(1)}^T x)^2 - (\gamma_{(2)}^T x)^2)]\right\}. \tag{7.6}$$

# 7 Statistical Modelling and Simulation Using the Fisher-Bingham Distribution

In this case the modal direction is given by $\boldsymbol{\gamma}_{(3)}$ and the major and minor axes by $\boldsymbol{\gamma}_{(1)}$ and $\boldsymbol{\gamma}_{(2)}$, respectively. We write the distribution as FB5$(\kappa, \beta, \Gamma)$ in terms of the concentration parameters $\kappa$, $\beta$, and the rotation matrix $\Gamma$.

A general orthogonal matrix $G$ can be decomposed as a product of two simpler matrices, $G = HK$, where

$$H = H(\delta, \eta) = \begin{bmatrix} \cos\delta\cos\eta & -\sin\eta & \sin\delta\cos\eta \\ \cos\delta\sin\eta & \cos\eta & \sin\delta\sin\eta \\ -\sin\delta & 0 & \cos\delta \end{bmatrix}, \quad K = K(\psi) = \begin{bmatrix} \cos\psi & -\sin\psi & 0 \\ \sin\psi & \cos\psi & 0 \\ 0 & 0 & 1 \end{bmatrix}. \quad (7.7)$$

The angles $\delta \in [0, \pi]$ and $\eta \in [0, 2\pi)$ represent the colatitude and longitude of the modal direction, and the angle $\psi \in [0, 2\pi)$ represents the rotation needed to align the major and minor axes with the coordinate axes, once the modal axis has been rotated to the north pole. Note that $K(\psi + \pi)$ changes the sign of the first two columns of $K(\psi)$ so that $\psi$ and $\psi + \pi$ determine the same two axes for $\boldsymbol{\gamma}_{(1)}$ and $\boldsymbol{\gamma}_{(2)}$. Hence $\psi$ can be restricted to the interval $[0, \pi)$. If $\boldsymbol{x}$ is a random unit vector following the density given by Eq. 7.6 and we set $\boldsymbol{y} = H^T \boldsymbol{x}$ and $\boldsymbol{z} = K^T \boldsymbol{y}$, so that $\boldsymbol{z} = G^T \boldsymbol{x}$, then $\boldsymbol{z}$ follows the standardized density given by Eq. 7.5.

Under high concentration, that is, for large $\kappa$ with $\lambda = 2\beta/\kappa$ held fixed, $0 \le \lambda < 1$, the FB5 distribution (Eq. 7.5) is approximately a bivariate normal distribution. In this case most of the probability mass is concentrated near $x_1 = 0$, $x_2 = 0$, $x_3 = 1$. In particular virtually all the mass lies in the northern hemisphere $x_3 > 0$, which we can represent by the orthogonal tangent coordinates, $x_1$ and $x_2$, with $x_1$ and $x_2$ small. Using the approximation $x_3 = \{1 - (x_1^2 + x_2^2)\}^{1/2} \approx 1 - \frac{1}{2}(x_1^2 + x_2^2)^2$, and noting $\{1 - (x_1^2 + x_2)^2\}^{-1/2} \approx 1$ in Eq. 7.3 yields the asymptotic formula for the density given by Eq. 7.5 (with respect to $dx_1\, dx_2$),

$$g(x_1, x_2) \propto \exp -\frac{\kappa}{2}\left\{(1 - \lambda)x_1^2 + (1 + \lambda)x_2^2\right\}. \quad (7.8)$$

That is, $\kappa^{1/2}x_1$ and $\kappa^{1/2}x_2$ are approximately independently normally distributed with 0 means and variances $1/(1 - \lambda)$ and $1/(1 + \lambda)$, respectively.

## 7.4 Estimation

Suppose we are given a set of data, $\{\boldsymbol{x}_i, i = 1, \ldots, n\}$ where each $\boldsymbol{x}_i = (x_{i1}, x_{i2}, x_{i3})^T$ is a unit 3-dimensional vector. Usually, the modal direction is unknown, so we need to estimate the orientation matrix $\Gamma$ as well as the concentration parameters $\kappa$ and $\beta$.

Estimation, or more specifically moment estimation, is described most easily by splitting the process into several stages.

1. Estimate the modal direction using the sample resultant vector. Rotate the data in three dimensions so that the modal direction points towards the north pole.

2. Project the data onto the tangent plane at the north pole, and compute the $2 \times 2$ matrix of second moments about the origin.
3. Rotate the data in the plane so that the second moment matrix is diagonal and use these diagonal values to estimate $\kappa$ and $\beta$.

Next we fill out this sketch, using a hat ˆ to indicate estimated quantities. Let $\bar{x} = \frac{1}{n} \sum x_i$ denote the sample mean vector of the data, also called the resultant vector, and write it in the form $\bar{x} = r_1 \bar{x}_0$, where the resultant length $r_1 = \{\bar{x}^T \bar{x}\}^{1/2}$ is the norm of $\bar{x}$ and where $\bar{x}_0$ is a unit vector pointing in the same direction as $\bar{x}$. In practice $0 < r_1 < 1$ with $r_1$ close to 1 for concentrated data. Write $\bar{x}_0 = \left( \sin \hat{\delta} \cos \hat{\eta}, \ \sin \hat{\delta} \sin \hat{\eta}, \ \cos \hat{\delta} \right)^T$ in polar coordinates, and define the orthogonal matrix $\hat{H} = H\left(\hat{\delta}, \hat{\eta}\right)$ as in Eq. 7.7. Define $y_i = \hat{H}^T x_i$, $i = 1, \ldots, n$ to be the rotated data. The mean vector of the $\{y_i\}$ now points towards the north pole $(0, 0, 1)^T$.

Define the $2 \times 2$ covariance matrix about the origin of the first two coordinates of the $\{y_i\}$ data,

$$S = \frac{1}{n} \sum_{i=1}^{n} \begin{bmatrix} y_{i1}^2 & y_{i1} y_{i2} \\ y_{i2} y_{i1} & y_{i2}^2 \end{bmatrix}.$$

Let $S = K_0 L K_0^T$ be a spectral decomposition of $S$ in terms of its eigenvalues and eigenvectors, where

$$K_0 = \begin{bmatrix} \cos \hat{\psi} & -\sin \hat{\psi} \\ \sin \hat{\psi} & \cos \hat{\psi} \end{bmatrix}$$

is a $2 \times 2$ rotation matrix and $L = \mathrm{diag}\,(l_1, l_2)$ is a diagonal matrix with $l_1 \geq l_2 > 0$. If the covariance matrix interpreted in terms of an ellipse, then the columns of $K_0$, the eigenvectors, represent the major and minor axes of the ellipse, and the eigenvalues $l_1$ and $l_2$ are proportional to the lengths of the major and minor axes. Note that $K_0$ is a submatrix of the $3 \times 3$ matrix $\hat{K} = K\left(\hat{\psi}\right)$ in Eq. 7.7 used in the standardization of the data.

Lastly estimate the concentration parameters $\kappa$ and $\beta$. Let $r_2 = l_1 - l_2$ measure the ellipticity of $S$. For concentrated data we can use the asymptotic normal result in Eq. 7.8 to get

$$\hat{\kappa} = (2 - 2r_1 - r_2)^{-1} + (2 - 2r_1 + r_2)^{-1}, \quad \hat{\beta} = \frac{1}{2}\left\{(2 - 2r_1 - r_2)^{-1} \right.$$

$$\left. - (2 - 2r_1 + r_2)^{-1}\right\}. \tag{7.9}$$

For data that are not highly concentrated, [365] developed a numerical algorithm based on a series expansion for the normalization constant $C(\kappa, \beta)$. This algorithm is straightforward to implement on a computer and a program in R is available from the author.

7  Statistical Modelling and Simulation Using the Fisher-Bingham Distribution    185

It is also possible to carry out estimation in a Bayesian framework. Assuming high concentration, the calculations can be done in closed form. The simplest strategy is to treat the data as approximately bivariate normally distributed and to use informative conjugate priors for the mean vector and covariance matrix. It is also possible to give the location vector $\gamma_{(3)}$ a non-informative prior uniform distribution on the sphere, but care is needed in the treatment of the remaining parameters.

## 7.5  Simulation

In modern Monte Carlo statistical methods, distributions such as FB5 are building blocks in a larger construction, and efficient algorithms are needed to simulate from such distributions. Here we give an exact simulation method with good efficiency properties for the whole range of $\kappa$ and $\beta$ values, $0 \leq 2\beta \leq \kappa$. In this case note that the exponent in (7.5), $\{\kappa \cos\theta + \beta \sin^2\theta(\cos^2\phi - \sin^2\phi)\}$, is a decreasing function of $\theta \in [0, \pi]$ for each $\phi$. (On the other hand, if $\beta > \kappa/2$, the density increases and then decreases in $\theta$ when $\phi = 0$.) The algorithm developed here was first set out in [367].

For the purposes of simulation it is helpful to use the equal area projection (Eq. 7.2) with coordinates $(u_1, u_2)$. For algebraic convenience, set $t_1 = u_1/2$, $t_2 = u_2/2$, so that $r^2 = t_1^2 + t_2^2 < 1$.

In $(t_1, t_2)$ coordinates, the probability density (with respect to $dt_1\, dt_2$ in the unit disk $t_1^2 + t_2^2 < 1$) takes the form

$$f(t_1, t_2) \propto \exp\left\{-2\kappa r^2 + 4\beta(r^2 - r^4)(\cos^2\phi - \sin^2\phi)\right\}$$
$$= \exp\left\{-2\kappa(t_1^2 + t_2^2) + 4\beta[1 - (t_1^2 + t_2^2)](t_1^2 - t_2^2)\right\}$$
$$= \exp\left\{-\frac{1}{2}[at_1^2 + bt_2^2 + \gamma(t_1^4 - t_2^4)]\right\}. \tag{7.10}$$

where the new parameters

$$a = (4\kappa - 8\beta), \quad b = (4\kappa + 8\beta), \quad \gamma = 8\beta \tag{7.11}$$

satisfy $0 \leq a \leq b$ and $\gamma \leq b/2$. Here we have used the double angle formulas, $\cos\theta = 1 - 2\sin^2(\theta/2)$, $\sin\theta = 2\sin(\theta/2)\cos(\theta/2)$.

Note that the density splits into a product of a function of $t_1$ alone and $t_2$ alone. Hence $t_1$ and $t_2$ would be independent except for the constraint $t_1^2 + t_2^2 < 1$. Our method of simulation, as sketched below, will be to simulate $|t_1|$ and $|t_2|$ separately by acceptance-rejection using a (truncated) exponential envelope, and then additionally to reject any values lying outside the unit disk. For a general background in acceptance-rejection sampling, see, for example [131].

The starting point for our simulation method is the simple inequality

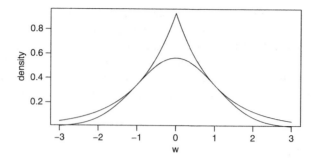

**Fig. 7.2** Acceptance-rejection simulation using Eq. 7.14. The *lower curve* is a Gaussian density. The *upper curve* is proportional to a double exponential density and always sits above the *lower curve*

$$\frac{1}{2}(\sigma|w| - \tau)^2 \geq 0 \qquad (7.12)$$

for any parameters $\sigma, \tau > 0$ and for all $w$. Hence

$$-\frac{1}{2}\sigma^2 w^2 \leq \frac{1}{2}\tau^2 - \sigma\tau|w|. \qquad (7.13)$$

After exponentiation, this inequality provides the basis for simulating a Gaussian random variable from a double exponential random variable by acceptance-rejection,

$$f(w) = (2\pi)^{-1/2} e^{-\frac{1}{2}w^2}, \quad g(w) = \frac{1}{2} e^{-|w|}, \quad f(w) \leq Cg(w), \qquad (7.14)$$

for all $w$, where $C = (2e/\pi)^{1/2} \approx 1.3$. Simulation from the bilateral exponential distribution is straightforward using the inverse method for $|w|$ [131]; set $w = s \log u$, where $u$ is uniform on $(0, 1)$, independent of $s$ which takes the values $\pm 1$ with equal probability. The constant $C$ gives the average number of simulations from $g$ needed to generate each simulation from $f$, and $1/C \approx 0.78$ is called the efficiency of the method. Figure 7.2 gives a plot of $f(w)$ and $Cg(w)$ in this case.

For our purposes some refinement of this simple method is needed. For $t_1$ we need to apply Eq. 7.13 twice, first with $\sigma = \gamma^{1/2}$, $\tau = 1$ and $w = t_1^2$, and second with $\sigma = (a + 2\gamma^{1/2})^{1/2}$, $\tau = 1$ and $w = t_1$, to get

$$-\frac{1}{2}(at_1^2 + \gamma t_1^4) \leq \frac{1}{2} - \frac{1}{2}(a + 2\gamma^{1/2})t_1^2$$

$$\leq c_1 - \lambda_1 |t_1| \qquad (7.15)$$

where

$$c_1 = 1, \quad \lambda_1 = (a + 2\gamma^{1/2})^{1/2}. \qquad (7.16)$$

# 7 Statistical Modelling and Simulation Using the Fisher-Bingham Distribution

**Table 7.1** Efficiencies of the acceptance-rejection simulation method of Sect. 7.5 for the FB5 distribution, based on 10,000 simulations for each entry

|  | | $\lambda$ | | | |
|---|---|---|---|---|---|
| $\kappa$ | 0.0 | 0.3 | 0.6 | 0.9 | 1.0 |
| 0.1 | 0.71 | 0.71 | 0.70 | 0.71 | 0.71 |
| 1 | 0.34 | 0.34 | 0.34 | 0.35 | 0.35 |
| 10 | 0.35 | 0.34 | 0.31 | 0.28 | 0.27 |
| 100 | 0.35 | 0.31 | 0.28 | 0.27 | 0.26 |
| 1000 | 0.35 | 0.30 | 0.26 | 0.25 | 0.26 |

To develop a suitable envelope for $t_2$ recall that $0 \leq 2\gamma \leq b$. To begin with suppose $b > 0$. From Eq. 7.13 with $\sigma = (b - \gamma)^{1/2}$, $\tau = (b/(b - \gamma))^{1/2}$, and $w = t_2^2$,

$$-\frac{1}{2}(bt_2^2 - \gamma t_2^4) \leq -\frac{1}{2}(b - \gamma)t_2^2 \leq c_2 - \lambda_2|t_2| \tag{7.17}$$

where

$$c_2 = b/\{2(b - \gamma)\} \leq 1, \quad \lambda_2 = b^{1/2}. \tag{7.18}$$

If $b = 0$ (and so $\gamma = 0$) then Eq. 7.17 continues to hold with $\lambda_2 = 0$ and $c_2 = 0$.

This construction is most effective for large $\kappa$. For small $\kappa$, say $\kappa \leq 1$, it is more efficient to use a simple uniform envelope on the square $|t_1| \leq 1$, $|t_2| \leq 1$. Table 7.1 summarizes some sample efficiencies for various values of $\kappa$ and $\beta$, with $\lambda = 2\beta/\kappa$, where the uniform envelope has been used if $\kappa \leq 1$. Note that the efficiencies range between about 25% and 75% for all choices of the parameters.

## 7.6 Describing a Protein Main Chain

Consider a sequence of vertices $\{v_i, i = 1, \ldots, n\}$ in $\mathbb{R}^3$ representing the $C\alpha$ trace of a protein. Edges can be defined by $e_i = v_i - v_{i-1}$. By assumption the $e_i$ all have unit size.

Two successive edges $e_{i-1}$ and $e_i$ determine a $3 \times 3$ orthogonal matrix $G = [g_{(1)}, g_{(2)}, g_{(3)}]$ defining a frame of reference at vertex $i$ as follows:

$$g_{(3)} = e_i, \quad g_{(1)} = e_{i-1} - (e_{i-1}^T e_i) e_i, \quad g_{(2)} = g_{(3)} \times g_{(1)}.$$

Our first model for the protein structure will be a third-order Markov process. Consider an FB5$(\kappa, \beta, R)$ distribution with fixed parameters, where $R$ is a $3 \times 3$ rotation matrix. Then given $v_i$, $v_{i+1} = v_i + e_{i+1}$ is simulated by letting

$$G^T e_{i+1} \sim \text{FB5}(\kappa, \beta, R).$$

where the different FB5 simulations are independent for each $i$. The process is third-order Markov because of the need to determine a frame of reference at each vertex.

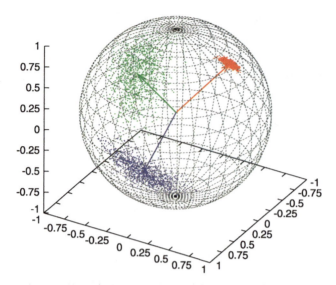

**Fig. 7.3** Simulation from three FB5 distributions; 1,000 samples from each are shown in *red*, *green* and *blue*, respectively. The mean directions are indicated by *arrows*. Figure taken from [275]

## 7.7 An HMM Protein Model

The previous model is a bit too simplistic to be useful in practice. Hence we let the parameters $(\kappa, \beta, R)$ vary according to a hidden Markov model (HMM) with a finite number of states. This model is discussed in more detail in Chap. 10, so we only give a short outline here. We call this HMM with discrete hidden nodes and observed FB5 nodes FB5HMM [275]. The discrete hidden nodes of the FB5HMM can be considered to model a sequence of fine-grained, discrete 'local descriptors'. These can be considered as fine grained extensions of helices, sheets and coils. The observed FB5 nodes translate these 'local descriptors' into corresponding angular distributions.

Hamelryck et al. [275] gives a thorough investigation of the FB5HMM model. Figure 7.3 is taken from that paper and shows a mixture of three FB5 distributions on the sphere with different location and concentration parameters. The three groups correspond to nodes which are typical representatives for different types of secondary protein structure: blue for coil, red for $\alpha$-helix, and green for $\beta$-strand.

# Part IV
# Shape Theory for Protein Structure Superposition

# Chapter 8
# Likelihood and Empirical Bayes Superposition of Multiple Macromolecular Structures

**Douglas L. Theobald**

## 8.1 Introduction

### 8.1.1 Overview

Superpositioning plays a fundamental role in current macromolecular structural analysis. By orienting structures so that their atoms match closely, superpositioning enables the direct analysis of conformational similarities and differences in three-dimensional Euclidean space. Superpositioning is a special case of Procrustes problems, in which coordinate vector sets are optimally oriented via rigid body rotations and translations. Optimal transformations are conventionally determined by minimizing the sum of the squared distances between corresponding atoms in the structures. However, the ordinary unweighted least-squares (OLS) criterion can produce inaccurate results when the atoms have heterogeneous variances (heteroscedasticity) or the atomic positions are correlated, both of which are common features of real data. In contrast, model-based probabilistic methods can easily allow for heterogeneous variances and correlations. Our likelihood treatment of the superposition problem results in more accurate superpositions and provides a framework for a full Bayesian analysis.

### 8.1.2 Superpositioning in Structural Biology

Superpositioning is a key technique in structural biology that enables the analysis of conformational differences among macromolecules with similar three dimensional

---

D.L. Theobald (✉)
Brandeis University, 415 South Street, Waltham, MA, USA
e-mail: dtheobald@brandeis.edu

T. Hamelryck et al. (eds.), *Bayesian Methods in Structural Bioinformatics*,
Statistics for Biology and Health, DOI 10.1007/978-3-642-27225-7_8,
© Springer-Verlag Berlin Heidelberg 2012

structures. The goal of macromolecular superpositioning methods is to orient two or more structures so that they are brought together as closely as possible in some optimal manner. Superpositioning is used routinely in the fields of NMR, X-ray crystallography, protein folding, molecular dynamics, rational drug design, and structural evolution [73, 191]. The interpretation of a superposition relies upon the validity of the estimated orientations, and hence accurate superpositioning tools are an essential component of modern structural analysis.

Throughout structural biology one is presented with different macromolecular structures that have biologically and physically significant conformational similarities. For instance, it is often possible to solve independent crystal structures of the same protein molecule, perhaps from different crystal forms. Similarly, NMR solution structure methods typically produce a family of distinct conformations for a single macromolecule, each of which is consistent with the experimental constraint data. These different structural solutions will generally differ in conformation, due either to experimental imprecision or to *bona fide* physical perturbations. In order to compare and analyze the conformations of these structures, it is of key interest to know where the structures are similar and where they are different. When determining and representing the structure of a macromolecule, in general the reference frame is arbitrary, since the absolute orientation usually has no particular physical significance. Even in cases where the reference frame may have some physical significance (for example, rotational and translational diffusion in an extended molecular dynamics trajectory), it is often desirable to distinguish global differences in molecular orientation from more local, internal conformational differences. The mathematical problem, then, lies in referring the structures to a common reference frame so that "real" conformational differences can be pinpointed.

### 8.1.3  Structure Alignment Versus Structure Superposition

The structural superposition problem is frequently confused with the structure-based alignment problem, a related, yet separate, bioinformatic challenge. The terms "alignment" and "superposition" are often conflated in common speech and even in the scientific literature, but here we draw a strict distinction. Performing a superposition requires an a priori one-to-one correspondence (a bijection) among the atoms in the different structures [73, 191, 248]. In many cases, the one-to-one correspondence is trivial, for instance when superpositioning multiple NMR models of the same protein or when superpositioning different conformations of identical proteins solved from independent crystal forms. On the other hand, for protein homologs with different numbers of residues or with different residue identities, we usually have no a priori knowledge of which amino acids correspond, especially when sequence similarity is low. If a sequence alignment is known, then it is possible to superposition the $\alpha$-carbons of homologous proteins, since a sequence alignment is a one-to-one map between the residues of each protein. Often, however, sequence similarity is so low as to prohibit reliable determination

## 8 Likelihood and Empirical Bayes Superposition

of a sequence alignment. Structure-based alignment methods attempt to determine the residue correspondences based on structural information alone. Determining a structure-based alignment is clearly a very different problem from finding the optimal orientations among the proteins, although superpositioning methods may be used as a component of a structural alignment algorithm. A Bayesian procedure for simultaneous determination of the alignment and superposition, using a Gaussian probability model similar to the superposition model presented here, is discussed in Chap. 9.

### 8.1.4 The Least-Squares Criterion for Optimal Superpositioning

Macromolecular structures are conventionally represented as ordered sets of three-dimensional vectors, one vector representing the Cartesian coordinates of each atom in the structure. In the classic molecular biology approach, the structures to be superpositioned are referred to a common reference frame using the conventional statistical optimization method of ordinary least-squares (OLS) [191]. For the OLS criterion, the molecules are rotated and translated in space to minimize the squared differences among corresponding atoms in the structures. In all common applications the rotations are rigid (orthogonal) and proper (without reflection) so as to preserve the interatomic distances and the chirality of the molecules. Thus the macromolecular superposition problem turns out to be a special case of general "Procrustes" problems, namely the least-squares orthogonal Procrustes problem for transforming rectangular matrices.

In 1905, Franz Boas (the "father of American anthropology") first formulated and provided the least-squares solution of a biological orthogonal Procrustes problem, for objectively comparing the two-dimensional morphological shapes of two human skulls. The analytical solution to the pair-wise least-squares orthogonal Procrustes problem for any number of dimensions was first given by John von Neumann in 1937 [744] (see Eq. 8.9), after which, evidently, it was largely forgotten for 15 years and then solved again independently many times, first by psychologists [252, 636] and later by researchers in the fields of photogrammetry[315], computer vision [731], and crystallography [356]. The pair-wise OLS solution was later extended to the more general case of superpositioning multiple structures simultaneously, again independently by different authors using a variety of iterative algorithms[149, 216, 248, 363, 397, 647, 707].

Why should the criterion of choice be the minimization of the average squared distance between structures? Naively, a more physically justifiable criterion might be to find the rotations and translations that bring the average of the absolute distances between the structures to a minimum. One reason is simply practical: as recounted above, the least-squares solution has a tractable analytic solution. However, the Gauss-Markov theorem also provides a well-known formal statistical justification for least-squares. Roughly speaking, the Gauss-Markov theorem states that, given certain general assumptions, the least-squares criterion renders estimates of unknown parameters that are optimal in the following specific statistical sense. A least-squares estimate is unbiased (i.e., on average it equals the true value of the parameter), and of all unbiased estimates the least-squares estimate has the least variance (i.e., on average it is closest to the true value, as measured by the mean squared deviation). For the guarantee of the Gauss-Markov theorem to be valid, two important requirements must be met: the data points must be uncorrelated, and they must all have the same variance (i.e., the data must be homoscedastic) [641].

In terms of a macromolecular superposition, the homoscedastic assumption stipulates that the corresponding atoms in the structures must have equal variances. However, the requirement for homogeneous variances is generally violated with macromolecular superpositions. For instance, the experimental precision attached to the atoms in a crystal structure vary widely as gauged by crystallographic temperature factors. Similarly, the variances of main chain atoms in reported superpositions of NMR models commonly range over three orders of magnitude. In comparisons of homologous protein domains, the structures deviate from each other with varying degrees of local precision: some atoms "superimpose well" and others do not. In molecular dynamics simulations, particular regions of a macromolecule may undergo relatively large conformational changes (a violation of the Eckart conditions [167], which have been used to provide a physical justification for using least-squares superpositions to refer different conformational states of a molecule to a common reference frame [404]). Ideally, the atomic positions should be uncorrelated, but this assumption is also violated for macromolecular superpositions. Adjacent atoms in a protein main chain covary strongly due to covalent chemical bonds, and atoms remote in sequence may covary due to other physical interactions. These theoretical problems are not simply academic. In practice, researchers usually perform an OLS superposition, identify regions that do not "superposition well", and calculate a new superposition in which the more variable regions have been subjectively excluded from the analysis. Different superpositions result depending on which pieces of data are discarded.

## 8.1.5  Accounting for Heteroscedasticity and Correlation in a Likelihood Framework

In previous work [712–714], we relaxed the Gauss-Markov assumptions in the superposition problem by allowing for heteroscedasticity and correlation within a likelihood framework [567] where the macromolecular structures are distributed normally (some theoretical work on such models appeared in [245]). In our non-isotropic likelihood treatment, superpositioning requires estimating five classes of parameters: (1) a mean structure, (2) a global covariance matrix describing the variance and correlations for each atom in the structures, (3) hierarchical parameters describing the covariance matrix, and, for each structure in the analysis, (4) a proper orthogonal rotation matrix and (5) a translation vector. Our ML method accounts

# 8 Likelihood and Empirical Bayes Superposition

for uneven variances and correlations by weighting by the inverse of the covariance matrix.

Estimation of the covariance matrix has historically been a significant impediment to a viable non-isotropic likelihood-based Procrustes analysis [159, 224, 242, 243, 430–432]. Simultaneous estimation of the sample covariance matrix and the translations is generally impossible. We permit joint identifiability by regularizing the covariance matrix using a hierarchical, empirical Bayes treatment in which the eigenvalues of the covariance matrix are considered as variates from an inverse gamma distribution.

In general, all the estimates of the unknown parameters are interdependent and cannot be solved for analytically. Furthermore, the smallest eigenvalues of the sample covariance matrix are zero due to colinearity imparted by the centering operation necessary to estimate the unknown translations. We treat these smallest eigenvalues as "missing data" using an Expectation-Maximization (EM) algorithm. For simultaneous estimation, we use iterative conditional maximization of the joint likelihood augmented by the EM algorithm. This method works very well in practice, with excellent convergence properties for the thousands of real cases analyzed to date.

## 8.2 A Matrix Gaussian Probability Model for the Macromolecular Superposition Problem

Consider $n$ structures ($\mathbf{X}_i$, $i = 1 \ldots n$), each with $k$ corresponding atoms (corresponding to "landmarks" in shape theory), where each structure is defined as a $k \times 3$ matrix holding $k$ rows of 3-dimensional coordinates for each atom. Following Goodall and Bose [244], we assume a probabilistic model for the superposition problem in which each macromolecular structure $\mathbf{X}_i$ is distributed according to a Gaussian probability density and is observed in a different unknown coordinate system [159, 241, 243, 245]. We allow heterogeneous variances and correlations among the atoms in the structures, as described by a $k \times k$ covariance matrix $\boldsymbol{\Sigma}$ for the atoms. For simplicity, we also assume that the variance about each atom is spherical (and that the covariance between two atoms is spherical). For our structural data, which is represented as a $k \times 3$ matrix, the appropriate Gaussian distribution is the matrix normal (Gaussian) distribution, designated by $\mathcal{N}_{k,d}(A, \boldsymbol{\Sigma}, \boldsymbol{\Xi})$, where $k$ and $d$ are the numbers of rows and columns in the variate matrices, respectively, $A$ is a $k \times d$ centroid or location matrix, $\boldsymbol{\Sigma}$ is a $k \times k$ covariance matrix for the rows of the variate matrices, and $\boldsymbol{\Xi}$ is a $d \times d$ covariance matrix for the columns of the variate matrices. Hence, each observed structure $\mathbf{X}_i$ can be considered to be an arbitrarily scaled, rotated, and translated zero-mean Gaussian matrix displacement $\mathbf{E}_i \sim \mathcal{N}_{k,d}(\mathbf{0}, \boldsymbol{\Sigma}, \mathbf{I}_3)$ of the mean structure $\boldsymbol{\mu}$

$$\mathbf{X}_i = (\boldsymbol{\mu} + \mathbf{E}_i)\, \mathbf{R}_i^T - \mathbf{1}_k \boldsymbol{\tau}_i^T, \tag{8.1}$$

where $\tau_i$ is a $3 \times 1$ column vector for the translational offset, $\mathbf{1}_k$ denotes the $K \times 1$ column vector of ones, and $\mathbf{R}_i$ is a proper, orthogonal $3 \times 3$ rotation matrix.

## 8.3 The Matrix Normal Likelihood Function for Multiple Superpositioning

The likelihood equation for the model given in Eq. 8.1 is obtained from a matrix normal distribution [136]. First, we define

$$\mathbf{E}_i = (\mathbf{X}_i + \mathbf{1}_k \tau_i^T)\mathbf{R}_i - \mu.$$

Then the corresponding likelihood is

$$L(\mathbf{R}, \tau, \mu, \Sigma | \mathbf{X}) = (2\pi)^{-\frac{3kn}{2}} |\Sigma|^{-\frac{3n}{2}} \exp\left(-\frac{1}{2} \sum_i^n \mathrm{tr}\left\{\mathbf{E}_i^T \Sigma^{-1}\mathbf{E}\right\}\right), \qquad (8.2)$$

and the log-likelihood is, up to a constant

$$\ell(\mathbf{R}, \tau, \mu, \Sigma | \mathbf{X}) = -\frac{3n}{2} \ln |\Sigma| - \frac{1}{2} \sum_i^n \mathrm{tr}\left\{\mathbf{E}_i^T \Sigma^{-1}\mathbf{E}\right\}. \qquad (8.3)$$

## 8.4 ML Superposition Solutions for Known Covariance Matrix

In order to find the maximum likelihood solution for the Gaussian model given in Eq. 8.1, the maximum likelihood estimates of the four classes of unknowns $(\mathbf{R}, \tau, \mu, \Sigma)$ must be determined jointly. In the following sections we provide the ML solutions for the translations, rotations, and mean form, assuming that the covariance matrix is known. In general the covariance matrix is not known and must be estimated as well. However, joint estimation of the translations and the covariance matrix presents certain challenges that will be given special attention (Sects. 8.5–8.8).

### 8.4.1 Conditional ML Estimates of the Translations $\tau$

For each of the $N$ structures $\mathbf{X}_i$, a translation vector must be estimated. The ML estimate of the optimal translation $\hat{\tau}_i$, conditional on the other nuisance parameters,

# 8 Likelihood and Empirical Bayes Superposition

is given in its most general form by

$$\hat{\tau}_{\text{gen},i} = \frac{\mathbf{R}_i \boldsymbol{\mu}^T \boldsymbol{\Sigma}^{-1} \mathbf{1}_k - \mathbf{X}_i^T \boldsymbol{\Sigma}^{-1} \mathbf{1}_k}{\mathbf{1}_k^T \boldsymbol{\Sigma}^{-1} \mathbf{1}_k} \tag{8.4}$$

Note that the denominator of Eq. 8.4 is a scalar.

In the superposition problem the absolute reference frame is not of interest; only relative orientations are needed. Consequently, the translations are identifiable only up to an arbitrary constant translation i.e., only the differences between the translations are identifiable). Thus we are free to arbitrarily fix the absolute reference frame, and it is customary to define from the outset the row-weighted centroid of the mean structure $\boldsymbol{\mu}$ to be at the origin: $\boldsymbol{\mu}^T \boldsymbol{\Sigma}^{-1} \mathbf{1}_k = \mathbf{0}$. This convention makes the translations uniquely identifiable and also neatly simplifies the ML estimates of the translations:

$$\hat{\tau}_i = -\frac{\mathbf{X}_i^T \boldsymbol{\Sigma}^{-1} \mathbf{1}_k}{\mathbf{1}_k^T \boldsymbol{\Sigma}^{-1} \mathbf{1}_k} \tag{8.5}$$

It will be convenient to define a centered structure $\check{\mathbf{X}}_i$ that has been translated with this vector so that its row-weighted center is at the origin:

$$\check{\mathbf{X}}_i = \mathbf{X}_i + \mathbf{1}_k \hat{\tau}_i^T \tag{8.6}$$

This row-weighted centering operation can also be concisely represented as premultiplication by a $k \times k$ square centering matrix $\mathbf{C}$:

$$\check{\mathbf{X}}_i = \mathbf{C} \mathbf{X}_i \tag{8.7}$$

where

$$\mathbf{C} = \mathbf{I} - \frac{\mathbf{1}_k \mathbf{1}_k^T \boldsymbol{\Sigma}^{-1}}{\mathbf{1}_k^T \boldsymbol{\Sigma}^{-1} \mathbf{1}_k} \tag{8.8}$$

The centering matrix $\mathbf{C}$ is also both singular and idempotent, since $\mathbf{C} \mathbf{1}_k = \mathbf{0}$ and $\mathbf{C} \mathbf{X}_i = \mathbf{C} \mathbf{C} \mathbf{X}_i$. These properties of the centering matrix will be important later for understanding the difficulties with estimation of the covariance matrix.

## 8.4.2 *Conditional ML Estimates of the Rotations* $\mathbf{R}$

The conditional ML estimates of the rotations are calculated using a singular value decomposition (SVD). Let the SVD of an arbitrary matrix $\mathbf{D}$ be $\mathbf{U} \boldsymbol{\Lambda} \mathbf{V}^T$. The optimal rotations $\hat{\mathbf{R}}_i$ are then estimated by

$$\boldsymbol{\mu}^T \boldsymbol{\Sigma}^{-1} \check{\mathbf{X}}_i = \mathbf{U} \boldsymbol{\Lambda} \mathbf{V}^T$$

$$\hat{\mathbf{R}}_i = \mathbf{V} \mathbf{P} \mathbf{U}^T \tag{8.9}$$

Rotoinversions are prevented by ensuring that the rotation matrix $\hat{\mathbf{R}}_i$ has a positive determinant: $\mathbf{P} = \mathbf{I}$ if $|\mathbf{V}||\mathbf{U}| = 1$ or $\mathbf{P} = \text{diag}(1, 1, -1)$ if $|\mathbf{V}||\mathbf{U}| = -1$. If the mean $\mu$ is row centered, as discussed in the previous section, then the estimate of the rotation is in fact independent of the translations, and it is strictly unnecessary to use the centered structure $\check{\mathbf{X}}_i$ in Eq. 8.9; one can just use $\mathbf{X}_i$ in the SVD, a fact which can simplify some superposition algorithms.

### 8.4.3 Conditional ML Estimate of the Mean Structure $\mu$

The mean structure is estimated as the arithmetic average of the optimally rotated and translated structures:

$$\hat{\mu} = \frac{1}{n} \sum_i^n \check{\mathbf{X}}_i \mathbf{R}_i \qquad (8.10)$$

The estimate of the mean structure $\hat{\mu}$ is independent of the covariance matrix $\Sigma$. The mean structure is also inherently centered, since

$$\hat{\mu} = \frac{1}{n} \sum_i^n \mathbf{C}\mathbf{X}_i \mathbf{R}_i$$

$$= \frac{1}{n} \mathbf{C} \left( \sum_i^n \mathbf{X}_i \mathbf{R}_i \right) \qquad (8.11)$$

It follows that $\mathbf{C}\hat{\mu} = \hat{\mu}$, due to the idempotency of the centering matrix.

## 8.5 Difficulties with the Covariance Matrix

We now turn to issues of identifiability with ML estimation of the covariance matrix. The conditional ML estimate of the covariance matrix from the likelihood function in Eq. 8.2 is simply the usual sample covariance matrix:

$$\hat{\Sigma}_s = \frac{1}{3n} \sum_i^n (\check{\mathbf{X}}_i \mathbf{R}_i - \hat{\mu})(\check{\mathbf{X}}_i \mathbf{R}_i - \hat{\mu})^T \qquad (8.12)$$

However, as is clear from the likelihood function 8.2, the Gaussian model requires that the covariance matrix be invertible, i.e., all its eigenvalues must be positive. The ML estimates of the translations and rotations also both require an invertible covariance matrix. These facts present three crucial difficulties for joint ML estimation of the parameters in the superposition problem.

## 8 Likelihood and Empirical Bayes Superposition

First, the ML estimate of the covariance matrix exists only when sufficient data are present to avoid rank deficiency, i.e., when $n \geq (k/3) + 1$ [165]. In practice, this condition is rarely satisfied due to limited data.

Second, the ML estimate of the covariance matrix requires centering the structures (see Eq. 8.5), which imparts a common linear constraint on the columns of the structure matrices. This can be seen by representing $\hat{\boldsymbol{\Sigma}}_s$ in terms of the centering matrices:

$$\hat{\boldsymbol{\Sigma}}_s = \frac{1}{3n} \sum_i^n (\mathbf{C}\mathbf{X}_i \mathbf{R}_i - \mathbf{C}\hat{\boldsymbol{\mu}})(\mathbf{C}\mathbf{X}_i \mathbf{R}_i - \mathbf{C}\hat{\boldsymbol{\mu}})^T$$

$$= \frac{1}{3n} \mathbf{C} \left[ \sum_i^n (\mathbf{X}_i \mathbf{R}_i - \hat{\boldsymbol{\mu}})(\mathbf{X}_i \mathbf{R}_i - \hat{\boldsymbol{\mu}})^T \right] \mathbf{C}^T \tag{8.13}$$

Thus, $\hat{\boldsymbol{\Sigma}}_s$ is both row and column centered, and since $\mathbf{C}$ is singular, the sample covariance matrix $\hat{\boldsymbol{\Sigma}}_s$ is also singular. Even with sufficient data, the sample covariance matrix is rank deficient, with at least one zero eigenvalue [431, 432], and therefore it is non-invertible.

Third, the displacements from the mean due to the covariance matrix $\boldsymbol{\Sigma}$ and the translations $\boldsymbol{\tau}_i$ are linearly entangled (see Eq. 8.1), and simultaneous estimation is not possible [242, 243, 431]. This last problem is particularly acute, since the unrestrained ML estimates of the translations $\boldsymbol{\tau}_i$ and of the covariance matrix $\boldsymbol{\Sigma}$ are strongly interdependent. Translations that closely superpose a given atom decrease the estimated variance, and a small variance in turn weights the translations to center on that atom. It is always possible to find translations that exactly superposition any selected atom, making its sample variance zero, the covariance matrix $\hat{\boldsymbol{\Sigma}}$ singular, and the likelihood infinite. The possibility of an infinite likelihood suggests that the covariance matrix should be regularized so that small eigenvalues are penalized. If all eigenvalues are constrained to be finite and positive, then the covariance matrix will be invertible and each of these problems solved.

## 8.6 An Extended Likelihood Function

In real applications the variances cannot take arbitrary values. For instance, the atoms in a macromolecule are linked by chemical bonds, and the atomic variances are similar in magnitude. Very small or large variances are improbable and physically unrealistic. Thus, simultaneous estimation of the covariance matrix and the translations can be enabled by restricting the variances to physically reasonable values.

To estimate the covariance matrix, we therefore adopt a hierarchical model where the eigenvalues ($\lambda_j$) of the covariance matrix $\boldsymbol{\Sigma}$ are inverse gamma distributed. The eigenvalues of the covariance matrix can be considered as variances with all correlations eliminated, e.g., the eigenvalues of a diagonal covariance matrix are the variances themselves. In Bayesian analysis, the inverse gamma distribution is commonly used as a proper conjugate prior for the variances in a Gaussian model [428]. For simplicity here we assume an inverse gamma distribution in which the shape parameter $\gamma$ is fixed at $\frac{1}{2}$:

$$p(\lambda_j) = \sqrt{\frac{\alpha}{\pi}} \lambda_j^{-\frac{3}{2}} e^{-\frac{\alpha}{\lambda_j}} \tag{8.14}$$

where $\alpha$ is the scale parameter. This distribution is also known as the Lévy distribution, which is one of the few analytically expressible stable distributions. It corresponds to a scaled inverse chi squared distribution with one degree of freedom, and thus has a straightforward Bayesian interpretation as a minimally informative, conjugate hierarchical prior. The inverse gamma distribution conveniently places a low probability on both small and large eigenvalues, with zero probability on zero-valued eigenvalues. The corresponding log-likelihood for the $K$ eigenvalues is (up to a constant)

$$\begin{aligned}
\ell(\alpha|\lambda) &= \frac{k}{2} \ln \alpha - \frac{3}{2} \sum_i^k \ln \lambda_j - \alpha \sum_i^k \frac{1}{\lambda_j} \\
&= \frac{k}{2} \ln \alpha - \frac{3}{2} \ln |\Sigma| - \alpha \operatorname{tr} \Sigma^{-1}
\end{aligned} \tag{8.15}$$

The complete joint log-likelihood $\ell_h$ for this hierarchical model (an *extended likelihood* [63, 567], also known as an *h-likelihood* [429] or *penalized likelihood* [255]) is then the sum of the "pure" log-likelihood from Eq. 8.3 and the log-likelihood of an inverse gamma distribution for the random eigenvalues (Eq. 8.15):

$$\ell_h = \ell(\mathbf{R}, \tau, \mu, \alpha | \mathbf{X}, \Sigma) = \ell(\mathbf{R}, \tau, \mu, \Sigma | \mathbf{X}) + \ell(\alpha|\lambda). \tag{8.16}$$

The full extended superposition log-likelihood $\ell_h$ is thus given by

$$\begin{aligned}
\ell_h = &-\frac{3N}{2} \ln |\Sigma| - \frac{1}{2} \sum_i^N \operatorname{tr} \left\{ \mathbf{E}_i^T \Sigma^{-1} \mathbf{E} \right\} \\
&+ \frac{K}{2} \ln \alpha - \frac{3}{2} \ln |\Sigma| - \alpha \operatorname{tr} \Sigma^{-1}.
\end{aligned} \tag{8.17}$$

The hierarchical model described by the likelihood in Eq. 8.17 is in fact identical to putting a diagonal inverse Wishart prior on the covariance matrix, with one degree of freedom and scale matrix is equal to $\alpha \mathbf{I}$.

# 8.7 Hierarchical Treatment of the Covariance Matrix $\Sigma$

## 8.7.1 Joint Maximization Over $\Sigma$ and All Parameters

Various pragmatic methods are available for likelihood inference using the extended likelihood function presented above. The first that we examine, presented in [712], is to treat the covariance matrix (with its associated eigenvalues) and the hyperparameter $\alpha$ as parameters of interest and maximize over them. This is appropriate, for instance, whenever the covariance matrix itself is considered informative and the correlation structure of an ensemble of molecular conformations is desired (e.g., [382, 714]).

The extended ML estimate $\hat{\Sigma}_h$ of $\Sigma$ is a linear function of the unrestricted conditional ML estimate $\hat{\Sigma}_s$ from Eq. 8.12:

$$\hat{\Sigma}_h = \frac{3n}{3n+3} \left( \frac{2\alpha}{3n} \mathbf{I} + \hat{\Sigma}_s \right) \tag{8.18}$$

In this ML hierarchical model, the point estimate of the inverse gamma parameter $\alpha$ is determined by the data, unlike when using a *bona fide* Bayesian prior. The $\hat{\Sigma}_h$ estimate can be viewed as a shrinkage estimate that contracts the eigenvalues of the covariance matrix to the mode of the inverse gamma distribution.

It will also be useful to specify the conditional ML estimate of the inverse gamma distributed eigenvalues $\hat{\Lambda}_h$ of the covariance matrix:

$$\hat{\Lambda}_h = \frac{3n}{3n+3} \left( \frac{2\alpha}{3n} \mathbf{I} + \hat{\Lambda}_s \right) \tag{8.19}$$

where $\hat{\Lambda}_s$ is the diagonal matrix of eigenvalues of the unrestricted sample covariance matrix $\hat{\Sigma}_s$, as determined by spectral decomposition of the covariance matrix (i.e., $\hat{\Sigma}_s = V \Lambda_s V^T$). This follows from the fact that maximum likelihood estimates are invariant to parameter transformations.

## 8.7.2 Maximization of the Marginal Likelihood

Alternatively, we can treat the covariance matrix as a nuisance parameter and integrate it out to give the marginal likelihood function:

$$L(\mathbf{R}, \tau, \mu, \alpha | \mathbf{X}) = \int_{\Sigma > 0} L(\mathbf{R}, \tau, \mu, \alpha | \mathbf{X}, \Sigma) \, d\Sigma \tag{8.20}$$

The marginal distribution (Eq. 8.20) has the form of a matrix Student-t density, which is difficult to treat analytically. However, the Expectation-Maximization algorithm can maximize the parameters of the marginal distribution indirectly [144,495,567] using the joint likelihood (Eq. 8.17), by substituting at each iteration the expected inverse covariance matrix conditional on the other current parameter estimates. The conditional expected inverse covariance matrix is

$$\mathbb{E}\left(\Sigma_{\mathrm{h}}^{-1}|\mathbf{R}, \tau, \mu, \alpha, \mathbf{X}\right) = \frac{3n+1}{3n}\left(\frac{2\alpha}{3n}\mathbf{I} + \hat{\Sigma}_s\right)^{-1} \tag{8.21}$$

where $\hat{\Sigma}_s$ is the sample covariance matrix from Eq. 8.12. Due to the rank deficiency of the sample covariance matrix, when using the marginal likelihood model it will generally be necessary to assume a diagonal structure (i.e., $\Sigma = \Lambda$) for the covariance matrix so that the expected inverses of the eigenvalues can be found easily:

$$\mathbb{E}\left(\Lambda_{\mathrm{h}}^{-1}|\mathbf{R}, \tau, \mu, \alpha, \mathbf{X}\right) = \frac{3n+1}{3n}\left(\frac{2\alpha}{3n}\mathbf{I} + \hat{\Lambda}_s\right)^{-1}. \tag{8.22}$$

## 8.8 Conditional Estimate of the Hyperparameter $\alpha$ of the Inverse Gamma Distribution

Note that there is an important complication with each of the estimates of the covariance matrix given above in Eqs. 8.18, 8.19, and 8.21. Namely, the smallest sample eigenvalues are nonidentifiable due to the rank degeneracy of the sample covariance matrix. Without special care for the rank degeneracy problem, then, the naive covariance estimates given in Eqs. 8.18, 8.19, and 8.21 are invalid (see, for example, the algorithm presented in [497], which results in degenerate solutions having arbitrary atoms perfectly superpositioned with zero variance). We deal with this problem by treating the missing eigenvalues with the EM algorithm. Recall that the sample covariance matrix has multiple zero eigenvalues, regardless of the number of structures used in the calculation. The sample covariance matrix is of maximum effective rank $k - 4$ (one linear translational constraint and three non-linear rotational constraints) and can be less when there are few structures (rank $= \min(3n - 7, k - 4)$). We treat these missing eigenvalues as missing data (from a left-truncated distribution) and estimate $\alpha$ conditional on the "observed" sample eigenvalues. In previous work, for simplicity we calculated the parameters of the inverse gamma distribution using the usual ML estimates but omitting the zero eigenvalues [713], a reasonable but inexact approximation. Here we give an exact solution, using an EM algorithm that determines the expected values for the inverses of the missing eigenvalues.

# 8 Likelihood and Empirical Bayes Superposition

In the following equations it is assumed that the eigenvalues are ordered from largest to smallest. The naive conditional ML estimate of $\alpha$ is given by

$$\hat{\alpha} = \frac{k}{2 \operatorname{tr} \boldsymbol{\Sigma}_{\mathrm{h}}^{-1}} = \frac{k}{2 \operatorname{tr} \boldsymbol{\Lambda}_{\mathrm{h}}^{-1}} = \frac{k}{2 \sum_i \lambda_i^{-1}} \tag{8.23}$$

Because the $m$ smallest eigenvalues are missing, the eigenvalue distribution is left-truncated. Hence, the EM estimate of $\alpha$ is given by

$$\hat{\alpha} = \frac{k}{2 \left( m \, \mathbb{E}\left(\lambda_{\mathrm{sm}}^{-1} | \alpha, \gamma, c\right) + \sum_i^{k-m} \lambda_i^{-1} \right)} \tag{8.24}$$

where $m$ is the number of missing eigenvalues, and $\mathbb{E}\left(\lambda_{\mathrm{sm}}^{-1} | \alpha, \gamma, c\right)$ is the expected value of the inverse of the $m$ smallest missing eigenvalues, conditional on the smallest observed eigenvalue $c$. The expected inverse of the smallest eigenvalues can be expressed analytically:

$$\mathbb{E}\left(\lambda_{\mathrm{sm}}^{-1} | \alpha, \gamma, c\right) = \frac{\Gamma\left(\gamma + 1, x\right)}{\hat{\alpha} \, \Gamma\left(\gamma, x\right)} \tag{8.25}$$

where $x = \hat{\alpha}/c$, $c$ is the smallest observed eigenvalue, $\gamma$ is the shape parameter of the inverse gamma distribution, and $\Gamma(a, s)$ is the (unnormalized) upper incomplete gamma function:

$$\Gamma(a, s) = \int_s^\infty t^{a-1} e^{-t} \, dt$$

for $a$ real and $s \geq 0$. Since we here assume that $\gamma = \frac{1}{2}$, Eq. 8.25 can be simplified:

$$\mathbb{E}\left(\lambda_{\mathrm{sm}}^{-1} | \alpha, c\right) = \frac{\Gamma\left(\frac{3}{2}, x\right)}{\hat{\alpha} \, \Gamma\left(\frac{1}{2}, x\right)} \tag{8.26}$$

$$= \frac{1}{2\hat{\alpha}} + \frac{e^{-x} \sqrt{x}}{\hat{\alpha} \sqrt{\pi} \, \mathrm{erfc}\left(\sqrt{x}\right)} \tag{8.27}$$

This EM algorithm, then, allows for valid estimation of the covariance matrix and its eigenvalues. Given a positive $\alpha$ parameter, the hierarchical model guarantees an invertible $\hat{\boldsymbol{\Sigma}}_{\mathrm{h}}$ by ensuring that all its eigenvalues (and variances) are positive, as can be seen from Eqs. 8.19 and 8.22. Hence the hierarchical model is sufficient to overcome all three of the difficulties with estimation of the covariance matrix enumerated above.

## 8.9 Algorithm

The conditional ML estimates given in the preceding sections must be solved simultaneously using a numerical maximization algorithm. We have developed the following iterative algorithm based on a Conditional Expectation-Maximization (EM) method [144, 567]. In brief:

1. **Initialize:** Set $\hat{\boldsymbol{\Sigma}} = \mathbf{I}$ and $\alpha = 0$. Randomly choose one of the observed structures to approximate the mean $\hat{\boldsymbol{\mu}}$.
2. **Translate:** For each structure, estimate the translation (Eq. 8.5) and center each $\mathbf{X}_i$ according to Eq. 8.6.
3. **Rotate:** Calculate each $\hat{\mathbf{R}}_i$ according to Eq. 8.9, and rotate each centered structure accordingly: $\mathbf{X}_i = \check{\mathbf{X}}_i \hat{\mathbf{R}}_i$.
4. **Estimate the mean:** Recalculate the average structure $\hat{\boldsymbol{\mu}}$ according to Eq. 8.10. Return to step 3 and loop to convergence.
5. **Calculate the sample covariance matrix (or sample variances):** Calculate $\hat{\boldsymbol{\Sigma}}_s$ from Eq. 8.12. If assuming a correlated model, spectrally decompose the sample covariance matrix $\hat{\boldsymbol{\Sigma}}_s$ to find the sample eigenvalues $\hat{\boldsymbol{\Lambda}}_s$. If assuming no correlations in the data, then this decomposition is unnecessary, since then the eigenvalues are the variances (the diagonal of $\hat{\boldsymbol{\Sigma}}_s$).
6. **Estimate the hyperparameter $\alpha$ of the inverse gamma distribution:** This step cycles until convergence between the following two substeps:

   (a) **Estimate eigenvalues conditional on current $\alpha$:** If using the joint likelihood model, modify the $k - m$ sample eigenvalues according to Eq. 8.19 to find $\hat{\boldsymbol{\Lambda}}_h$. Otherwise, if using the marginal likelihood model, find the expected inverse eigenvalues with Eq. 8.22.
   (b) **Estimate $\alpha$ conditional on current eigenvalues:** Find the ML estimate of the inverse gamma scale parameter for the current eigenvalues using Eq. 8.24. In the first iteration, $\mathbb{E}(\lambda_{sm}^{-1})$ can simply be omitted and $k$ replaced with $k - m$. In subsequent iterations, starting values can be provided by the parameter values from the previous iteration.

7. **Estimate the covariance matrix $\hat{\boldsymbol{\Sigma}}_h$:** If assuming a fully correlated $\boldsymbol{\Sigma}$, find $\hat{\boldsymbol{\Sigma}}_h$ by modifying $\hat{\boldsymbol{\Sigma}}_s$ according to Eq. 8.18. Otherwise $\hat{\boldsymbol{\Sigma}}_h = \hat{\boldsymbol{\Lambda}}_h$ (or $\mathbb{E}(\boldsymbol{\Sigma}_h^{-1}) = \mathbb{E}(\boldsymbol{\Lambda}_h^{-1})$ for the marginal likelihood model), which has already been determined.
8. **Loop:** Return to step 2 and loop until convergence.

When assuming that the variances are all equal and that there are no correlations (i.e., when $\boldsymbol{\Sigma} \propto \mathbf{I}$), then the above algorithm is equivalent to the classic least-squares algorithm for simultaneous superpositioning of multiple structures [149, 216, 363, 647]. Examples of ML superpositions, with the corresponding OLS superpositions, are presented in Figs. 8.1 and 8.2.

8 Likelihood and Empirical Bayes Superposition

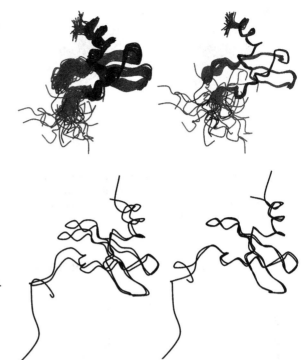

**Fig. 8.1** At *left* is a least-squares superposition of 30 NMR models from PDB ID 2SDF. At *right* is the maximum likelihood superposition, which allows each atom to have a different variance. All superpositions were performed using only the $\alpha$-carbon atoms

**Fig. 8.2** At *left* is the pairwise, least-squares superposition of two NMR models from PDB ID 2SDF. At *right* is the maximum likelihood superposition. Note the close correspondence with Fig. 8.1 despite substantially less data. Both superpositions were performed using only the $\alpha$-carbon atoms

## 8.10 Performance of ML Superpositioning with Simulated Data

We have investigated the ability of the ML superposition method to determine known parameters from a family of simulated protein conformations. A data set of 400 randomly perturbed protein structures was generated, assuming a matrix normal distribution with a known mean and covariance matrix. Each structure was additionally randomly translated and rotated.

ML superpositioning of this dataset provided estimates of the mean structure, the covariance matrix, and the original superposition (before arbitrary rotational and translational transformation). Results are shown in Fig. 8.3, which shows the close correspondence between the ML superposition and the "true" superposition. The LS superposition, in contrast, is markedly inaccurate.

### 8.10.1 Accuracy of ML Estimates of the Covariance and Correlation Matrices

We performed two different simulation analyses to gauge the ability of the ML method to accurately determine the superposition covariance matrices. Two families of protein structures, each with a unique conformation, were generated randomly

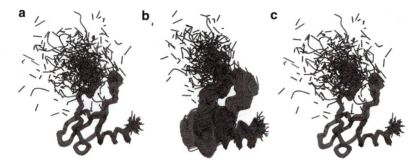

**Fig. 8.3** Superpositions of a simulated ensemble of protein structures. (**a**) In *grey* is the true superposition, before random rotation and translation. (**b**) The least squares superposition. (**c**) A maximum likelihood superposition using the inverse gamma hierarchical matrix normal model (Modified from [712])

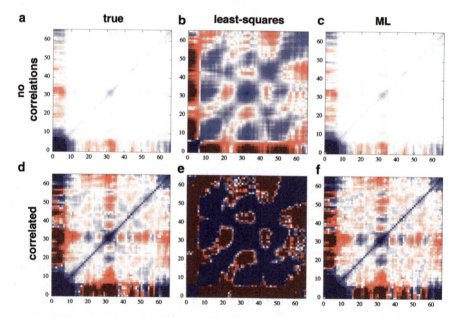

**Fig. 8.4** Comparison of true and estimated covariance matrices, by both LS and ML superpositioning methods. In these plots, *blue* indicates positive covariance whereas *red* indicates negative covariance; *white* indicates no covariance. The *top* row (panels **a**, **b**, and **c**) shows results from a simulation with very weak correlations. The *bottom* row (**d**, **e**, and **f**) shows results from a simulation with strong correlations. The true covariance structure is shown in the left-most column (panels **a** and **d**). Covariance matrices were estimated using the least-squares criterion (middle panels **b** and **e**) and the maximum likelihood method (right-most panels **c** and **f**)

8  Likelihood and Empirical Bayes Superposition                                    207

assuming a matrix Gaussian distribution with known mean and known covariance matrices. The simulations were based on two different covariance matrices. The first has extremely weak correlations and uneven variances that ranged over three orders of magnitude (Fig. 8.4a). The second has the same variances but contained strong correlations/covariances (Fig. 8.4d). The covariance matrices were then estimated from these simulations using both least-squares (Fig. 8.4b and e) and ML (Fig. 8.4c and f).

The ML estimate of the covariance matrix is considerably more accurate than the least-squares estimate (Fig. 8.4). The least-squares estimate is markedly biased and shows a strong artifactual pattern of correlation (Fig. 8.4b and e). The ML estimate, in contrast, is nearly visually indistinguishable from the matrix assumed in the simulations (Fig. 8.4c and f).

## 8.11  Implementation in THESEUS

The ML superpositioning method detailed above has been implemented in the command-line UNIX program THESEUS [712, 713], which is freely available on the web (http://www.theseus3d.org/). THESEUS operates in two different modes: (1) a mode for superpositioning structures with identical sequences (such as multiple NMR models of the same protein) and (2) an "alignment mode", which superpositions proteins based on an assumed alignment (a sequence alignment must be provided by the user). THESEUS will additionally calculate the principal components of the estimated covariance matrix (or the corresponding correlation matrix). THESEUS takes as input a set of standard PDB formatted structure coordinate files [45, 46]). Each principal component is written to the temperature factor field of two output files: (1) a PDB file of the ML superposition (where each structure is given as a different MODEL) and (2) a PDB file of the mean structure. Principal components can be viewed and analyzed with macromolecular visualization programs, such as PyMOL [143], RasMol [627], or MolScript [394], that can color structures by temperature factor.

## 8.12  A Full Bayesian Extension

The likelihood analysis described above does not address the uncertainty in the estimates of the parameters. Bayesian analysis, however, can provide posterior distributions for the parameters and can also incorporate prior knowledge from other data (e.g., crystallographic B-factors or NMR order parameters). Due to the close philosophical relationship between likelihood and Bayesian methods, the likelihood treatment described above provides a natural theoretical foundation for a full Bayesian treatment of the multiple superposition problem.

For a Bayesian analysis we assume that the $\Sigma, \mu, \mathbf{R}, \tau$ parameters are independent, so that the posterior joint distribution of the parameters is given by

$$p(\Sigma, \mu, \mathbf{R}, \tau | \mathbf{X}) \propto p(\mathbf{X} | \Sigma, \mu, \mathbf{R}, \tau) \, p(\Sigma) \, p(\mu) \, p(\mathbf{R}) \, p(\tau) \qquad (8.28)$$

where $p(\mathbf{X} | \Sigma, \mu, \mathbf{R}, \tau)$ is the likelihood given in Eq. 8.2, and $p(\Sigma)$, $p(\mu)$, $p(\mathbf{R})$, and $p(\tau)$ are the prior distributions. To enable a fully heteroskedastic, correlated model we adopt a hierarchical, diagonal inverse Wishart prior for $\Sigma$ in which the scale matrix of the inverse Wishart is proportional to $\phi\mathbf{I}$:

$$p(\Sigma) \propto p(\Sigma | \phi) \, p(\phi | \alpha) \, p(\alpha) \qquad (8.29)$$

In this Bayesian formulation we use standard conjugate priors, and thus the conditional distributions of all of the parameters except the rotations have convenient analytical representations. The posterior for the rotations belongs to matrix von Mises-Fisher distribution, which has a normalization constant with no known analytical form. We have solved the MAP estimates for this Bayesian superposition model and have developed a hybrid Gibbs-Metropolis sampling algorithm to approximate the joint posterior distribution of the parameters [481, 710, 711]. Future versions of our THESEUS software will include the option of performing a Bayesian superposition analysis based on this methodology.

**Acknowledgements** Much of this methodology was initially developed with Deborah S. Wuttke at the University of Colorado at Boulder. I thank Phillip Steindel, Thomas Hamelryck, Kanti Mardia, Ian Dryden, Colin Goodall, and Subhash Lele for helpful comments and criticism. This work was supported by NIH grants 1R01GM094468 and 1R01GM096053.

# Chapter 9
# Bayesian Hierarchical Alignment Methods

**Kanti V. Mardia and Vysaul B. Nyirongo**

## 9.1 Protein Structure Alignment and Superposition

This chapter considers the problem of matching configurations of biological macro-molecules when both alignment and superposition transformations are unknown. *Alignment* denotes correspondence – a bijection or mapping – between points in different structures according to some objectives or constraints. *Superposition* denotes rigid-body transformations, consisting of translations and rotations, that bring whole or partial configurations together, typically in Euclidean space, as closely as possible and according to some objectives or constraints. Further details are given in Chap. 8.

The main objective of alignment is maximizing the number of corresponding points. Particular constraints in aligning biological macromolecules such as proteins include imposing sequence order and avoiding gaps between corresponding points. In fact, most applications require strict sequence order, respecting the rules of sequence alignment. Structure-based alignment methods may allow limited exceptions to sequence order, which are then penalized in the objective function. Here, we do not include sequence order constraints.

An additional objective would be to preferably align points with the same or similar attributes. When aligning atoms, amino acids or other building blocks in biological macromolecules, such attributes may include atom type, amino acid type, charge, ligand binding properties and other physical or chemical properties.

---

K.V. Mardia (✉)
Department of Statistics, School of Mathematics, The University of Leeds, Leeds, West Yorkshire, LS2 9JT, UK
e-mail: K.V.Mardia@leeds.ac.uk

V.B. Nyirongo
Statistics Division, United Nations, New York, NY 10017, USA
e-mail: nyirongov@un.org

T. Hamelryck et al. (eds.), *Bayesian Methods in Structural Bioinformatics*, Statistics for Biology and Health, DOI 10.1007/978-3-642-27225-7_9, © Springer-Verlag Berlin Heidelberg 2012

Given the alignment, the main objective of superposition is typically to minimize the root mean square deviation (RMSD) between corresponding points [159, 357]. The problem of superposition of proteins *given* their alignment is extensively discussed in Chap. 8.

However, in many applications both alignment and superposition are unknown [235, 236, 254, 370, 477, 549, 723]. In that case, matching implies optimizing both the alignment and the superposition, subject to all the relevant constraints. This is called the *unlabeled rigid body transformation problem* in statistical shape theory. Alternative names for the same problem include the *form transformation problem* and the *size-and-shape transformation problem*.

This chapter mainly considers the Bayesian alignment model introduced in [254], henceforth called ALIBI, which stands for Alignment by Bayesian Inference. ALIBI includes full Bayesian inference for all parameters of interest in the unlabeled rigid-body transformation problem. ALIBI can additionally take into account attributes of points in the configurations. The chapter also highlights methods that extend ALIBI for matching multiple configurations and incorporate information on bonded atoms. Other related shape analysis models and methods are also highlighted.

## 9.2   Form and Shape

Two objects have the same *form* or *size-and-shape* if they can be translated and rotated onto each other so that they match exactly. In other words, the objects are rigid-body transformations of each other [159]. Quantitatively, form is all the geometrical information remaining when orientation and translation are filtered out. On the other hand, *shape* is the geometrical information remaining when transformations in general – including orientation, translation and additionally scaling – are filtered out. *Landmark points* on an object can be used to describe or represent its form or shape. Two objects with the same form are equivalent up to translation and rotation. That is, they not only have the same shape, but also the same size. Thus, shape analysis deals with configurations of points with some invariance.

## 9.3   Labelled Shape Analysis

Landmarks in *labeled shape analysis* are supposed to be uniquely defined for similar objects [333]. Thus, for labeled shape analysis, landmark matrices and alignments are known and given. Figure 9.1 is an example of a set of pre-specified landmarks that describe shape. Shown in the figure is an outline and three landmarks for a fossil, microscopic planktonic foraminiferan, *Globorotalia truncatulinoides*, found in the ooze on the ocean bed. Given a set of fossils, a basic application is estimating

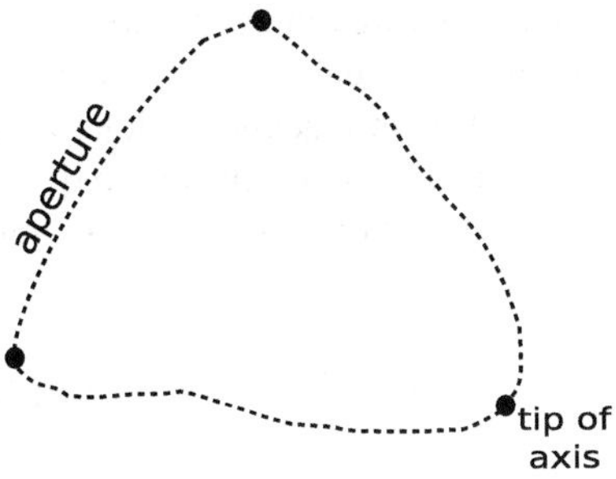

# 9 Bayesian Hierarchical Alignment Methods

**Fig. 9.1** An outline of a microfossil of a microscopic planktonic foraminiferan found in the ooze on the ocean bed. Shown are three landmarks selected on the outline [159] (Reproduced and adapted by permission of the Institute of Mathematical Studies)

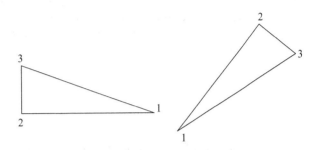

**Fig. 9.2** Triangles with the same form and labels. Figure 9.3 shows the same triangles but labeled arbitrarily

an average shape and describing shape variability.[1] Now consider a simple example: a case of triangles in Fig. 9.2. In shape space, these triangles are equivalent under rotation and translation. In reflection shape analysis, the equivalence class also includes reflected triangles. For $m$ fixed landmarks and $d$ dimensions, let $\mathbf{X}$ be the $(m \times d)$ configuration matrix with $m$ rows as landmarks $\mathbf{x}_i, i = 1, \ldots, m$. The standard model for rigid transformation is (see for example [159], p. 88)

$$\mathbf{X}_i = (\boldsymbol{\mu} + \mathbf{E}_i)\mathbf{R}_i + \mathbf{1}_k \boldsymbol{\tau}_i^T$$

where the $\mathbf{E}_i$ are zero mean $m \times d$ independent random error matrices, $\boldsymbol{\mu}$ is an $m \times d$ matrix representing the mean configuration, $\mathbf{R}_i$ are rotation matrices and $\boldsymbol{\tau}_i$ are translation vectors. For $\mathbf{E}_i$ there are two common modelling assumptions. First, $\mathbf{E}_i$ can be isotropic matrix normal, which implies that the errors are independent and have the same variability. Second, $\mathbf{E}_i$ can have independent rows that are multivariate normal, but each row has the same covariance matrix $\boldsymbol{\Sigma}$. In that case, the errors are independent between landmarks, but allow some common correlation between the dimensional coordinates for each landmark.

The main focus of labeled shape analysis is to estimate $\boldsymbol{\mu}$ and carry out standard hypotheses. This can be done in shape space. Procrustes analysis is the basic tool for exploratory work in shape analysis. Also, for rigid body transformations, we can work directly on distances between landmarks in each configuration so the nuisance

---

[1] See p. 15 in [159] for other applications of shape analysis in biology.

parameters are removed a priori. For similarity shape, one needs some modification to allow for scaling. For Bayesian methods in labeled shape analysis see Chap. 8 and [712]. In machine vision, *active shape models* play a key role [119, 120]. In these models, $\mu$ is allowed to dynamically deform within constraints learned from a training set of previous configurations.

## 9.4 Unlabeled Shape Analysis

In unlabeled shape analysis, the labeling is unknown. For example, the triangles in Fig. 9.3 have the same form even after allowing for the six label permutations of the vertices. The matching solution is the set of pairs $(1, 2')$, $(3, 1')$ and $(2, 3')$. The pairwise matching can be represented by the *matching* or *permutation matrix* **M**:

$$\mathbf{M} = \begin{pmatrix} 0 & 1 & 0 \\ 0 & 0 & 1 \\ 1 & 0 & 0 \end{pmatrix}.$$

Another frequent feature of unlabeled form or shape analysis is that of *partial matching*. Consider a pair of configurations in Fig. 9.4 in which points schematically

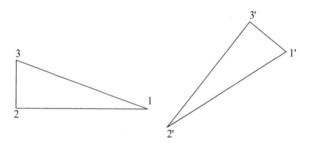

**Fig. 9.3** Triangles with the same form but arbitrary labels

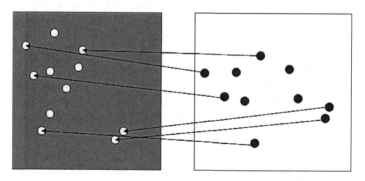

**Fig. 9.4** Alignment of two configurations of points schematically representing atomic positions in two proteins

# 9 Bayesian Hierarchical Alignment Methods

**Fig. 9.5** Alignment and superimposition of the configurations shown in Fig. 9.4

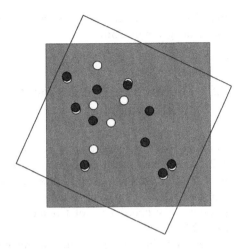

represent atomic positions in two proteins. Only a few points match. That is, the corresponding points and the number of matches in addition to the transformation are unknown beforehand. Figure 9.5 shows configurations after a superposition that involves rotating the second frame clockwise.

The aim of unlabeled shape analysis is somewhat different from labeled shape analysis. The aim of unlabeled shape analysis is to estimate the matching matrix **M**, while in labeled shape analysis the aim is to estimate $\mu$. Note that in practice, specific coordinate representations of labeled shape have been useful; they include Bookstein coordinates [67] and Procrustes tangent coordinates [366]. For a set of configurations close to $\mu$, specific coordinate representations could be useful even for unlabeled shape analysis.

A basic problem in unlabeled shape analysis is to find matching points. Suppose we have two configurations **X** and **Y**, for example representing the atomic positions in two proteins. Matching can be represented by a matching matrix **M**:

$$\mathbf{M}_{jk} = \begin{cases} 1 & \text{if the } j\text{th point in } \mathbf{X} \text{ corresponds to the } k\text{th point in } \mathbf{Y}, \\ 0 & \text{otherwise.} \end{cases}$$

Given the mapping matrix **M**, it is straight-forward using least squares estimation to calculate the transformation parameters in order to bring the two configurations into superposition or optimal alignment [159, 357]. Also, see Chap. 8 for a Bayesian approach.

In unlabeled shape analysis, **M** is unknown a priori, and transformation parameters need to be concurrently estimated and used for estimating **M**. Certainly this makes unlabeled shape analysis more challenging than labeled shape analysis. A hierarchical model for inference on both the matching matrix **M** and transformation parameters [254] is presented in Sect. 9.5. An instructional approach on using the model is given in Sects. 9.5–9.8. Practical approaches with illustrations using the model are given in [256].

## 9.5 Hierarchical Models for Pairs of Configurations: ALIBI Model

Consider two configurations, $\mathbf{X} = \{\mathbf{x}_j, j = 1, 2, \ldots, m\}$ and $\mathbf{Y} = \{\mathbf{y}_k, k = 1, 2, \ldots, n\}$ that consist of sets of coordinates in $d$-dimensional space, $\mathbb{R}^d$. That is, $\mathbf{x}_j$ and $\mathbf{y}_k$ are the $j$th and $k$th atoms in $\mathbf{X}$ and $\mathbf{Y}$ respectively. The points are arbitrarily labeled for identification only. In practice, these points could be atomic positions, say coordinates of $C\alpha$ atoms in protein structures or atoms in active sites [236].

Represent the transformation bringing the configurations into alignment by $\mathbf{x}_j = \mathbf{R}\mathbf{y}_k + \boldsymbol{\tau}$ for $\mathbf{M}_{jk} = 1$ where $\mathbf{R}$ is a rotation matrix and $\boldsymbol{\tau}$ is a translation vector. Also, assume the presence of Gaussian noise $\mathcal{N}_d(0, \sigma^2 \mathbf{I}_d)$ in the atomic positions for $\mathbf{x}_j$ and $\mathbf{y}_k$. Let $r_1(\mathbf{x}_j, k) = \mathbf{x}_j - \mathbf{R}\mathbf{y}_k - \boldsymbol{\tau}$. The posterior distribution for the model is

$$p(\mathbf{M}, \mathbf{R}, \boldsymbol{\tau}, \sigma | \mathbf{X}, \mathbf{Y}) \propto \text{prior} \times \text{likelihood}$$

$$= |\mathbf{R}|^n p(\mathbf{R}) p(\boldsymbol{\tau}) p(\sigma) \times \prod_{j,k} \mathbf{M}_{jk} \kappa \frac{\phi\left(r_1(\mathbf{x}_j, k)/\sigma\sqrt{2}\right)}{(\sigma\sqrt{2})^d} \quad (9.1)$$

where $p(\mathbf{R})$, $p(\boldsymbol{\tau})$ and $p(\sigma)$ denote prior distributions for $\mathbf{R}$, $\boldsymbol{\tau}$ and $\sigma$; $|\mathbf{R}|$ is the Jacobian of the transformation from the space of $\mathbf{X}$ into the space of $\mathbf{Y}$ and $\phi(\cdot)$ is the standard normal probability density function. $\kappa$ measures the tendency a priori for points to be matched and can depend on attributes of $\mathbf{x}_j$ and $\mathbf{y}_k$; for example, the atom types when matching protein structures [254].

For two-dimensional configurations such as protein gels, $d = 2$ and for three-dimensional configurations such as active sites, $d = 3$. We describe the motivation of the model and the mathematical derivation in the next section.

### 9.5.1 Motivation and Statistical Model Derivation

Suppose that the data generating mechanism can be modelled by a homogeneous spatial Poisson process with rate $\lambda$ over a region in $\mathbb{R}^d$ with volume $v$. A realization of a spatial homogeneous Poisson point process is a random pattern of points in $d$-dimensional space. Points occur within a given region in a completely random manner, independently of each other in the region. A realization of any point in a

# 9 Bayesian Hierarchical Alignment Methods

region does not affect the occurrence of any other point in its neighborhood. However, the probability of points overlapping is zero. The number of points in a given disjoint subregion is random and independently follows the Poisson distribution with rate $\lambda v$ where $v$ is volume of the region. The intensity of points from a homogeneous spatial Poisson process does not vary over the region and points are expected to be uniformly located [128].

We suppose that $N$ locations spanning a super-population over a region of volume $v$ are realized. Each point may belong to $\mathbf{X}$, $\mathbf{Y}$, neither, or both with probabilities $p_x$, $p_y$, $1 - p_x - p_y - \rho p_x p_y$, and $\rho p_x p_y$ respectively, where $\rho$ is the tendency of configuration points to match a priori [135, 254]. Under this model, the number of locations which belong to $\mathbf{X}$, $\mathbf{Y}$, neither and both are independent Poisson random variables with counts $m - L$, $n - L$, $N - m - n + L$ and $L$ where $L = \sum_{jk} \mathbf{M}_{jk}$. Thus, both point sets $\mathbf{X}$ and $\mathbf{Y}$ are regarded as noisy observations on subsets of a set of true locations $\boldsymbol{\mu} = \{\boldsymbol{v}_i\}$, where we do not know the mappings from $j$ and $k$ to $i$. There may be a geometrical transformation between the coordinate systems of $\mathbf{X}$ and $\mathbf{Y}$, which may also be unknown. The objective is to make model-based inference about these mappings, and in particular to make probability statements about matching: which pairs $(j, k)$ correspond to the same true location?

Conditional on $m$ and $n$, $L$ follows a hypergeometric distribution, which belongs to the exponential family:

$$p(L|m, n, D) = \frac{KD^L}{(m - L)!(n - L)!L!} \tag{9.2}$$

where $K$ is a normalizing constant dependent on $n$, $m$ and $D = \rho/\lambda v$. Henceforth, we let $D \propto \kappa$ assuming $v$ constant. The distribution is derived from the joint distribution for counts $m - L$, $n - L$, $N - m - n + L$ and $L$.

Denote the mappings between $\boldsymbol{\mu}$ and that of the data $\mathbf{X}$ and $\mathbf{Y}$ by indexing arrays $\xi = \{\xi_j\}$ and $\eta = \{\eta_k\}$. Thus $\mathbf{x}_j$ corresponds to a noisy observation of $\boldsymbol{v}_{\xi_j}$. Similarly, $\mathbf{y}_k$ corresponds to a noisy observation of $\boldsymbol{v}_{\eta_k}$. We model $\mathbf{x}_j$ and $\mathbf{y}_k$ conditional on $\boldsymbol{\mu}$, $\xi_j$ and $\eta_k$ by $\mathcal{N}(\mathbf{0}, \sigma^2 \mathbf{I}_d)$. Furthermore, we assume the following geometrical transformations:

$$\mathbf{x}_j = \boldsymbol{v}_{\xi_j} + \boldsymbol{\varepsilon}_j \tag{9.3}$$

and

$$\mathbf{R}\mathbf{y}_k + \boldsymbol{\tau} = \boldsymbol{v}_{\eta_k} + \varepsilon_k \tag{9.4}$$

where $\varepsilon_j, \varepsilon_k \sim \mathcal{N}(0, \sigma^2 \mathbf{I}_d)$ for $j = 1, 2, \ldots, m$ and $k = 1, 2, \ldots, n$. $\mathbf{R}$ and $\boldsymbol{\tau}$ are the rotation matrix and translation vector to geometrically transform $\mathbf{Y}$ into the coordinate system of $\boldsymbol{\mu}$. However $\mathbf{X}$ is in the same coordinate system as $\boldsymbol{\mu}$. Thus $\mathbf{Y}$ corresponds to $\mathbf{X}' = \mathbf{R}\mathbf{Y} + \boldsymbol{\tau}$ in the coordinate system of $\mathbf{X}$. All $\boldsymbol{\varepsilon}_j$ and $\boldsymbol{\varepsilon}_k$ are independent of $\boldsymbol{\mu}$ and each other.

216            K.V. Mardia and V.B. Nyirongo

### 9.5.2   The Likelihood

Thus we have

$$\mathbf{x}_j \sim \mathcal{N}_d(\boldsymbol{\nu}_{\xi_j}, \sigma^2 \mathbf{I}_d) \qquad \text{and} \qquad \mathbf{R}\mathbf{y}_k + \boldsymbol{\tau} \sim \mathcal{N}_d(\boldsymbol{\nu}_{\eta_k}, \sigma^2 \mathbf{I}_d).$$

Integrating over $\boldsymbol{\mu}$ and assuming that $v$ is large enough relative to the supports of densities for $\mathbf{x}_j - \boldsymbol{\nu}_{\xi_j}$ and $\mathbf{R}\mathbf{y}_k + \boldsymbol{\tau} - \boldsymbol{\nu}_{\eta_k}$, the likelihood is

$$p(\mathbf{X}, \mathbf{Y} | \mathbf{M}, \mathbf{R}, \boldsymbol{\tau}, \sigma) \propto |\mathbf{R}|^n \prod_{j,k} \mathbf{M}_{jk} \left( \kappa \frac{\phi \left( r_1(\mathbf{x}_j, k)/\sigma \sqrt{2} \right)}{(\sigma \sqrt{2})^d} \right) \qquad (9.5)$$

where $|\mathbf{R}|$ is the Jacobian of the transformation from the coordinate system of $\mathbf{Y}$ into the coordinate system of $\mathbf{X}$.

Although the model was motivated with a spatial homogeneous Poisson process with points expected to be uniformly located and no inhibition distance between points, none of the parameters for the process are in the joint model of interest; $N$, $v$ and $\boldsymbol{\mu}$ are integrated out. This allows the model to be applicable to many patterns of configurations, including biological macromolecules with some given minimum distances between points due to steric constraints.

### 9.5.3   Prior Distributions

We use conditionally conjugate priors for $\boldsymbol{\tau}, \sigma^{-2}$ and $\mathbf{R}$ so

$$\boldsymbol{\tau} \sim \mathcal{N}_d(\boldsymbol{\nu}_\tau, \sigma_\tau^2 \mathbf{I}_d); \qquad \sigma^{-2} \sim G(\alpha, \beta); \qquad \mathbf{R} \sim F_{d \times d}(\mathbf{F}_0)$$

where $F_{d \times d}(\mathbf{F}_0)$ is the matrix Fisher distribution on which more details are given in Sect. 9.5.4. $G(\alpha, \beta)$ is a gamma distribution. $\mathbf{M}$ is assumed uniform a priori, and conditional on $L$

$$p(\mathbf{M}|L) \propto \kappa^L.$$

### 9.5.4   The Matrix Fisher Distribution

The matrix Fisher distribution plays an important role in the ALIBI model. This distribution is perhaps the simplest non-uniform distribution for rotation matrices [157, 254, 474]. The distribution takes an exponential form, i.e. for a matrix $\mathbf{R}$:

$$p(\mathbf{R}) \propto \exp \left\{ \text{tr}(\mathbf{F}_0^T \mathbf{R}) \right\}$$

where the matrix $\mathbf{F}_0$ is a parameter.

9 Bayesian Hierarchical Alignment Methods

In practice, an uninformative uniform distribution on $\mathbf{R}$, corresponding to a matrix Fisher distribution with $\mathbf{F}_0 = \mathbf{0}$, is found to be an adequate prior distribution.

For $d = 2$, $\mathbf{R}$ has a von Mises distribution. An arbitrary rotation matrix $\mathbf{R}$ can be written as

$$\mathbf{R} = \begin{pmatrix} \cos\theta & -\sin\theta \\ \sin\theta & \cos\theta \end{pmatrix}$$

where $\theta \in (0, 2\pi)$. A uniformly distributed choice of $\mathbf{R}$ corresponds to

$$p(\mathbf{R}) \propto 1.$$

with $p(\mathbf{R}) \propto \exp(\mathrm{tr}(F_0^T A))$, where a (non-unique) choice for $F_0$ is

$$\mathbf{F}_0 = \kappa/2 \begin{pmatrix} \cos\nu & -\sin\nu \\ \sin\nu & \cos\nu \end{pmatrix}.$$

Hence, the distribution for $\theta$ is

$$p(\theta) \propto \exp(\kappa \cos(\theta - \nu)) = \exp(\kappa \cos\nu \cos\theta + \kappa \sin\nu \sin\theta) \qquad (9.6)$$

which is a von Mises distribution .

For $d = 3$, it is useful to express $\mathbf{R}$ in terms of Euler angles. Euler angles are one of several ways of specifying the relative orientation or rotation of a configuration in three-dimensional Euclidean space. That is, an orientation described by the rotation matrix $\mathbf{R}$ is decomposed into a sequence of three elementary rotations around the $x-$, $y-$ and $z-$axes. Each elementary rotation is simply described by a corresponding Euler angle, say $\theta_{12}$, $\theta_{13}$ and $\theta_{23}$ quantifying rotations around the $z-$, $y-$ and $x-$axes.

$\mathbf{R}$ is a product of elementary rotations

$$\mathbf{R} = \mathbf{R}_{12}(\theta_{12})\mathbf{R}_{13}(\theta_{13})\mathbf{R}_{23}(\theta_{23}). \qquad (9.7)$$

The first two matrices represent longitude and co-latitude, respectively, while the third matrix represents the rotation of the whole frame; see for example [254]. For $i < j$, let $\mathbf{R}_{ij}(\theta_{ij})$ be the $3 \times 3$ matrix with $a_{ii} = a_{jj} = \cos\theta_{ij}$, $-a_{ij} = a_{ji} = \sin\theta_{ij}, a_{rr} = 1$ for $r \neq i, j$ and other entries 0.

## 9.6 Posterior Distribution and Inference

The full posterior distribution is given by Eq. 9.1. Inference is done by sampling from conditional posterior distributions for transformation parameters and variance ($\mathbf{R}$, $\tau$ and $\sigma$) and updating the matching matrix $\mathbf{M}$ using the Metropolis-Hastings algorithm. As discussed in Chap. 2, the Metropolis-Hastings algorithm is a Markov chain Monte Carlo method used for obtaining random samples from a distribution where direct sampling is difficult. Details on updating $\mathbf{M}$ are given in Sect. 9.6.2.

With $p(\mathbf{R}) \propto \exp(\text{tr}(\mathbf{F}_0^T \mathbf{R}))$ for some matrix $\mathbf{F}_0$, the posterior has the same form with $\mathbf{F}_0$ replaced by

$$\mathbf{F} = \mathbf{F}_0 + \mathbf{S}^\dagger \tag{9.8}$$

where

$$\mathbf{S}^\dagger = 1/2\sigma^2 \sum_{j,k} \mathbf{M}_{jk}(\mathbf{x}_j - \boldsymbol{\tau})\mathbf{y}_k^T. \tag{9.9}$$

In the three-dimensional case the joint full conditional density for Euler angles is

$$\propto \exp[\text{tr}\{\mathbf{F}^T \mathbf{R}\}] \cos \theta_{13}$$

for $\theta_{12}, \theta_{23} \in (-\pi, \pi)$ and $\theta_{13} \in (-\pi/2, \pi/2)$. The cosine term arises since the natural dominating measure corresponding to the uniform distribution of rotations has volume element $\cos \theta_{13} \mathbf{d}\theta_{12} \mathbf{d}\theta_{13} \mathbf{d}\theta_{23}$ in these coordinates.

From the representation in Eq. 9.7

$$\text{tr}\{\mathbf{F}^T \mathbf{R}\} = a_{12} \cos \theta_{12} + b_{12} \sin \theta_{12} + c_{12}$$
$$= a_{13} \cos \theta_{13} + b_{13} \sin \theta_{13} + c_{13}$$
$$= a_{23} \cos \theta_{23} + b_{23} \sin \theta_{23} + c_{23}$$

where

$$a_{12} = (\mathbf{F}_{22} - \sin \theta_{13} \mathbf{F}_{13}) \cos \theta_{23} + (-\mathbf{F}_{23} - \sin \theta_{13} \mathbf{F}_{12}) \sin \theta_{23} + \cos \theta_{13} \mathbf{F}_{11}$$
$$b_{12} = (- \sin \theta_{13} \mathbf{F}_{23} - \mathbf{F}_{12}) \cos \theta_{23} + (\mathbf{F}_{13} - \sin \theta_{13} \mathbf{F}_{22}) \sin \theta_{23} + \cos \theta_{13} \mathbf{F}_{21}$$
$$a_{13} = \sin \theta_{12} \mathbf{F}_{21} + \cos \theta_{12} \mathbf{F}_{11} + \sin \theta_{23} \mathbf{F}_{32} + \cos \theta_{23} \mathbf{F}_{33}$$
$$b_{13} = (- \sin \theta_{23} \mathbf{F}_{12} - \cos \theta_{23} \mathbf{F}_{13}) \cos \theta_{12}$$
$$\quad +(- \sin \theta_{23} \mathbf{F}_{22} - \cos \theta_{23} \mathbf{F}_{23}) \sin \theta_{12} + \mathbf{F}_{31}$$
$$a_{23} = (\mathbf{F}_{22} - \sin \theta_{13} \mathbf{F}_{13}) \cos \theta_{12} + (- \sin \theta_{13} \mathbf{F}_{23} - \mathbf{F}_{12}) \sin \theta_{12} + \cos \theta_{13} \mathbf{F}_{33}$$
$$b_{23} = (-\mathbf{F}_{23} - \sin \theta_{13} \mathbf{F}_{12}) \cos \theta_{12} + (\mathbf{F}_{13} - \sin \theta_{13} \mathbf{F}_{22}) \sin \theta_{12} + \cos \theta_{13} \mathbf{F}_{32}.$$

The $c_{ij}$ are ignored as they are combined into the normalizing constants. Thus, the full conditionals for $\theta_{12}$ and $\theta_{23}$ are von Mises distributions. The posterior for $\theta_{13}$ is proportional to

$$\exp[a_{13} \cos \theta_{13} + b_{13} \sin \theta_{13}] \cos \theta_{13}.$$

In the two-dimensional case the full conditional distribution for $\theta$ is of the same von Mises form (Eq. 9.6), with $\kappa \cos \nu$ updated to $(\kappa \cos \nu + \mathbf{S}_{11}^\dagger + \mathbf{S}_{22}^\dagger)$, and $\kappa \sin \nu$ to $(\kappa \sin \nu - \mathbf{S}_{12}^\dagger + \mathbf{S}_{21}^\dagger)$. The von Mises distribution is the conjugate prior for rotations for the spherical Gaussian error distributions (Model 9.1).

## 9.6.1 MCMC Implementation for Transformation Parameters

Gibbs sampling is used for inference about $\mathbf{R}$ in two-dimensions. For example, assuming a von Mises prior distribution on the rotation angle $\theta$, one can use the Best/Fisher algorithm ([474], p. 43) to sample from the full conditional for $\theta$, which is also a von Mises distribution.

For $d = 3$, Gibbs sampling and the Metropolis-Hastings algorithm are used. Each of the generalized Euler angles $\theta_{ij}$ is updated conditioning on the other two angles and the other variables $(\mathbf{M}, \tau, \sigma, \mathbf{X}, \mathbf{Y})$ entering the expression for $\mathbf{F}$. Note that in practice $\sigma_\tau^2$ is taken large and $\mathbf{R}$ uniform to preserve shape invariance. Euler angles $\theta_{12}$ and $\theta_{23}$ are updated by Gibbs sampling since their full conditionals are von Mises distributions. It is difficult to sample directly from the conditional for $\theta_{13}$ so a random walk Metropolis algorithm with a uniformly distributed perturbation is used to update it.

## 9.6.2 The Matching Matrix

For updating $\mathbf{M}$ conditionally, we need some new ways. The full posterior for the matrix $\mathbf{M}$ is not required, and is actually complex. The matching matrix $\mathbf{M}$ is updated, respecting detailed balance, using Metropolis-Hastings moves that only propose changes to a few entries: the number of matches $L = \sum_{j,k} \mathbf{M}_{jk}$ can only increase or decrease by one at a time, or stay the same. The possible changes are as follows:

(a) Adding a match, which changes one entry $\mathbf{M}_{jk}$ from 0 to 1;
(b) Deleting a match, which changes one entry $\mathbf{M}_{jk}$ from 1 to 0;
(c) Switching a match, which simultaneously changes one entry from 0 to 1 and another in the same row or column from 1 to 0.

The proposal proceeds as follows. First a uniform random choice is made from all the $m + n$ data points $\mathbf{x}_1, \mathbf{x}_2, \ldots, \mathbf{x}_m, \mathbf{y}_1, \mathbf{y}_2, \ldots, \mathbf{y}_n$. Suppose without loss of generality, by the symmetry of the set-up, that an $\mathbf{X}$ is chosen, say $\mathbf{x}_j$. There are two possibilities: either $\mathbf{x}_j$ is currently matched, in that there is some $k$ such that $\mathbf{M}_{jk} = 1$, or not, in that there is no such $k$. If $\mathbf{x}_j$ is matched to $\mathbf{y}_k$, with probability $p^*$ we propose to delete the match, and with probability $1 - p^*$ we propose switching it from $\mathbf{y}_k$ to $\mathbf{y}_{k'}$, where $k'$ is drawn uniformly at random from the currently unmatched $\mathbf{Y}$ points. On the other hand, if $\mathbf{x}_j$ is not currently matched, we propose to add a match between $\mathbf{x}_j$ and a $\mathbf{y}_k$, where again $k$ is drawn uniformly at random from the currently unmatched $\mathbf{Y}$ points. We give MCMC implementation in Sect. 9.6.3.

220                                                                 K.V. Mardia and V.B. Nyirongo

## 9.6.3  MCMC Implementation for the Matching Matrix

The matching matrix $\mathbf{M}$ is updated, respecting detailed balance, using Metropolis-Hastings moves that propose adding, deleting or switching a match (See Sect. 9.6.2). The proposal proceeds as follows: first a uniform random choice is made from all the $m + n$ data points $\mathbf{x}_1, \mathbf{x}_2, \ldots, \mathbf{x}_m, \mathbf{y}_1, \mathbf{y}_2, \ldots, \mathbf{y}_n$. Suppose without loss of generality, by the symmetry of the set-up, that an $\mathbf{x}$ is chosen, say $\mathbf{x}_j$. There are two possibilities: either $\mathbf{x}_j$ is currently matched ($\exists k$ such that $\mathbf{M}_{jk} = 1$) or not (there is no such $k$). If $\mathbf{x}_j$ is matched to $\mathbf{y}_k$, with probability $p^*$ propose *deleting* the match, and with probability $1 - p^*$ propose *switching* it from $\mathbf{y}_k$ to $\mathbf{y}_{k'}$, where $k'$ is drawn uniformly at random from the currently unmatched $\mathbf{Y}$ points. On the other hand, if $\mathbf{x}_j$ is not currently matched, propose *adding* a match between $\mathbf{x}_j$ and a $\mathbf{y}_k$, where again $k$ is drawn uniformly at random from the currently unmatched $\mathbf{Y}$ points. The acceptance probabilities for these three possibilities are derived from the joint model in Eq. 9.1. In each case the proposed new matching matrix $\mathbf{M}'$ is only slightly perturbed from $\mathbf{M}$ so that the ratio $p(\mathbf{M}', \tau, \sigma | \mathbf{x}, \mathbf{y}) / p(\mathbf{M}, \tau, \sigma | \mathbf{X}, \mathbf{Y})$ has only a few factors. Taking into account also the proposal probabilities, whose ratio is $(1/n_u) \div p^*$, where $n_u = \#\{k \in 1, 2, \ldots, n : \mathbf{M}_{jk} = 0 \forall j\}$ is the number of unmatched $\mathbf{Y}$ points in $\mathbf{M}$; the acceptance probability for adding a match $(j, k)$ is

$$\min \left\{ 1, \frac{\kappa \phi \left( r_1(\mathbf{x}_j, k)/\sigma \sqrt{2} \right) p^* n_u}{(\sigma \sqrt{2})^d} \right\} . \tag{9.10}$$

Similarly, the acceptance probability for switching the match of $\mathbf{x}_j$ from $\mathbf{y}_k$ to $\mathbf{y}_{k'}$ is

$$\min \left\{ 1, \frac{\phi \left( r_1(\mathbf{x}_j, k')/\sigma \sqrt{2} \right)}{\phi \left( r_1(\mathbf{x}_j, k)/\sigma \sqrt{2} \right)} \right\} . \tag{9.11}$$

For deleting the match $(j, k)$ the acceptance probability is

$$\min \left\{ 1, \frac{(\sigma \sqrt{2})^d}{\kappa \phi \left( r_1(\mathbf{x}_j, k)/\sigma \sqrt{2} \right) p^* n_u'} \right\} , \tag{9.12}$$

where $n_u' = \#\{k \in 1, 2, \ldots, n : \mathbf{M}'_{jk} = 0 \forall j\} = n_u + 1$. Along with just one of each of the other updates, including for $\mathbf{R}, \tau$ and $\sigma$, several moves updating $\mathbf{M}$ per sweep are made, since the changes effected to $\mathbf{M}$ are so modest. The method bypasses the reversible jump algorithm [253] as $\mu$ is integrated out.

## 9.6.4  Concomitant Information

Attributes of points in the configurations can be taken into account. For example, atom type, charge or binding affinity are some of the attributes which can be incorporated in ALIBI. We call such attributes *concomitant information*. Henceforth these attributes are considered as *point colors*. When the points in each configuration are 'colored', with the interpretation that like-colored points are more likely to be matched than unlike-colored ones, it is appropriate to use a modified likelihood that allows us to exploit such information. Let the colors for the $\mathbf{x}_j$ and $\mathbf{y}_k$ points be $\{r_j^x, j = 1, 2, \ldots, m\}$ and $\{r_k^y, k = 1, 2, \ldots, n\}$ respectively. The hidden-point model is augmented to generate the point colors, as follows. Independently for each hidden point, with probability $(1 - p_x - p_y - \rho p_x p_y)$ we observe points belonging to neither $\mathbf{X}$ nor $\mathbf{Y}$, as before. With probabilities $p_x \pi_r^x$ and $p_y \pi_r^y$, respectively, we observe points only in $\mathbf{X}$ or $\mathbf{Y}$ with color $r$ from an appropriate finite set. With probability

$$\rho p_x p_y \pi_r^x \pi_s^y \exp(\gamma I [r = s] + \delta I [r \neq s]), \tag{9.13}$$

where $I[\cdot]$ is an indicator function, we observe a point in $\mathbf{X}$ colored $r$ and a point in $\mathbf{Y}$ colored $s$. Our original likelihood is equivalent to the case $\gamma = \delta = 0$, where colors are independent and so carry no information about matching. If $\gamma$ and $\delta$ increase, then matches are more probable, a posteriori, and, if $\gamma > \delta$, matches between like-colored points are more likely than those between unlike-colored ones. The case $\delta \to -\infty$ allows the prohibition of matches between unlike-colored points, a feature that might be adapted to other contexts such as the matching of shapes with given landmarks.

In the implementation of the modified likelihood with concomitant information, the Markov chain Monte Carlo acceptance ratios have to be modified accordingly. For example, if $r_j^x = r_k^y$ and $r_j^x \neq r_{k'}^y$, then Eq. 9.10 has to be multiplied by $\exp(-\gamma)$ and Eq. 9.11 by $\exp(\delta - \gamma)$.

Other, more complicated, coloring distributions where the log probability can be expressed linearly in entries of $\mathbf{M}$ can be handled similarly. For example, continuous concomitant information on van der Waals radii or partial charges can be incorporated in the model. Pairwise continuous functions can be used for continuous concomitant information instead of an indicator function. In Eq. 9.13, $\gamma I [r = s] + \delta I [r \neq s]$ would be replaced by, say $\gamma (r_j^x, r_k^y)$ where $r_j^x$ and $r_k^y$ are van der Waals radii or partial charges for atoms $\mathbf{x}_j$ and $\mathbf{y}_k$ respectively.

## 9.7 Sensitivity Analysis

A sensitivity analysis on model parameters, the parameters of the homogeneous Poisson process, and the use of concomitant information was carried out in [254] and [549]. ALIBI is not sensitive to homogeneous Poisson process parameters such as the volume of the region $v$. ALIBI is also not sensitive to model parameters, including $\kappa$. Although an assumption of a hidden homogeneous Poisson process was made to formulate the model, the algorithm is not sensitive to this assumption. The algorithm can thus match configurations where distances between points are constrained by sterical considerations, such as is the case for atoms in molecules.

## 9.8 Extensions: Multiple Configurations and Refinement of ALIBI

The model in Eq. 9.1 can be extended for matching configurations with bonded points and also for matching multiple configurations simultaneously. Furthermore, the model can be used for refining matching solution found by heuristic but fast methods, such as approaches based on graph theory [15,235,477,635]. We consider extensions to matching bonded points and multiple configurations in Sects. 9.8.1 and 9.8.2 below.

### 9.8.1 Bonded Points ALIBI

The model in Eq. 9.1 can be extended for matching bonded points in a configuration [477, 549]. Bonded points in a configurations are dependent. This is motivated by the requirement in bioinformatics to prefer matching amino acids with similar orientation. Thus, we can take into account relative orientation of side chains by using $C\alpha$ and $C\beta$ atoms in matching amino acids. The positions of these two covalently bonded atoms in the same amino acid are strongly dependent.

We use the example of superimposing a potential active site, called the *query site*, on a true active site, called the *functional site*. Let $\mathbf{y}_{1k}$ and $\mathbf{x}_{1j}$ denote coordinates for $C\alpha$ atoms in the query and the functional site. We denote $C\beta$ coordinates for the query and functional site by $\mathbf{y}_{2k}$ and $\mathbf{x}_{2j}$ respectively. Thus $\mathbf{x}_{1j}$ and $\mathbf{x}_{2j}$ are dependent. Similarly, $\mathbf{y}_{1k}$ and $\mathbf{y}_{2k}$ are dependent. We take into account the position of $\mathbf{y}_{2k}$ by using the conditional distribution given the position of $\mathbf{y}_{1k}$. Given $\mathbf{x}_{1j}, \mathbf{y}_{1k}$, it is plausible to assume that

$$p(\mathbf{x}_{1j}, \mathbf{x}_{2j}, \mathbf{y}_{1k}, \mathbf{y}_{2k}) = p(\mathbf{x}_{1j}, \mathbf{y}_{1k})p(\mathbf{x}_{2j}, \mathbf{y}_{2k}|\mathbf{x}_{1j}, \mathbf{y}_{1k}),$$
$$\mathbf{x}_{2j}|\mathbf{x}_{1j} \sim \mathcal{N}(\mathbf{x}_{1j}, \sigma_*^2\mathbf{I}_3),$$
$$\mathbf{R}\mathbf{y}_{2k}|\mathbf{y}_{1k} \sim \mathcal{N}(\mathbf{R}\mathbf{y}_{1k}, \sigma_*^2\mathbf{I}_3)$$

or that the displacement is

$$\mathbf{x}_{2j} - \mathbf{R}\mathbf{y}_{2k}|(\mathbf{x}_{1j}, \mathbf{y}_{1k}) \sim \mathcal{N}(\mathbf{x}_{1j} - \mathbf{R}\mathbf{y}_{1k}, 2\sigma_*^2\mathbf{I}_3).$$

We assume for "symmetry" that $p(\mathbf{x}_{2j} - \mathbf{R}\mathbf{y}_{2k}|\mathbf{x}_{1j}, \mathbf{y}_{1k})$ depends only on the displacement as in the likelihood in Eq. 9.1. Let $r_2(\mathbf{x}_{2j}, k) = \mathbf{x}_{2j} - \mathbf{x}_{1j} - \mathbf{R}(\mathbf{y}_{2k} - \mathbf{y}_{1k})$. Thus $\phi(.)$ in Eq. 9.1 is replaced by $\phi(.) \times \phi(r_2(\mathbf{x}_{2j}, k)/\sigma_*\sqrt{2})$ for the new full likelihood. Now the likelihood becomes

$$\mathcal{L}(\vartheta) \propto |\mathbf{R}|^n \prod_{j,k} \mathbf{M}_{jk} \left( \kappa \frac{\phi(r_1(\mathbf{x}_j, k)/\sigma\sqrt{2}) \times \phi(r_2(\mathbf{x}_{2j}, k)/\sigma_*\sqrt{2})}{(\sigma\sqrt{2})^d} \right) \tag{9.14}$$

where $\vartheta = \{\mathbf{M}, \mathbf{R}, \boldsymbol{\tau}, \sigma\}$.

# 9 Bayesian Hierarchical Alignment Methods

Some probability mass for the distribution of $\mathbf{x}_{2j}|\mathbf{x}_{1j}$ and $\mathbf{Ry}_{2k}|\mathbf{y}_{1k}$ is unaccounted for because there are distance constraints between $\mathbf{x}_{2j}$ and $\mathbf{x}_{1j}$ and also between $\mathbf{y}_{2k}$ and $\mathbf{y}_{1k}$. Thus $\mathbf{x}_{2j} - \mathbf{x}_{1j}$ is not isotropic. In theory, a truncated distribution is required. However, using the Gaussian distribution is not expected to affect the performance of the algorithm because the relative matching probabilities for $\mathbf{x}_{2j} - \mathbf{x}_{1j}$ vectors are unaffected.

### 9.8.1.1 Posterior Distributions and Computations

The additional term in the new full likelihood does not involve $\tau$. Hence, the posterior and updating of $\tau$ is unchanged. The full conditional distribution of $\mathbf{R}$ is

$$p(\mathbf{R}|\mathbf{M}, \tau, \sigma, \mathbf{X}_1, \mathbf{Y}_1, \mathbf{X}_2, Y_2) \propto$$
$$|\mathbf{R}|^{2n} p(\mathbf{R}) \times \prod_{j,k} \mathbf{M}_{jk} \phi \left( \frac{r_1(\mathbf{x}_j, k)}{\sigma \sqrt{2}} \right) \times \phi \left( \frac{r_2(\mathbf{x}_{2j}, k)}{\sigma_* \sqrt{2}} \right). \tag{9.15}$$

Thus

$$p(\mathbf{R}|\mathbf{M}, \tau, \sigma, \mathbf{X}_1, \mathbf{Y}_1, \mathbf{X}_2, Y_2) \propto p(\mathbf{R}) \times \exp \left\{ \operatorname{tr} \left( (\mathbf{B}^\dagger + \mathbf{B}^{\dagger\dagger})\mathbf{R} \right) \right\}$$

where

$$\mathbf{B}^\dagger = 1/2\sigma^2 \sum_{j,k} \mathbf{M}_{jk} \mathbf{y}_{1k} (\mathbf{x}_{1j} - \tau)^T$$

and

$$\mathbf{B}^{\dagger\dagger} = 1/2\sigma_*^2 \sum_{j,k} \mathbf{M}_{jk} (\mathbf{y}_{2k} - \mathbf{y}_{1k})(\mathbf{x}_{2j} - \mathbf{x}_{1j})^T.$$

Similar to Eq. 9.8, with $p(\mathbf{R}) \propto \exp(\operatorname{tr}(\mathbf{F}_0^T \mathbf{R}))$ for some matrix $\mathbf{F}_0$, the full conditional distribution of $\mathbf{R}$ – given data and values for all other parameters – has the same form with $\mathbf{F}_0$ replaced by

$$\mathbf{F} = \mathbf{F}_0 + \mathbf{S}^\dagger + \mathbf{S}^{\dagger\dagger} \tag{9.16}$$

where $\mathbf{S}^\dagger$ is defined in Eq. 9.9 and

$$\mathbf{S}^{\dagger\dagger} = 1/2\sigma_*^2 \sum_{j,k} \mathbf{M}_{jk} (\mathbf{x}_{2j} - \mathbf{x}_{1j})(\mathbf{y}_{2k} - \mathbf{y}_{1k})^T. \tag{9.17}$$

That is, the matrix Fisher distribution is a conjugate prior for $\mathbf{R}$.

### 9.8.1.2 Updating M

Similar to Expression 9.10, the acceptance probability for adding a match $(j, k)$ is

$$\min\left\{1, \frac{\kappa\phi(r_1(\mathbf{x}_j,k)/\sigma\sqrt{2})p^*n_u}{(\sigma\sqrt{2})^d} \times \frac{\phi(r_2(\mathbf{x}_{2j},k)/\sigma_*\sqrt{2})}{(\sigma_*\sqrt{2})^d}\right\}.$$

Similarly, the acceptance probability for switching the match of $\mathbf{x}_j$ from $\mathbf{y}_k$ to $\mathbf{y}_{k'}$ is

$$\min\left\{1, \frac{\phi(r_1(\mathbf{x}_j,k')/\sigma\sqrt{2})}{\phi(r_1(\mathbf{x}_j,k)/\sigma\sqrt{2})} \times \frac{\phi(r_2(\mathbf{x}_{2j},k')/\sigma_*\sqrt{2})}{\phi(r_2(\mathbf{x}_{2j},k)/\sigma_*\sqrt{2})}\right\}$$

and for deleting the match $(j,k)$ is

$$\min\left\{1, \frac{(\sigma\sqrt{2})^d}{\kappa\phi(r_1(\mathbf{x}_j,k)/\sigma\sqrt{2})p^*n_u} \times \frac{(\sigma_*\sqrt{2})^d}{\phi(r_2(\mathbf{x}_{2j},k)/\sigma_*\sqrt{2})}\right\}.$$

### 9.8.2 Multiple ALIBI

The pairwise model in Eq. 9.1 is extended to matching multiple configurations simultaneously [482, 616]. Suppose there are $C$ point configurations in a set $\mathcal{X} = \{\mathbf{X}^{(1)}, \mathbf{X}^{(2)}, \ldots, \mathbf{X}^{(C)}\}$, such that $\mathbf{X}^{(c)} = \{\mathbf{x}_j^{(c)}, j = 1, 2, \ldots, n_c\}$, where $\mathbf{x}_j^{(c)}$ is in $\mathbb{R}^d$ and $n_c$ is the number of points in the configuration $\mathbf{X}^{(c)}$. As in the pairwise case, assume the existence of a set of "hidden" points $\boldsymbol{\mu} = \{\boldsymbol{v}_i\} \subset \mathbb{R}^d$ underlying the observations. Considering only rigid body transformations, the multiple-configuration model is :

$$\mathbf{R}^{(c)}\mathbf{x}_j^{(c)} + \boldsymbol{\tau}^{(c)} = \boldsymbol{v}_{\xi_j^{(c)}} + \boldsymbol{\varepsilon}_j^{(c)}. \quad \text{for } j = 1, 2, \ldots, n_c, \quad c = 1, 2, \ldots, C. \quad (9.18)$$

The unknown transformation $\mathbf{R}^{(c)}\mathbf{X}^{(c)} + \boldsymbol{\tau}^{(c)}$ brings the configuration $\mathbf{X}^{(c)}$ back into the same frame as the $\boldsymbol{\mu}$-points, and $\xi^{(c)}$ is a labeling array linking each point in configuration $\mathbf{X}^{(c)}$ to its underlying $\boldsymbol{\mu}$-point. As before, the elements within each labeling array are assumed to be distinct. In this context a match can be seen as a set of points $\mathcal{X}_j^I = \{\mathbf{x}_{j_1}^{(i_1)}, \mathbf{x}_{j_2}^{(i_2)}, \ldots, \mathbf{x}_{j_k}^{(i_k)}\}$ such that $\xi_{j_1}^{(i_1)} = \xi_{j_2}^{(i_2)} = \ldots = \xi_{j_k}^{(i_k)}$. Define the set $I = \{i_1, \ldots, i_k\}$ as a set of $k$ matching configurations on the $j$th match. A set of all matches is denoted by $J_I = \{j : j = 1, \ldots, L_I\}$. We also consider sets $I = \{c\}$ for $c = 1, \ldots, C$ as sets for a type of matching whereby all points in the configuration are essentially unmatched, that is, $|I| = 1$.

#### 9.8.2.1 The Likelihood

Let $\mathbf{R}^{(c)}\mathbf{x}_j^{(c)} + \boldsymbol{\tau}^{(c)} \sim \mathcal{N}_d\left(\boldsymbol{v}_{\xi_j^{(c)}}, \sigma^2\mathbf{I}_d\right)$ for $c = 1, 2 \ldots, C$ and $j = 1, 2, \ldots, n_c$. Key steps for integrating out $\boldsymbol{v}_{\mathbf{x}_j^{(c)}}$s are given in [470]. Furthermore, let $\kappa_I$ measure

# 9 Bayesian Hierarchical Alignment Methods

the tendency a priori for points to be matched in the $I$th set. The posterior model has the form

$$\mathbf{p}\left(\mathcal{A}, \mathcal{M} \mid \mathcal{X}\right) \propto \prod_{c=1}^{C} \left\{ p(\mathbf{R}^{(c)}) p(\boldsymbol{\tau}^{(c)}) \left|\mathbf{R}^{(c)}\right|^{n_c} \right\}$$

$$\times \prod_{I} \prod_{J_I} \kappa_I \frac{\exp\left\{-1/2\sigma^2 \gamma_A\left(\mathcal{X}_j^I\right)\right\}}{|I|^{d/2} (2\pi\sigma^2)^{d(|I|-1)/2}},$$

where $\mathcal{A} = \left\{\mathbf{R}^{(1)}, \mathbf{R}^{(2)}, \dots \mathbf{R}^{(C)}\right\}$ and $\mathcal{M} = \left\{\mathbf{M}^{(1)}, \mathbf{M}^{(2)}, \dots \mathbf{M}^{(C)}\right\}$ is the matching array and

$$\gamma_A\left(\mathcal{X}_j^I\right) = \sum_{k=1}^{|I|} \left|\left|\mathbf{R}^{(i_k)} \mathbf{x}_{j_k}^{(i_k)} - \mathbf{c}\right|\right|^2$$

with $\mathbf{c} = 1/|I| \sum_{k=1}^{|I|} \mathbf{R}^{(i_k)} \mathbf{x}_{j_k}^{(i_k)}$ and $||\cdot||$ denotes the Euclidean norm.

### 9.8.2.2 Prior Distributions

Prior distributions for the $\boldsymbol{\tau}^{(c)}, \mathbf{R}^{(c)}$, and $\sigma^2$ are identical to those in Eq. 9.1. For $c = 1, 2, \dots, C$,

$$\boldsymbol{\tau}^{(c)} \sim \mathcal{N}_d\left(\boldsymbol{\nu}^{(d)}, \sigma_c^2 \mathbf{I}_d\right); \qquad \sigma^{-2} \sim G(\alpha, \beta); \qquad \mathbf{R}^{(c)} \sim F_{d \times d}(\mathbf{F}_c).$$

$\mathbf{M}^{(c)}$ is assumed uniform a priori, and conditional on $L$ and $I$

$$p(\mathbf{M}^{(c)}|L, I) \propto (\kappa_I)^{L_I}.$$

That is, here the prior distribution for the matches $\mathcal{M}$ also assume that $\boldsymbol{\mu}$-points follow a Poisson process with constant rate $\lambda$ over a region of volume $v$. Each point in the process gives rise to a certain number of observations, or none at all. Let $q_I$ be the probability that a given hidden location generates an $I$-match. Then, consider the following parameterization:

$$q_I = \rho_I \cdot \prod_{c \in I} q_{\{c\}},$$

where $\rho_I$ is the tendency a priori for points to be matched in the $I$th set and $\rho_I = 1$ if $|I| = 1$. Define $L_I$ as the number of $I$-matches contained in $\mathcal{M}$, and assume the conditional distribution of $\mathcal{M}$ given the $L_I$ number is uniform. The prior distribution for the matches can be expressed as

$$p(\mathcal{M}) \propto \prod_I \left( \frac{\kappa_I}{\nu^{|I|-1}} \right)^{L_I},$$

where $\kappa_I = \rho_I / \lambda^{|I|-1}$. This is a generalization of the prior distribution for the matching matrix $\mathbf{M}$ [616]. Similar to the role of $\kappa$ in the pairwise model (Eq. 9.1), $\kappa_I$ measures the matching tendency a priori for $I$-matches. If $\kappa_I \gg \kappa_{I'}$ then one would see more $I$-matches than $I'$-matches. For example, for $C = 3$, if $\kappa_{\{1,2\}}$, $\kappa_{\{1,3\}}$ and $\kappa_{\{2,3\}}$ are much larger than $\kappa_{\{1,2,3\}}$, then one would mostly see pairwise matches and fewer triple matches.

### 9.8.2.3 Identifiability Issues

There is symmetry between configurations if $\mathbf{R}^{(c)}$ are uniformly distributed a priori. It is then true that the relative rotations $\left( \mathbf{R}^{(c_1)} \right)^T \cdot \mathbf{R}^{(c_2)}$ are uniform and independent for $c_2 \neq c_1$ and fixed $c_1$. So without loss of generality, the identifiability constraint is imposed by fixing $\mathbf{R}^{(1)}$ to the identity transformation. This corresponds to assuming that the first data configuration lies in the same space as the hidden point locations, similar to the pairwise model (Eq. 9.1).

## 9.9 Other Statistical Methods

ALIBI is connected to the mixture model formulation for matching configurations [368], combinatorial algorithms minimizing the RMSD and also to the method used to score similarity between matches; the Poisson Index [135]. We briefly outline the connections between the Bayesian hierarchical alignment method on one hand and combinatorial, mixture model and Poisson Index approaches on the other hand. Comments on some other methods are given further in Sect. 9.10.

### 9.9.1 Connection with Combinatorial Algorithms

A strong relationship between combinatorial algorithms minimizing the RMSD and the Bayesian approach with a Gaussian error model has been noted [477]. There is a connection between maximizing the joint posterior (Eq. 9.1) and minimizing the RMSD, which is defined by:

$$\text{RMSD}^2 = Q/L, \text{ where } Q = \sum_{j,k} \mathbf{M}_{jk} ||r_1(\mathbf{x}_j, k)||^2, \tag{9.19}$$

# 9 Bayesian Hierarchical Alignment Methods

and $L = \sum_{j,k} \mathbf{M}_{jk}$ denotes the number of matches. The RMSD is the focus of study in combinatorial algorithms for matching. In the Bayesian formulation, the log likelihood (with uniform priors) is proportional to

$$\text{constant} - 2 \left( \sum \mathbf{M}_{jk} \right) \ln \sigma + \left( \sum \mathbf{M}_{jk} \right) \ln \rho - Q/2\sigma^2 \sqrt{2}.$$

The maximum likelihood estimate of $\sigma$ for a given matching matrix $\mathbf{M}$ is the same as the RMSD which is the least squares estimate. The RMSD is a measure commonly used in bioinformatics, although the joint uncertainty in the RMSD and the matrix $\mathbf{M}$ is difficult to appreciate outside the Bayesian formulation.

Sometimes a purely algorithmic approach that minimizes the RMSD is justified, but it has the underlying assumption of normality. The situation is not very different when using likelihood estimators versus the RMSD. Thus, in the context of the statistical methodology, the choice to base the combinatorial objective function on the squared Euclidean distance is equivalent to deciding on using Gaussian errors in the probability model. In the Bayesian approach, the objective function is a probability distribution – the joint distribution of all unknowns given the data – while in the combinatorial approach, the objective function is some measure of mismatch. Typically, the RMSD is minimized under some constraints. These constraints are required as it is trivial to minimize the RMSD to zero by matching a single point [15, 99, 221, 222, 235, 239, 311, 379, 396, 635, 655, 686, 723, 749, 766]. Thus the two objective functions are mathematically very closely related.

## 9.9.2 EM Algorithm

The problem can be recast as mixture model estimation. This might suggest considering maximization of the posterior or likelihood using the EM algorithm [254]. In the EM formulation, the "missing data" are $\mathbf{M}_{jk}$s. The EM algorithm alternates between finding $\mathbb{E}[\mathbf{M}_{jk}|\mathbf{X}, \mathbf{Y}]$ at current values of $\mathbf{R}$, $\tau$ and $\sigma$ and maximizing the log-posterior or likelihood, with $\mathbf{M}_{jk}$ replaced by $\mathbb{E}[\mathbf{M}_{jk}|\mathbf{X}, \mathbf{Y}]$. Here $\mathbf{P} = \mathbb{E}[\mathbf{M}_{jk}|\mathbf{x}, \mathbf{y}]$ is a matrix with probabilities of matching. Thus, the assumption that a point can only be matched with at most one other point is dropped in this framework [254, 368]. However $\sum_k \mathbb{E}[\mathbf{M}_{jk}|\mathbf{x}, \mathbf{y}]$ is not bound between zero and one. We constrain $\sum_k \mathbb{E}[\mathbf{M}_{jk}|\mathbf{x}, \mathbf{y}] + \mathbf{P}_{j0} = 1$ (as the model is formulated conceptualizing that a larger configuration "represents" the "population" or $\boldsymbol{\mu}$) whereby $\mathbf{P}_{j0}$ denotes the probability of not matching the $j$th point. Thus a *coffin bin* is used to indicate unmatched points in the $\mathbf{X}$ configuration.

We have $\mathbf{P}_{j0} = 1 - \sum_k \mathbf{P}_{jk}$, which allows matching different sizes of configurations. Alternatively some probability model can be specified for the coffin bin; for example, $\mathbf{x}_j \sim \mathcal{N}(\boldsymbol{\nu}_0, \sigma_0^2 \mathbf{I}_d)$ where $\sigma_0^2$ is large. The approach that specifies a separate distribution for the coffin bin gives flexibility in modeling non-matches, but requires repeatedly re-normalizing the matching matrix $\mathbf{P}$. Thus, for maximizing the posterior the E-step involves calculating $\mathbf{P}_{jk} = \mathbf{W}_{jk}/(1 + \mathbf{W}_{jk})$, where

$$\mathbf{W}_{jk} = \kappa \times \phi(r_1(\mathbf{x}_j, k)/\sigma\sqrt{2}). \tag{9.20}$$

The M-step involves maximizing

$$\ln\{|\mathbf{R}|^n p(\mathbf{R})p(\tau)p(\sigma)\} + \sum_{j,k} \mathbf{P}_{jk} \ln \mathbf{W}_{jk} \tag{9.21}$$

over $\mathbf{R}$, $\tau$ and $\sigma$ for given $\mathbf{P}_{jk}$.

In case of the maximum likelihood estimation approach, Eq. 9.20 is replaced by $\mathbf{W}_{jk} = \phi(r_1(\mathbf{x}_j, k)/\sigma\sqrt{2})$ and Expression 9.21 by $\sum_{j,k} \mathbf{P}_{jk} \ln \mathbf{W}_{jk}$.

### 9.9.3 The Poisson Index

There is a need to assess the significance of biological molecule matches. One approach is to use an index quantifying the "goodness" of matches, such as the Tanimoto Index [767]. A statistical model-based index [135], the *Poisson index* (PI) has been proposed based on the Poisson model [254]. The main advantage of using PI over similar indices such as the Tanimoto Index is that the PI is based on an intuitive model, and has a natural probability distribution.

The PI model assumes a super-population $\mu$ from which all the points are drawn randomly within a given volume. Then, two subsets of the points are sampled from this population with a probability structure which imposes similarity and dissimilarity characteristics. This distribution can be used to obtain $p$-values to indicate whether the match could have occurred "by chance". In matching two configurations $\{\mathbf{x}_j, j = 1, 2, \ldots, m\}$ and $\{\mathbf{y}_k, k = 1, 2, \ldots, n\}$ with $n \geq m$, the $p$-value is the tail probability of finding a match as good as or better than the observed $L = L_{\text{obs}}$ matches given $m, n$ and $D = \rho/(\lambda v)$. Thus the $p$-value is

$$PI = \sum_{L=L_{\text{obs}}}^{m} p(L|m, n, D) \tag{9.22}$$

where $p(L|m, n, D)$ is given by Eq. 9.2. Details on obtaining maximum likelihood estimates for $D$ are given in [135].

## 9.10 Discussion

In this chapter, we have mainly considered the ALIBI model of Green and Mardia [254] and its extensions. However there are some other Bayesian alignment methods in the literature [764]. As suggested by [158], these Bayesian alignment methods mainly fall into two classes: those involving marginalization and those involving maximization of the transformation parameters. In the maximization approach, the joint distribution is maximized over the nuisance parameters – via a Procrustes alignment using the matched points – and inference for the matching parameter $\mathbf{M}$ is then performed conditionally. On the other hand, the marginalization approach integrates out nuisance parameters, as in our case.

Embedding Procrustes analysis in maximization methods effectively assumes that the Procrustes alignment is "correct" and uncertainty in geometric alignment is not explicitly modelled [765]. Chapter 8 gives an empirical Bayesian approach for superposition which allows variance heterogeneity and correlations among points. ALIBI uses marginalization and is fully Bayesian, as transformation parameters are included in the model and the transformation uncertainty is explicitly modelled. ALIBI integrates out hidden point locations, thereby avoiding significant computational penalties. Furthermore, ALIBI is symmetrical with respect to the configurations compared to the Procrustes analysis embedding approach which treats one configuration as the reference. That is, ALIBI treats matching configurations $\mathbf{X}$ and $\mathbf{Y}$ symmetrically, while Procrustes analysis regresses one on to the other.

Maximization methods [130, 160, 607, 632] filter out arbitrary transformations of the original data by statistical inference in the tangent space, for example using a linear approximation to the size-and-shape space. These methods use a MAP estimator for matching, after estimating transformation parameters which are considered nuisance parameters. Additionally, a non-Bayesian approach [370] uses the EM algorithm; first, the likelihood is maximized over the transformation parameters given expected values of the unknown matching labels in the M–Step; then, the expectation of the labels is taken with respect to the resulting maxima of the parameters in the E-Step. That is, the labels are treated as missing data; a coffin bin is introduced for unmatched landmarks.

There are also other approximate or hybrid methods designed to significantly improve the computational capability for matching more rapidly a large number of configurations in a database [477, 607, 632]. In [632], the Procrustes analysis approach is made faster by using a geometric hashing algorithm for matching. In [607], a profile likelihood for matching is used. The use of a profile likelihood is highly efficient for local sampling, and sufficient for matching configurations with very similar transformations. However, the approach may not perform as well when multiple alternative alignments with distinct transformations exist. Therefore, an additional step that proposes global moves to solve or improve on the matching is added. The additional step is used independent of the transformation problem. Additionally, this method can also model sequence gaps in matching. An algorithm based on ALIBI is also proposed for refining matches [477]. This algorithm searches for alternative matching solutions that are better in terms of the RMSD and the number of matched points than solutions found by deterministic methods[15, 99, 221, 222, 235, 239, 311, 379, 396, 635, 655, 686, 723, 749, 766]. That is, the algorithm finds better alternative solutions, or just returns the solution of the deterministic method if no such better solution is found. An excellent review of deterministic methods is found in [607].

Some important issues in Bayesian alignment have been highlighted in the literature; in particular that the uniform prior for the matching matrix $\mathbf{M}$ would be strongly biased towards matching a higher number of points [469, 765].

Some initial comparisons between the fully Bayesian and approximate approaches have been done [158]. For example, the Procrustes analysis embedding approach in [160] often gets stuck in local modes. However, for small variability, both the full Bayesian and the Procrustes analysis embedding approach lead to similar results [158]. Further review of various methods has been given in [256]. More work is needed to compare the performance of different approaches in practical situations.

**Acknowledgements** We are thankful to Professor Peter Green and Dr. Yann Ruffieux for many useful discussions on ALIBI based approaches. Our thanks also go to Chris Fallaize and Zhengzheng Zhang for their helpful comments.

# Part V
# Graphical Models for Structure Prediction

# Chapter 10
# Probabilistic Models of Local Biomolecular Structure and Their Applications

**Wouter Boomsma, Jes Frellsen, and Thomas Hamelryck**

## 10.1 Introduction

In 1951, before the first experimental determination of a complete protein structure, Corey and Pauling predicted that certain typical local structural motifs would arise from specific hydrogen bond patterns [121]. These motifs, referred to as $\alpha$-*helices* and $\beta$-*sheets*, were later confirmed experimentally, and are now known to exist in almost all proteins. The fact that proteins display such strong local structural preferences has an immediate consequence for protein structure simulation and prediction. The efficiency of simulations can be enhanced by focusing on candidate structures that exhibit realistic local structure, thus effectively reducing the conformational search space. Probabilistic models that capture local structure are also the natural building blocks for the development of more elaborate models of protein structure. This chapter will explore some of the ways in which this idea can be exploited in structural simulations and predictions. Although the chapter focuses primarily on proteins, the described modeling approaches are quite generally applicable. This is illustrated by a probabilistic model of a different biomolecule, namely RNA.

---

W. Boomsma
Dept. of Astronomy and Theoretical Physics, Lund University, Lund, Sweden
e-mail: wouter@thep.lu.se

Dept. of Biomedical Engineering, DTU Elektro, Technical University of Denmark,
Lyngby, Denmark

J. Frellsen · T. Hamelryck
Bioinformatics Centre, Dept. of Biology, University of Copenhagen, Copenhagen, Denmark
frellsen@binf.ku.dk; thamelry@binf.ku.dk

T. Hamelryck et al. (eds.), *Bayesian Methods in Structural Bioinformatics*,
Statistics for Biology and Health, DOI 10.1007/978-3-642-27225-7_10,
© Springer-Verlag Berlin Heidelberg 2012

## 10.2 Modeling Local Structure

The set of recurring structural motifs found by Corey and Pauling has been extended substantially in the last decades. In addition to $\alpha$-helices and $\beta$-sheets, the list of known motifs now includes $\beta$-turns, $\beta$-hairpins, $\beta$-bulges and N-caps and C-caps at the ends of helices. These motifs have been studied extensively, both experimentally and by knowledge-based approaches, revealing their amino acid preferences and structural properties [327].

Starting in the late 1980s, attempts were made to automate the process of detecting local structural motifs in proteins, using the increasing amount of publicly available structural data. Several groups introduced the concept of a structural *building block*, consisting of a short fragment spanning between four and eight residues. The blocks were found using various clustering methods on protein fragments derived from the database of solved structures [326, 610, 732].

At the time, the low number of available solved structures severely limited the accuracy of the local structure classification schemes. However, the *sequence* databases contained much more information and grew at a faster rate. This fact motivated an alternative approach to local structure classification. Instead of clustering known structures and analyzing amino acid preferences of these structures, the idea was to find patterns in sequence space first, and only then consider the corresponding structural motifs [277]. This approach was later extended to a clustering approach that simultaneously optimized both sequence and structure signals, leading to the I-sites fragment library [93].

### 10.2.1 Fragment Assembly

While most of the earlier studies focused primarily on *classification* of local structure, there were also examples of methods using local structural motifs directly in the prediction of protein structure. The methods were based on assembling fragments of local structure to form complete structures. This technique is called *fragment assembly*, and was already in 1986 proposed by Jones and Thirup in X-ray crystallography as a way to construct models from density maps [351].

In the field of protein structure simulation, the term fragment assembly has been used in two different contexts. One definition describes fragment assembly as a technique in which a forcefield is used to energy-minimize small fragments of fixed length, and subsequently merge these fragments into a complete structure. The alternative definition covers methods where structural fragments are *extracted* from the database of known structures, and subsequently merged in various ways during simulation to construct candidate structures. While the first method was explored in several early studies [657, 664, 735], the second approach is now most common. It is also this second approach that is most closely related to the modeling of local structure, and will therefore be the definition of fragment assembly used in this chapter.

In 1994, Bowie and Eisenberg presented the first complete fragment assembly method for *ab initio* protein structure prediction. For a given target sequence, all nine-residue sequence segments were extracted and compared with all fragments in their library, using a sequence-structure compatibility score. Longer fragments (15–25 residues) were identified using sequence alignment with the protein data bank (PDB) [45]. Using various fragment-insertion techniques, an initial population of candidate structures was generated. To minimize the energy, an evolutionary algorithm was designed to work on the angular degrees of freedom of the structures, applying small variations (mutations) on single candidate structures, and recombinations on pairs of candidate structures (cross-over). The study reported remarkably good results for small helical proteins [74].

In 1997, two other studies made important contributions to the development of fragment assembly based techniques. Jones presented a fragment assembly approach based on fragments with manually selected supersecondary structure, and demonstrated a correct prediction of a complete protein target from the second CASP experiment [350]. The second study was by Baker and coworkers, who presented the first version of their Rosetta protein structure prediction method, inspired by the Bowie and Eisenberg study. This method included a knowledge-based energy function and used multiple sequence alignments to select relevant fragments [658]. Reporting impressive results on a range of small proteins, the paper made a significant impact on the field, and remains heavily cited. The Rosetta method itself has consistently remained among the top-performing participants in CASP experiments – a testament to the great potential of incorporating local structural information into the conformational search strategy.

Although fragment assembly has proven to be an extremely efficient tool in the field of protein structure prediction, the technique has several shortcomings. An obvious concern is how to design a reasonable scheme for merging fragments. Either an overlap of fragments will occur, requiring an averaging scheme for the angular degrees of freedom in that region, or the fragments are placed side-by-side, which introduces an angle configuration at the boundary that is not generally present in the fragment library. This might seem like a minor issue, and it can be remedied by a smoothing scheme that adjusts unrealistic angles. However, in the context of a Markov chain Monte Carlo (MCMC) simulation, the boundary issue is symptomatic for a more fundamental problem. In principle, using fragment assembly to propose candidate structures corresponds to an implicit proposal distribution. However, as the boundary issue illustrates, this implicit distribution is not well-defined; when sampling, angles that should have zero probability according to the fragment library do occur in boundary regions. The introduction of a smoothing strategy improves the angles in the boundary region, but the non-reversibility of this additional step constitutes another problem for the detailed balance property (see Chap. 2). The construction of a reversible fragment library is possible, but comes at the expense of extending the library with artificial fragments [112].

## 10.2.2 Rotamers

Fragment libraries address the problem of conformational sampling of the main chain of proteins, at the expense of sacrificing the continuous nature of the conformational space. However, one also needs to address the problem of conformational sampling of the side chains of the amino acids, in order to obtain a protein conformation in atomic detail. If one assumes ideal bond lengths and bond angles, the conformation of an amino acid side chain is parameterized by a set of zero to four dihedral angles, called the $\chi$ angles (see Fig. 10.1). As is the case for the main chain, current methods typically solve the problem by discretizing the conformational space using conformations derived from experimental structures. These conformations are called *rotamers*, which stands for *rotational isomer* [108, 162, 163, 451, 588]. Collections of rotamers used in modelling the side chains are called *rotamer libraries*. These libraries are usually compiled from experimentally determined, high resolution protein structures by clustering the side chain conformations. Rotamer libraries with large number of rotamers are a good approximation of the conformational space, but are also computationally challenging [773]. In addition, side chain conformations are strongly dependent on the structure of the main chain at that position. Capturing this dependency using discretized models requires huge amounts of data, especially in the case of amino acids with many dihedral angles.

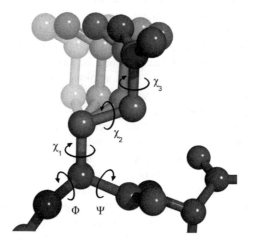

**Fig. 10.1** Dihedral angles in glutamate. Dihedral angles are the main degrees of freedom for the main chain and the side chain of an amino acid. The $\phi$ and $\psi$ angles describe the main chain conformation, while a number of $\chi$ angles describe the side chain conformation. The number of $\chi$ angles varies between zero for alanine and glycine and four for arginine and lysine. The figure shows a *ball-and-stick* representation of glutamate, which has three $\chi$ angles. The *light gray* conformations in the background illustrate a rotation around $\chi_1$. The figure was made using PyMOL (http://www.pymol.org) (Figure adapted from Harder et al. [290])

As in the case of the main chain, the discretization of the conformational space is also inherently problematic. The hydrophobic core of proteins typically consist of many tightly packed side chains. It is far from trivial to handle such dense systems using rotamer libraries. The problem is that very small differences in side chain conformations can lead to large differences in energy. By not considering possible conformations that fall in between rotamers, one might miss energetically favorable conformations. In practice, these inherent shortcomings of rotamer libraries are addressed using various heuristics [257].

## 10.3 Requirements for an Appropriate Probabilistic Model

The previous section illustrates the difficulty of incorporating a non-probabilistic model in MCMC simulations. We will now show that it is possible to formulate probabilistic models that provide rigorous alternatives to fragment and rotamer libraries. Before proceeding, however, we define a set of simple properties required in order for a model to function optimally in the context of an MCMC simulation:

**Probabilistic interpretation**
The model should be interpretable as a probability distribution, which is necessary to allow correct integration with MCMC simulations.

**Continuous space**
We require a detailed exploration of the conformational space in a continuous way, avoiding the usual discretizations.

**Amino acid dependent sampling**
It should be possible to model protein structure corresponding to a specific amino acid sequence, rather than simply the average structural behavior of a protein.

**Non-redundant representation**
The probabilistic model should allow sampling and constructing well-defined structures in an unambiguous way.

The last point is perhaps the most important: the only way to ensure consistent sampling from the model is through a non-redundant representation. We will illustrate this point in the next section.

## 10.4 Probabilistic Models for the Protein Main Chain

One important aspect of the design of a probabilistic model is the underlying representation of the modelled entity – in our case the local structure of a biomolecule. An early approach that is worth highlighting in this context is that of Hunter and States from 1992, on the development of a Bayesian classifier for local protein structure [326]. Their model was trained on a dataset of 53 proteins, from which all fragments of a specified length were extracted, and translated into vectors of atomic coordinates in a local reference frame, which were then modeled using normal distributions. A Bayesian scheme was used to automatically determine the number of clusters supported by the data. For each cluster obtained in this fashion, an analysis of secondary structure and amino acid preferences was conducted. In addition, the Markov transition probabilities between clusters were calculated, effectively turning the method into a hidden Markov model (HMM). The method was designed for classification purposes, not for sampling. However, for the sake of our discussion on optimal designs of probabilistic models, it is important to understand *why* such a model is not useful in the context of simulation. The problem lies in the choice of representation of local structure. First, since the basic unit of the model is a structure fragment, it is faced with similar problems as the fragment assembly technique. Second, and more importantly, the *representation* of the method prevents consistent sampling of the fragments themselves, since the sampling of atom positions according to Gaussian distributions will tend to violate the strong stereochemical constraints proteins have on bond lengths and bond angles.

In 1999, Camproux et al. presented an HMM of local protein structure [96]. Much along the lines of the work by Hunter and States, their model represented a protein chain as a number of overlapping fragments. Fragments of length four were used, where the internal structure of a fragment was captured through the distances between the $C\alpha$ atoms. The sequential dependencies along the chain were modeled by a Markov chain, where each state in the HMM corresponded to specific parameters for a four-dimensional Gaussian distribution. While this approach proved quite useful for classification purposes, from a sampling perspective, their representation suffers from the same problems as the Hunter and States model. The representation of neighboring HMM states are overlapping, failing to satisfy the non-redundancy requirement, and even within a single state, the representation is not consistent, since four-valued vectors can be sampled that do not correspond to any three-dimensional conformation of atoms. De Brevern, Etchebest and Hazout presented a similar model in 2000, using a representation of overlapping five-residue long fragments [139]. In this method, the eight internal dihedral angles of each fragment were used as degrees of freedom, thereby solving the problem of internal consistency within a fragment. However, also this model has a redundant structural representation, and was therefore not ideal for simulation purposes. Several variations of these types of models haven been proposed, with similar representations [35, 97, 175].

In 1996, Dowe et al. [156] proposed a model using only the $(\phi, \psi)$ angles of a single residue to represent states in structural space. The angles were modeled using the von Mises distribution, which correctly handled the inherent periodicity of the angular data. While their original method was primarily a clustering algorithm in $(\phi, \psi)$ space, the approach was extended to an HMM in a later study [168]. Since $(\phi, \psi)$ angles only contain structural information of one residue at a time, this representation is non-redundant. Although the model does not incorporate amino acid sequence information, thereby violating one of our requirements, this work was the first to demonstrate how angular distributions can be used to elegantly model local protein structure.

The methods described above are all fundamentally geometrical, in that they do not take amino acid or secondary structure information into account directly in the design of the model. For simulation and prediction purposes, it is important that models can be conditioned on any such available input information. The first model to rigorously solve this problem was the HMMSTR method by Bystroff, Thorsson and Baker, from 2004 [94]. Their model can be viewed as a probabilistic version of the I-sites fragment library described previously. Although it was trained on fragments, the model avoids the representation issues mentioned for some of the earlier methods. The sequential dependency in HMMSTR is handled exclusively by the HMM, and emitted symbols at different positions are independent given the hidden sequence. The model was formulated as a multi-track HMM that simultaneously models sequence, secondary structure, supersecondary structure and dihedral angle information. Based on this flexible model architecture, the authors identified a wide range of possible applications for the model, including gene finding, secondary structure prediction, protein design, sequence comparison and dihedral angle prediction, and presented impressive results for several of these applications. Unfortunately, for the purpose of protein simulation or prediction, HMMSTR had one significant drawback. The $(\phi, \psi)$ dihedral angle output was discretized into a total of eleven bins, representing a significant limitation on the structural resolution of the model.

Recently, several models have been proposed specifically with MCMC simulation in mind, designed to fulfill the requirements presented in the previous section. These models differ in the level of detail used in the representation of the protein main chain. Two common choices are the $C\alpha$-only representation and the full-atom representation, illustrated in Fig. 10.2. The $C\alpha$-only representation is more coarse-grained, involving fewer atoms, and can thus, in principle, lead to more efficient simulations. The full-atom representation more closely reflects the underlying physics, and it is often easier to formulate force fields in this representation. The corresponding models for these representations are the FB5HMM model [275] and the TORUSDBN model [68], respectively. The structure of the two models is similar. They are both dynamic Bayesian networks (DBNs), consisting of a Markov chain of hidden states emitting amino acid, secondary structure, and angle-pairs representing the local structure. The main difference between the models is the angular output given by the representation.

### 10.4.1 The $C\alpha$ Representation and the FB5 Distribution

In the $C\alpha$-only scenario, each residue is associated with one pseudo-bond angle $\theta \in [0°, 180°)$ and a dihedral angle $\tau \in [-180°, 180°)$, assuming fixed bond lengths (Fig. 10.2a). Each pair of these angles corresponds to a position on the

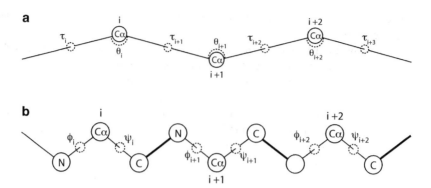

**Fig. 10.2** Two different coarse-grained representations of the protein main chain. In the full-atom model, all atoms in the main chain are included, while the $C\alpha$ representation consists only of $C\alpha$ atoms and pseudo-bonds connecting them. The nature of the degrees of freedom in these representations is different, giving rise to two different angular distributions. (**a**) $C\alpha$-only representation and (**b**) heavy-atom-only main chain representation

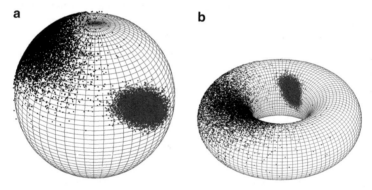

**Fig. 10.3** Examples of the angular distributions used to model the angular preferences of proteins. (**a**) samples from two FB5 distributions and (**b**) samples from two bivariate von Mises distributions

sphere.[1] To capture the periodicity of the angular degrees of freedom, a protein in the $C\alpha$ representation is therefore naturally modeled as a sequence of points on the sphere, or, equivalently, a sequence of unit vectors. As we saw in Chap. 7, the FB5 distribution is the equivalent of a bivariate Gaussian on the sphere, and is therefore a natural choice when modeling data of this type. Figure 10.3a shows samples drawn from two FB5 distributions, with different means and concentration parameters. Using a mixture – a convex combination – of these distributions, it is possible to model the more complex distribution resulting from the $(\theta, \tau)$ preferences of proteins.

---

[1]This follows directly from the definition of the spherical coordinate system. Note that the sphere is the two-dimensional surface of the three-dimensional ball.

## 10.4.2 The Full-Atom Main Chain Representation and the Bivariate Von Mises Distribution

The full-atom main chain representation is usually also associated with two degrees of freedom: the $\phi$ and $\psi$ angles (Fig. 10.2b). Both of these are dihedral angles, ranging from $-180°$ to $180°$. All bond lengths and bond angles are in this representation typically assumed to be fixed. The dihedral angle $\omega$ around the bond connecting the $C$ and $N$ atoms usually has the value $180°$ (*trans* state), with an occasional value of $0°$ (*cis* state).

The $(\phi, \psi)$ dihedral angles are well known in the biochemistry literature, where the scatter plot of $(\phi, \psi)$ is referred to as the Ramachandran plot, often used to detect errors in experimentally determined structures. The fact that both angular degrees of freedom span $360°$ has the effect that $(\phi, \psi)$ pairs should be considered as points on the torus. The bivariate Gaussian equivalent on this manifold is the bivariate von Mises distribution. Chapter 6 describes different variants of this distribution, and demonstrates the efficient parameter estimation and sampling techniques that have recently been developed for it. In the TORUSDBN model, the cosine variant was used. Figure 10.3b shows samples from two distributions of this type.

## 10.4.3 Model Structure

The FB5HMM and TORUSDBN models are single-chain DBNs[2] [68, 275]. The sequential signal in the chain is captured by a Markov chain of hidden nodes, each node in the chain representing a residue at a specific position in a protein chain (Fig. 10.4). The hidden node can adopt a fixed number of states. Each of these states corresponds to a distinct emission probability distribution over angles ($\mathbf{x}$), amino acids ($\mathbf{a}$) and secondary structure ($\mathbf{s}$). The amino acid and secondary structure nodes are simple discrete distributions, while the angular distribution is either the FB5 distribution or the bivariate von Mises (cosine variant). In addition, the TORUSDBN model has a discrete *trans/cis* node ($\mathbf{c}$), determining whether the $\omega$ angle is $180°$ or $0°$, respectively.

For ease of reference, we denote the set of observable nodes at position $i \in \{1, \ldots, n\}$ by $o_i$, where

$$o_i = \begin{cases} \{a_i, x_i, s_i\} & \text{FB5HMM} \\ \{a_i, x_i, s_i, c_i\} & \text{TORUSDBN} \end{cases} \tag{10.1}$$

---

[2]The models can equally well be considered as multi-track HMMs, but the graphical formalism for DBNs is more convenient for the TORUSDBN and FB5HMM models, since they use fully connected transition matrices (see Chap. 1).

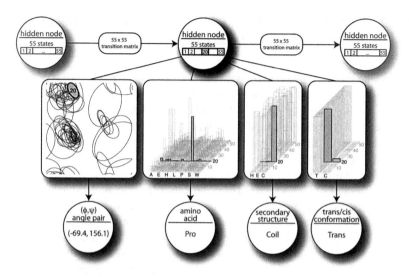

**Fig. 10.4** The TORUSDBN model. The *circular nodes* represent stochastic variables, while the *rectangular boxes* along the *arrows* illustrates the nature of the conditional probability distribution between them. The lack of an arrow between two nodes denotes that they are conditionally independent. Each hidden node has 55 states and the transition probability are encoded in a 55 × 55 matrix. A hidden node emits angle pairs, amino acid information, secondary structure labels (H:helix, E:strand, C:coil) and *cis/trans* information (C:cis, T:trans). The arbitrary hidden node value 20 is highlighted to demonstrate how the hidden node value controls which mixture component is chosen (Figure adapted from Boomsma et al. [68])

Given the structure of the model, the joint probability for a sequence of observables $\mathbf{o} = (o_1, \ldots, o_n)$ can be written as a sum over all possible node sequences $\mathbf{h} = (h_1, \ldots, h_n)$

$$p(\mathbf{o}) = \sum_{\mathbf{h}} p(\mathbf{h}) p(\mathbf{o}|\mathbf{h})$$

$$= \sum_{\mathbf{h}} p(h_1) \left( \prod_{o \in O_1} p(o|h_1) \right) \prod_{i>1} p(h_i|h_{i-1}) \left( \prod_{o' \in O_i} p(o'|h_i) \right)$$

where $n$ denotes the length of the protein. As the number of possible hidden node sequence grows exponentially with the protein length, calculating this sum directly is generally intractable. Instead, this is normally done using the forward algorithm [62].

The emission nodes (**a**, **x**, **s**, and **c**) can each be used either for input or output. When input values are available, the corresponding emission nodes are fixed to these specific values, and the node is referred to as an observed node. The emission nodes for which no input data is available will be sampled during simulation. For example, in the context of structure prediction simulations, the amino acid sequence and perhaps a predicted secondary structure sequence will be available as input, while the angle node will be repeatedly resampled to create candidate structures.

# 10 Probabilistic Models of Local Biomolecular Structure and Their Applications 243

## 10.4.4 Sampling

Sampling from these types of models involves two steps: First, the hidden node sequence is sampled conditioned on the *observed* emission nodes, after which, in the second step, values for the *unobserved* emission nodes are sampled given the sampled hidden node sequence.

In order for a model to be useful for simulation, it must be possible to resample a subsequence of the chain. This can be efficiently done using the *forward-backtrack algorithm* [102] – not to be confused with the more well known *forward-backward algorithm*. More precisely, we wish to resample a subsequence of hidden nodes from index $s$ to index $t$ while taking into account the transition at the boundaries. Given observed values of $o_s, \ldots o_t$, and hidden values of $h_{s-1}$ and $h_{t+1}$ at the boundaries, the probability distribution for the last hidden value can be written as

$$p(h_t|o_s, \ldots, o_t, h_{s-1}, h_{t+1}) = \frac{p(h_t, o_s \ldots, o_t, h_{s-1}, h_{t+1})}{p(o_s \ldots, o_t, h_{s-1}, h_{t+1})}$$
$$\propto p(o_s, \ldots, o_t, h_{s-1}, h_t) p(h_{t+1}|h_t) . \qquad (10.2)$$

The first factor can be efficiently calculated using the forward algorithm[3] [594]. The second is simply given by the transition matrix of the model. Equation 10.2 thus represents a discrete distribution over $h_t$, from which a value can be sampled directly (after normalizing). The key insight is that the situation for $h_{t-1}$ is equivalent, this time conditioned on $h_t$ at the boundary. For $h_{t-1}$, the calculation will involve the factor $p(o_s, \ldots, o_{t-1}, h_{s-1}, h_{t-1})$, which is available from the same forward matrix as before. The entire sampling procedure can thus be reduced to a single forward pass from $s$ to $t$, followed by a backtrack phase from index $t$ to $s$, sampling values based on Eq. 10.2.

Once a new hidden node (sub)sequence has been sampled, values for the unobserved nodes are sampled directly from the emission distributions corresponding to the hidden state at each position. For instance, the angles-pairs at a given position are sampled according to the bivariate von Mises or FB5 distribution component associated to the current hidden node value at that position.

## 10.4.5 Estimation

When training models of the type described above, a method is needed that can deal with the hidden (latent) variables in the model. Without these hidden variables, the maximum likelihood estimate of the parameters of the model would simply

---

[3]This requires that the probability of $h_{s-1}$ is included, by taking it into consideration when filling in the first column of the forward matrix (position s).

be the observed frequencies in the dataset. However, the data contains no direct information on the distribution of hidden values. The expectation maximization (EM) algorithm is a common maximum likelihood solution to this problem [144, 219]. It is an iterative procedure where in each iteration, parameters are updated based on estimated frequencies given the parameters of the previous iteration. The algorithm is guaranteed to produce estimates of non-decreasing likelihood, but occasionally gets trapped in a local optimum, failing to find the global likelihood maximum. Several variants of the EM algorithm exist. In cases where large amounts of data are available, a stochastic version of EM algorithm can be an appealing alternative known to avoid convergence to local optima [223, 544]. It is an iterative procedure consisting of two steps: (1) draw samples of the hidden node sequence given the data and the parameter estimates obtained in the previous iteration, (2) update the parameters in the model as if the model was fully observed, using the current transition and emission frequencies. Just like standard EM, the two steps are repeated until the algorithm convergences. EM algorithms with a stochastic E-step come in two flavors [62, 223]. In Monte Carlo EM (MC-EM), a large number of samples is generated in the E-step, while in Stochastic EM (S-EM) only one sample is generated for each hidden node [103, 223, 544]. Accordingly, the E-step is considerably faster for S-EM than for MC-EM. Furthermore, S-EM is especially suited for large datasets, while for small datasets MC-EM is a better choice.

The sampling in step (1) could be implemented using the forward-backtrack algorithm described previously. However, it turns out that a less ambitious sampling strategy may be sufficient. For the training of the models presented above, a single iteration of Gibbs sampling was used to fill-in the hidden node values.[4] Since this approach avoids the full dynamic programming calculation, it speeds up the individual cycles of the stochastic EM algorithm and was found to converge consistently.

The model design of TORUSDBN and FB5HMM gives rise to a single hyperparameter that is not automatically updated by the described procedure: the *hidden node size*, which is the number of states that the hidden node can adopt. This parameter was optimized by estimating a range of models with varying hidden node size, and evaluating the likelihood. The best model was selected based on the Bayesian information criterion (BIC) [638]. The optimal values were 55 states for TORUSDBN, and 75 for FB5HMM.

### 10.4.6 Model Evaluation

Both the TORUSDBN and FB5HMM models have been demonstrated to successfully reproduce the local structural properties of proteins [68, 275]. In particular,

---

[4]In random order, all hidden nodes were resampled based upon their current left and right neighboring h values and the observed emission values at that residue.

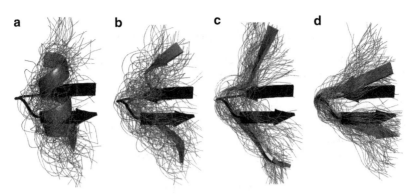

**Fig. 10.5** Samples from TORUSDBN with various types of input on a hairpin fragment: (**a**) no input, (**b**) sequence input, (**c**) predicted secondary structure, (**d**) sequence and predicted secondary structure. The *black structure* represents native, the cloud of *light grey* structures are 100 samples from the model. In *dark grey*, we highlight the centroid of this cloud as a representative structure (the structure with lowest RMSD to all other samples in the cloud) (Figure adapted from Boomsma et al. [68])

the marginal distribution of angle pairs – marginalized over all positions and amino acid inputs – is indistinguishable from the distribution found in naturally occurring proteins. This marginalized distribution conceptually corresponds to the Ramachandran plot of TORUSDBN samples. Figure 10.5 illustrates the types of local structure that can be expected as varying degrees of input are given to the model. In the example, when sampling without input data, the model produces random structures. When sequence information is available, the model narrows the angle distributions to roughly the correct secondary structure, but fails to identify the correct boundaries. The distributions are further sharpened when providing a predicted secondary structure signal. Finally, when both sequence and predicted secondary structure are provided, the hairpin region becomes confined to allow only samples compatible with the characteristic sharp turn.

### 10.4.7 Detailed Balance

We now investigate how models such as the ones described above can be used in MCMC simulations. In particular, we will consider the move corresponding to the resampling of a subsequence of the chain using the forward-backtrack method, discussed above in Sect. 10.4.4.

Let $\mathbf{h}_{s:t}$ and $\mathbf{x}_{s:t}$ denote a resampled subsequence of hidden nodes and angle-pairs, respectively, while $\mathbf{h}_{\overline{s:t}}$ and $\mathbf{x}_{\overline{s:t}}$ refers to the hidden nodes and angle-pairs that are not resampled. The probability of proposing a move from angle-pair and hidden node sequence $(\mathbf{x}, \mathbf{h})$ to a new angle-pair and hidden node sequence $(\mathbf{x}', \mathbf{h}')$ is given by

$$p(\mathbf{x}, \mathbf{h} \to \mathbf{x}', \mathbf{h}') = p(\mathbf{x}'_{s:t}|\mathbf{h}'_{s:t})p(\mathbf{h}'_{s:t}|\mathbf{h}_{\overline{s:t}}) .$$

To reduce unnecessary notational complexity, we will go through the argument without conditioning on amino acid information or secondary structure. However, the derivations for these cases are identical.

Now, assume that we want to sample from a given target distribution, $\pi(\mathbf{x})$, over the angle-pair sequence. As our sample space also includes the hidden node sequence, we need to choose a joint target distribution $\pi(\mathbf{x}, \mathbf{h})$ that has marginal distribution $\pi(\mathbf{x})$. A natural way of selecting this distribution is

$$\pi(\mathbf{x}, \mathbf{h}) = \pi(\mathbf{x})p(\mathbf{h}|\mathbf{x}) ,$$

where $p(\mathbf{h}|\mathbf{x})$ is the conditional distribution of the hidden node sequence given the angle sequence according to the local model. In order to obtain a detailed balance in the Markov chain, the Metropolis-Hastings algorithm prescribes that the acceptance probability is given by

$$
\begin{aligned}
\alpha(\mathbf{x}, \mathbf{h} \to \mathbf{x}', \mathbf{h}') &= \min\left(1, \frac{\pi(\mathbf{x}', \mathbf{h}')p(\mathbf{x}', \mathbf{h}' \to \mathbf{x}, \mathbf{h})}{\pi(\mathbf{x}, \mathbf{h})p(\mathbf{x}, \mathbf{h} \to \mathbf{x}', \mathbf{h}')}\right) \\
&= \min\left(1, \frac{\pi(\mathbf{x}')p(\mathbf{h}'|\mathbf{x}')p(\mathbf{x}_{s:t}|\mathbf{h}_{s:t})p(\mathbf{h}_{s:t}|\mathbf{h}_{\overline{s:t}})}{\pi(\mathbf{x})p(\mathbf{h}|\mathbf{x})p(\mathbf{x}'_{s:t}|\mathbf{h}'_{s:t})p(\mathbf{h}'_{s:t}|\mathbf{h}_{\overline{s:t}})}\right) .
\end{aligned}
\tag{10.3}
$$

Using the conditional independence structure encoded in the DBN, the conditional distribution $p(\mathbf{h}|\mathbf{x})$ can be rewritten in terms of the resampled subsequences,

$$p(\mathbf{h}|\mathbf{x}) = \frac{p(\mathbf{x}_{s:t}|\mathbf{h}_{s:t})p(\mathbf{x}_{\overline{s:t}}|\mathbf{h}_{\overline{s:t}})p(\mathbf{h}_{s:t}|\mathbf{h}_{\overline{s:t}})p(\mathbf{h}_{\overline{s:t}})}{p(\mathbf{x})} .$$

By inserting this expression into Eq. 10.3, the acceptance probability simply reduces to

$$\alpha(\mathbf{x}, \mathbf{h} \to \mathbf{x}', \mathbf{h}') = \min\left(1, \frac{\pi(\mathbf{x}')p(\mathbf{x})}{\pi(\mathbf{x})p(\mathbf{x}')}\right) .
\tag{10.4}$$

This means that we have to sum over all hidden sequences in the network, and accordingly calculate the full forward array, to find the acceptance probability for each proposed move. Since the transition matrix in an HMM only has limited memory, the acceptance probability can be well approximated by only calculating the probability in a window around the changed sequence. For a window size of $w \geq 0$, the acceptance probability becomes

$$\alpha(\mathbf{x}, \mathbf{h} \to \mathbf{x}', \mathbf{h}') \approx \min\left(1, \frac{\pi(\mathbf{x}')p(\mathbf{x}_{(s-w):(t+w)}|\mathbf{h}_{s-w-1}, \mathbf{h}_{t+w+1})}{\pi(\mathbf{x})p(\mathbf{x}'_{(s-w):(t+w)}|\mathbf{h}'_{s-w-1}, \mathbf{h}'_{t+w+1})}\right) .
\tag{10.5}$$

10 Probabilistic Models of Local Biomolecular Structure and Their Applications      247

In protein structure prediction studies using fragment assembly, this acceptance criterion is often omitted. This corresponds to the assumption that the target distribution, $\pi(\mathbf{x})$, factorizes into a global, $p_G(\mathbf{x})$, and a local, $p_L(\mathbf{x})$, contribution

$$\pi(\mathbf{x}) = p_G(\mathbf{x})\,p_L(\mathbf{x}) \tag{10.6}$$

where the local part $p_L(\mathbf{x})$ could be given by our model of local structure. It follows from Eq. 10.5 that the acceptance probability in this case simply becomes

$$\alpha(\mathbf{x}, \mathbf{h} \to \mathbf{x}', \mathbf{h}') = \min\left(1, \frac{p_G(\mathbf{x}')\,p_L(\mathbf{x}')\,p_L(\mathbf{x})}{p_G(\mathbf{x})\,p_L(\mathbf{x})\,p_L(\mathbf{x}')}\right) = \min\left(1, \frac{p_G(\mathbf{x}')}{p_G(\mathbf{x})}\right).$$

The acceptance probability is simply given by the ratio of the global energy in the new and the old configuration. If it is true that the target distribution can be written as a product of a local and a global term, we can thus remove the local contribution from the energy evaluation, and instead sample from it, thereby increasing the acceptance rate. Note that this is indeed the case for distributions constructed using the reference ratio method, described in Chaps. 4 and 3.

## 10.5 BARNACLE: A Probabilistic Model of RNA Conformational Space

The probabilistic models discussed so far only describe the main chain of proteins, omitting a parameterization of the major degrees of freedom in the protein side chain. In this section we will consider a probabilistic model of another biomolecule, RNA, which incorporates all major degrees of freedom in a single model. This model is called BARNACLE [197], which loosely stands for *Bayesian network model of RNA using circular distributions and maximum likelihood estimation*. In the next section, we will show that a more complete model can also be constructed for proteins by combining TORUSDBN with an additional model, BASILISK.

The BARNACLE model is conceptually related to the protein models presented previously. However, it is not a trivial extension of these, as the RNA main chain contains many more relevant degrees of freedom than the protein main chain. Furthermore, if a similar model design was to be applied to RNA, this would require a higher dimensional multivariate von Mises distribution (see Chap. 6), which was not available when the model was developed. Instead, a very different model design was used.

### 10.5.1 Description of the BARNACLE Model

An RNA molecule is comprised of a sugar-phosphate main chain with a nitrogen-containing base attached to each sugar ring. If we assume that all bond angles and bond lengths are fixed to idealized values, the geometry of an RNA molecule can be characterized by the remaining free dihedral angles. It can be shown that seven dihedral angles per residue are sufficient to describe RNA in atomic detail [197]. These are the six dihedral angles $\alpha$ to $\zeta$ that describes the course of the main chain and the $\chi$-angle, which describes the dihedral angle around the bond connecting the sugar ring and the base, as depicted in Fig. 10.6.

The BARNACLE model was expressed as a DBN that can capture the marginal distribution of each of the seven dihedral angle and the local dependencies between the angles (Fig. 10.7). The DBN has one slice per angle and each slice consists of three stochastic variables: $d_j$, $h_j$ and $x_j$. The angle identifier $d_j$ is a bookkeeping variable that specifies which of the seven angles are described in the given slice,

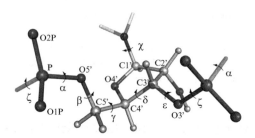

**Fig. 10.6** The dihedral angles in an RNA fragment. The fragment is shown in *ball-and-stick* representation and the dihedral angles are placed on the central bond of the four consecutive atoms that defined the dihedral angle. For clarity, the base is only shown partially. The six dihedral angles $\alpha$ to $\zeta$ describe the course of the main chain, while the $\chi$-angle is the rotation of the base relative to the sugar ring (Figure adapted from Frellsen et al. [197])

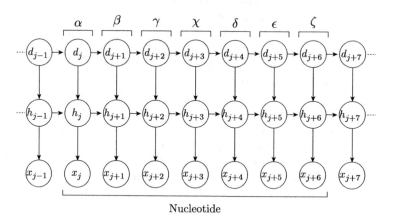

**Fig. 10.7** The BARNACLE dynamic Bayesian network. Nine consecutive slices of the network are shown, where the seven central slices describes the angles in one nucleotide. Each slice contains three variables: an angle identifier, $d_j$, a hidden variable, $h_j$, and an angular variable, $x_j$. The angle identifier, $d_j$, controls which type of angle ($\alpha$ to $\zeta$) is described by the given slice, while the value of the angle is represented by the variable $x_j$. The hidden node describes the dependencies between all the angles along the sequence (Figure adapted from Frellsen et al. [197])

10 Probabilistic Models of Local Biomolecular Structure and Their Applications    249

while the actual angle values are represented by the stochastic variable $x_j$, which takes values in the interval $[0, 2\pi)$. Finally, $h_j$ is a hidden variable that is used for modeling the dependencies between the angles and for constructing a mixture of the angular output. Both the angle identifier, $d_j$, and the hidden variable, $h_j$, are discrete variables, and the conditional distributions $p(d_j|d_{j-1})$ and $p(h_j|h_{j-1}, d_j)$ are described by conditional probability tables (see Chap. 1). The conditional distribution of the angular variable, $p(x_j|h_j)$, is specified by an independent univariate von Mises distribution for each value of $h_j$ (see Chap. 6). The joint probability distribution of $n$ angles in an RNA molecule encoded by this model is given by

$$p(x_1, \ldots, x_n|d_1, \ldots, d_n) = \sum_{\mathbf{h}} p(h_1|d_1)p(x_1|h_1) \prod_{j=2}^{n} p(h_j|h_{j-1}, d_j)p(x_j|h_j),$$

where the sum runs over all possible hidden node sequences $\mathbf{h} = (h_1, \ldots, h_n)$. In the model, it is assumed that the angle identifiers are always specified in the order $\alpha, \beta, \gamma, \chi, \delta, \epsilon, \zeta, \alpha, \beta$ and so forth.

A full sequence of angles, $\mathbf{a} = (x_1, \ldots, x_n)$, can be sampled from the model by hierarchical sampling (see Chap. 1). First a sequence of identifier values are specified by the user, that is

$$d_1 = \alpha, d_2 = \beta, d_3 = \gamma, d_4 = \chi, d_5 = \delta, d_6 = \epsilon, d_7 = \zeta, d_8 = \alpha, \ldots, d_n = \zeta.$$

Then a sequence of hidden states are sampled conditioned upon the values of the identifier variables, and subsequently a sequence of angles are sampled conditioned on the sampled hidden values. When the model is used in an MCMC simulation, new structures can be proposed by resampling a subsequence of angles. This can be done by the forward-backtrack algorithm, using an approach similar to what is described for the protein models in Sects. 10.4.4 and 10.4.7. Note, that although amino acid dependent sampling was one of the four requirements for an appropriate probabilistic model of protein structure (see Sect. 10.3), BARNACLE does not incorporate nucleotide information. In the case of RNA the influence of the sequence on the local structure is more subtle, and presumably better captured by adding a model of nonlocal interactions.

The parameters of the BARNACLE model were estimated using stochastic EM [223, 544] and the optimal number of hidden states was selected using the Akaike information criterion (AIC) [91], see Sect. 1.7.3. As training data the RNA05 dataset from the Richardson lab [531] was used, containing 9486 nucleotides from good quality X-ray structures. The optimal number of hidden states was twenty.

## 10.5.2 Evaluation of BARNACLE

Analysis of BARNACLE shows that it describes the essential properties of local RNA structure [197]. The model accurately captures both the marginal distribution of the individual seven dihedral angles and the joint distribution of dihedral angle pairs. In order to have a simple continuous baseline, BARNACLE was compared to a mixture model, where the seven angles are modeled as independent mixtures of von Mises distributions. An example of this is illustrated in Fig. 10.8, where the marginal distribution of the $\alpha$-angle in BARNACLE is compared to both the distribution observed in the data set and the distribution according to the mixture model. This figure shows that BARNACLE is on par with the mixture model for the individual angles. However, in contrast to the mixture model, BARNACLE also captures the length distribution of helical regions correctly (Fig. 10.9), and the model is consistent with a previously published rotameric description of the RNA main chain [531]. Finally, the model has been tested in MCMC simulations, using a simple geometric base pair potential based on secondary structure. A comparison with the FARNA method by Das and Baker [133] shows that this approach readily generates state-of-the-art quality decoys for short RNA molecules.

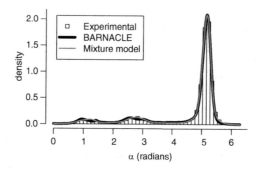

**Fig. 10.8** The marginal distribution of the $\alpha$-angle in RNA. The distribution of the training data is shown as a *histogram*, and the density function for BARNACLE is shown as a *black line*. For comparison, the density function according to a simple mixture model is shown as a *gray line* (Figure adapted from Frellsen et al. [197])

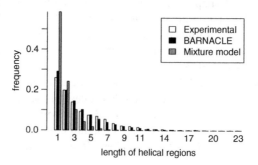

**Fig. 10.9** Histogram of the lengths of helical regions in RNA. The distribution in the training data and in data sampled from BARNACLE are shown in *white* and *black*, respectively. For comparison, the length distribution in data sampled from a mixture model is shown in *gray* (Figure adapted from Frellsen et al. [197])

## 10.6 BASILISK: A Probabilistic Model of Side Chain Conformations

TORUSDBN constitutes a probabilistic model of the protein main chain on a local length scale. As such, it makes it possible to sample protein-like main chain conformations that include the positions of the $C$, $C\alpha$, $C\beta$, $O$ and $N$ atoms, together with their associated hydrogen atoms. However, TORUSDBN does not provide information on the conformations of the side chains attached to the amino acids. This information can be provided by a second probabilistic model, called BASILISK [290]. BASILISK loosely stands for ***Ba****yesian network model of **si**de chain conformations estimated by maximum **li**kelihood*. As there is a strong correlation between the side chain conformation of an amino acid and its local main chain conformation, BASILISK also includes information on the main chain. For sampling purposes, the two models are used in concert in the following way. First, one samples a main chain conformation from TORUSDBN. Then, the side chain conformations are sampled from BASILISK, conditioned on the previously sampled main chain conformation. This is done one amino acid at a time, for each sequence position. TORUSDBN combined with BASILISK constitutes the first rigorous, generative probabilistic model of protein structure in atomic detail, and in continuous space.

### 10.6.1 Model Structure of BASILISK

From a statistical point of view, modeling the side chain is similar to modeling the main chain; the challenge consists in modeling a sequence of dihedral angles. Each amino acid type has a fixed, small number of such dihedral angles, ranging from zero for glycine and alanine, over one for serine and valine, to four for arginine and lysine. These angles are labelled $\chi_1$ to $\chi_4$. In total, 18 types amino acid types need to be modeled. As we also want to capture the dependency on the main chain, the model also includes nodes that represent the $(\phi, \psi)$ angle pair for that amino acid. In principle, 18 different models could be formulated and trained; one for each amino acid type excluding glycine and alanine. However, we decided to include all 18 amino acid types in a single probabilistic model. This approach is known as *multitask* or *transfer learning* in the field of machine learning and has several advantages [101, 556]. As the same set of distributions is used to model all amino acids, it leads to a lower amount of free parameters. Moreover, it makes "knowledge transfer" possible during training between amino acids with similar conformational properties. Finally, for rotamer libraries, one needs to determine the optimal number of rotamers for each amino acid type separately; in our approach, only the size of the hidden node needs to be determined.

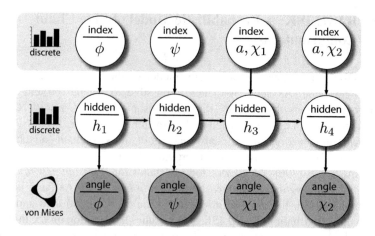

**Fig. 10.10** The BASILISK dynamic Bayesian network. The network shown represents an amino acid with two $\chi$ angles, such as for example leucine. In the case of leucine, the DBN consists of four slices: two slices for the $\phi$, $\psi$ angles, followed by two slices for the $\chi$ angles. Sampling a set of $\chi$ angles is done as follows. First, the values of the input nodes (*top row*) are set to bookkeeping indices that determine both the amino acid type and dihedral angle position. For example, in the case of leucine, the first two indices denote $\phi$ and $\psi$ angles, followed by two indices that denote the $\chi_1$ and $\chi_2$ angles of leucine. In the next step, the hidden node values (*middle row*, discrete nodes) are sampled conditioned upon the observed nodes. These observed nodes always include the index nodes (*top row*, discrete nodes), and optionally also the $\phi$ and $\psi$ nodes (first two nodes in the *bottom row*) if the sampling is conditioned on the main chain. Finally, a set of $\chi$ angles is drawn from the von Mises nodes (*bottom row*), whose parameters are specified by the sampled values of the hidden nodes (Figure adapted from Harder et al. [290])

BASILISK was formulated as a DBN whose structure (see Fig. 10.10) is very similar to the structure of BARNACLE, the model of local RNA structure that is described in the previous section [197]. Each slice in the DBN represents a single dihedral angle using the von Mises distribution [474] as child node. The first two slices represent the main chain angles $\phi$ and $\psi$; they are included to make it possible to sample the side chain angles conditional on the main chain conformation. The third and subsequent slices represent the dihedral angles of the side chain itself. As in the case of BARNACLE, bookkeeping input nodes specify which angle is modeled at that position in the DBN. The input nodes for the first two slices indicate that these slices concern the $\phi$ and $\psi$ main chain angles, without specifying any amino acid type. Specifying the type is superfluous, as the model is exclusively used for sampling conditional upon the values of the $\phi$ and $\psi$ values. It is TORUSDBN that provides a generative model for these angles. For the subsequent slices, the bookkeeping input nodes not only specify the $\chi$ angles that are represented, but also the amino acid type. Let us consider the example of leucine, which has two dihedral angles $\chi_1$ and $\chi_2$. The first two input nodes simply indicate that the $\phi$ and $\psi$ main chain angles are modeled. The subsequent two input nodes indicate that the two associated slices represent the $\chi_1$ and $\chi_2$ values of a leucine residue.

10 Probabilistic Models of Local Biomolecular Structure and Their Applications 253

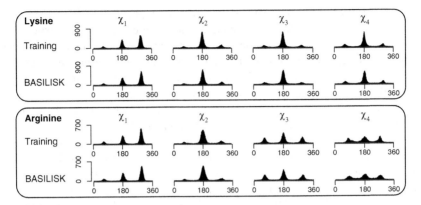

**Fig. 10.11** Univariate histograms of the $\chi$ angles for lysine (*top*) and arginine (*bottom*). The histograms marked "Training" represent the data set used for training BASILISK. The histograms marked "BASILISK" represent samples generated using BASILISK. For each amino acid, all histograms are plotted on the same scale. The X-axis indicates the value of the dihedral angles in degrees, while the Y-axis denotes the number number of occurrences (Figure adapted from Harder et al. [290])

BASILISK was trained using data from 1,703 high quality crystal structures. The optimal number of hidden node states was determined using the Akaike information criterion (AIC) [91], see Sect. 1.7.3, and resulted in a model with 30 states.

## 10.6.2 Evaluation of BASILISK

Extensive analysis indicates that BASILISK is an excellent generative probabilistic model of the conformational space of amino acid side chains. This includes reproduction of the univariate and pairwise histograms of the dihedral angles, concurrence with a standard rotamer library, capturing the influence of the protein main chain and filling in the side chain conformations on a given main chain [290]. As an illustration, Fig. 10.11 compares the marginal angular distributions of the training set with those of BASILISK samples for arginine and lysine. We chose arginine and lysine because they are the only amino acids with four $\chi$ angles, and are the most challenging amino acids to model. As the figure indicates, the histograms of the individual $\chi$ angles in samples generated from BASILISK are in excellent agreement with the training data.

## 10.7 Conclusions

Fragment libraries were a major step forward in the exploration of the conformational space of proteins. However, their *ad hoc* nature leaves much to be desired for many applications. The powerful combination of graphical models and directional statistics makes it possible to formulate efficient and elegant probabilistic models of biomolecular structure that do not suffer from the disadvantages associated with discretizing the conformational space. Already, other groups have explored variants of the models described in this chapter, including conditional random fields [757, 794, 795][5] and Dirichlet process mixtures [434, 435].

The probabilistic models described in this chapter only model biomolecular structure on a *local* length scale. Global, *nonlocal* features such as hydrogen bond networks or hydrophobic cores are not captured by these models. However, these models are still of great significance. First, they can be used for efficient conformational sampling, in combination with any arbitrary energy function. Second, the models can be used as prior distributions describing protein structure in general. Recently, TORUSDBN and BASILISK provided the prior distribution in inferential structure determination from NMR data, with impressive results [552]. Finally, these models can be used as starting points for the development of rigorous probabilistic models of biomolecular structure that capture both local and nonlocal features. The mathematical machinery to do this was recently developed by us and is discussed in Chap. 4. In conclusion, we expect that probabilistic models will find widespread application in structural bioinformatics, including prediction, inference from experimental data, simulation and design.

**Acknowledgements** The authors acknowledge funding by the Danish Research Council for Technology and Production Sciences (FTP, project: Protein structure ensembles from mathematical models, 274-09-0184) and the Danish Council for Independent Research (FNU, project: A Bayesian approach to protein structure determination, 272-08-0315).

---

[5]It should be noted that these articles erroneously state that the models described in this chapter only capture the dependencies between neighboring residues. Obviously, the presence of a Markov chain of hidden nodes actually enforces dependencies along the whole sequence. In practice, such Markov chains do have a finite memory.

# Chapter 11
# Prediction of Low Energy Protein Side Chain Configurations Using Markov Random Fields

**Chen Yanover and Menachem Fromer**

## 11.1 Introduction

The task of predicting energetically favorable amino acid side chain configurations, given the three-dimensional structure of a protein main chain, is a fundamental subproblem in computational structural biology. Specifically, it is a key component in many protocols for de novo protein folding, homology modeling, and protein-protein docking. In addition, fixed main chain protein design can be cast as a generalized version of the side chain placement problem. For all these purposes, the objective of pursuing low energy side chain configurations is equivalent to finding the most probable assignments of a corresponding Markov random field. Consequently, this problem can be addressed using message-passing probabilistic inference algorithms, such as max-product belief propagation (BP) and its variants. In this chapter, we review the inference techniques that have been successfully applied to side chain placement, discuss their current limitations, and outline promising directions for future improvements.

## 11.2 Biological Background

Full comprehension of biological protein function, such as enzymatic activity, interaction affinity, and specificity, can be achieved only through the consideration

---

C. Yanover (✉)
Fred Hutchinson Cancer Research Center, Seattle, WA, USA
e-mail: cyanover@fhcrc.org

M. Fromer
The Hebrew University of Jerusalem, Jerusalem, Israel
e-mail: fromer@cs.huji.ac.il

T. Hamelryck et al. (eds.), *Bayesian Methods in Structural Bioinformatics*,
Statistics for Biology and Health, DOI 10.1007/978-3-642-27225-7_11,
© Springer-Verlag Berlin Heidelberg 2012

of atomic protein structure. Despite major advances in experimental techniques, solving protein structures is still a labor intensive, and often unattainable, task. Consequently, of the millions of protein sequences documented [709], a three-dimensional structure has of yet been experimentally determined for only a tiny fraction (less than 1%). In an attempt to fill this gap, scientists have sought methods to computationally model and predict the structures of proteins.

Here, we focus on the *side chain placement problem* and demonstrate how it can play a role in providing more accurate representations of natural phenomena such as protein folding. Moreover, it equips researchers with tools to selectively modify natural protein structures and functions and even design proteins with novel functionalities.

## 11.2.1 Side Chain Placement

A protein is composed of a linear sequence of amino acids. Each amino acid contains an amino group (-NH$_2$), a carboxyl group (-COOH), and a side chain group, all of which are connected to a central carbon atom. During its formation, a protein is successively polymerized, so that the carboxyl group at the end of the polymer is joined to an amino group of the newly incorporated amino acid. The resulting series of carbon, nitrogen, and oxygen atoms is termed the *protein main chain* (Fig. 11.1, top, left). The side chain of each amino acid residue is what confers its unique character (e.g., hydrophobic, hydrophilic, charged, polar, etc.), among the 20 naturally ubiquitous amino acids. Amino acid side chains are free to assume continuous spatial conformations (Fig. 11.1, top, right) in the context of the complete protein structure, i.e., when interacting with one another. Note that such interactions often involve amino acids that are distant from one another in the primary amino acid sequence, yet spatially proximal in the three-dimensional protein structure.

As noted above, despite high-throughput genomic and metagenomic sequencing projects that have already documented millions of protein sequences [709], structural data for these proteins is still relatively scarce. And, this is despite seemingly exponential growth in the world-wide protein structure repository [44]. Thus, in order to more fully understand and predict the functions of these proteins, there is a pressing need to formulate and develop methods that predict the structures of proteins [400, 401].

One approach to tackling this general *protein structure prediction*, or *protein folding*, problem is to predict the relative compatibility of a particular protein sequence with previously determined protein structures in a procedure known as *fold recognition*. It is reasonable to expect that newly discovered sequences will adopt previously known folds, since this was often empirically found to be the case, even when protein sequences with novel structures were actively pursued [436]. Historically, fold recognition was performed using a sequence *threading* procedure [75], where a protein sequence is computationally placed onto a fixed structural

11 Prediction of Low Energy Protein Side Chain Configurations 257

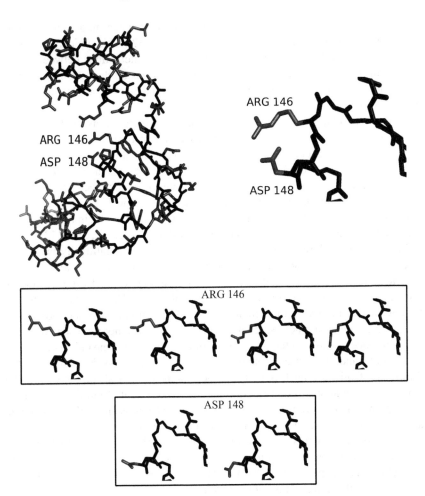

**Fig. 11.1** *Top, left*: Protein main chain (*colored black*) and side chains (*colored gray*), depicted for a sequential protein fragment of 85 amino acids (PDB id: 1AAY). *Top, right*: Zoom-in for a single pair of positions, ARG 146 and ASP 148. *Bottom*: A few main chain-dependent rotamers for each of these amino acid residues (Taken from the Dunbrack rotamer library [163])

protein scaffold. This scaffold is represented using a one-dimensional biochemical profile to which the fit of the one-dimensional protein sequence is evaluated. This low-resolution procedure can be used to quickly scan a database of tens of thousands of structures for the candidate structures most compatible with a query amino acid sequence.

The process of protein *side chain placement* is a refinement of the simple threading procedure described above. More accurate predictions are made by accounting for the degrees of freedom and interactions of the amino acid side chains. This higher resolution structural modeling is characterized by a contact map that explicitly considers the pairs of positions that are deemed sufficiently proximal in space to be capable of interacting. Based on this contact map, pairwise amino acid-amino acid interaction energies are computed [352]. This typically entails the use of atomic-level energy functions to quantify the interactions between pairs of side chains of the query amino acids placed at positions on the scaffold, where the interaction energies are based on atom distances and relative angles in the three-dimensional space [246].

More generally, the need for accurate and rapid predictions of the side chain configurations of a protein, given the fixed spatial conformation of its main chain, arises in diverse structural applications, ranging from de novo protein structure prediction to protein docking. In these scenarios, side chain configurations are typically optimized for many related low resolution structures, yielding a large number of local minima of a coarse-grained energy function. Subsequently, searches initialized from each of these alternative minima are used to obtain low energy side chain placements that correspond to high resolution structural models [134].

To make the high resolution placement procedure computationally feasible, the side chain of an amino acid at a particular position is typically permitted to adopt only a small number of dihedral angle conformations, termed *rotamers* (see the lower panels of Fig. 11.1 for some example rotamers). These rotamers are taken from a discrete library of energetically favorable empirical side chain observations [163]. Thus, the amino acid side chain conformations are not represented continuously in space but are limited for the sake of computational convenience.

Notwithstanding the use of rotamer-discretized amino acid side chains, a seeming drawback of this approach is that a naive search through all possible side chain placements would still require iteration over an exponential number of combinations, even if there were just two side chain conformations allowed for each amino acid at each position; see Sect. 11.3.1 for details on the computational difficulty of this class of problems. Due to this inherent complexity, numerous and diverse algorithms have been devised to perform this task accurately [147, 237, 247, 374, 448, 578] and quickly [95, 98, 148, 349, 381, 389, 405, 800]. In Sect. 11.4, some major archetypes of these algorithms are detailed.

## 11.2.2 Protein Design

The goal of constructing a protein sequence capable of performing a target biological function is termed *protein design*. Since the search for a protein sequence with a specific function without further constraints presents a highly formidable challenge, the design problem is typically confined to the pursuit of a sequence that adopts a target three-dimensional structure [406], expecting that it will possess the function(s) determined by this structure. This objective is the reverse of that of protein folding (in which the target structure of a particular sequence is sought), so that this task is also known as the *inverse protein folding* problem.

11 Prediction of Low Energy Protein Side Chain Configurations

Interestingly, despite the seemingly reverse nature of the two problems, the protein design problem can be formalized as a generalization of the side chain placement problem (Sect. 11.2.1), which itself is a subtask of protein folding. Whereas in side chain placement each protein position is modeled with the side chain conformers of a *single* amino acid, in protein design the side chain conformers of *multiple* amino acids are modeled at each design position. Thus, historically, all algorithms previously devised for side chain placement were simply transferred to the domain of protein design. Although this makes sense on a theoretical level, there were, however, a number of practical problems with this approach.

Foremost of these is the fact that there are many more rotamers in the per-position rotamer set. In fact, when considering all design positions together, there are exponentially more possible rotamer combinations than there are for side chain placement. Thus, clearly, algorithms that managed to successfully approach low energy solutions for the protein side chain placement domain would not necessarily successfully scale up to the larger protein design cases. Therefore, numerous adaptations were made to the side chain placement algorithms in order to enable their feasibility. Some of these were guaranteed to provide optimal solutions (albeit with no guarantee of terminating within a short, or non-exponential, time) [247, 374, 425, 448, 578], while others were more heuristic in nature but were found to fare reasonably in practice [298, 389, 405].

## 11.2.3 Atomic Energy Functions for Side Chain Placement and Design

The exponential growth in the world-wide protein structure data bank (PDB) [44] – to currently include tens of thousands of structures – has deepened scientific understanding of the biophysical forces that dominate in vivo protein folding. Specifically, this large number of experimentally determined protein structures has enabled the deduction of "knowledge-based" energy terms, which have facilitated more accurate structural predictions [670]. A weighted sum of such statistically derived and physically realistic energy terms is what constitutes most modern energy functions.

In particular, the energy function used by the successful SCWRL program [98] combines a (linear approximation of a) van der Waals repulsive term, with main chain-dependent rotamer probabilities estimated from a large set of representative protein structures. Other programs, such as Rosetta [405] and ORBIT [132], include additional energy terms to consider attractive steric, hydrogen bond, and electrostatic interactions, as well as interactions with the surrounding solvent (i.e., the water molecules).

## 11.3 Formalization of Side Chain Placement and Protein Design

The input to the side chain placement problem consists of a three-dimensional protein main chain structure, the $N$ sequence positions to be modeled, the permitted rotamer set of the respective amino acid(s) at each position, and an atomic energy function. Formally, we denote by $Rots_i$ the set of all possible rotamers at position $i$; let $\mathbf{r} = (r_1, \ldots, r_N)$ denote an assignment of rotamers for all $N$ positions. For a given pairwise atomic energy function, the energy of assignment $\mathbf{r}$, $E(\mathbf{r})$, is the sum of the interaction energies between:

1. Rotamer $r_i \in Rots_i$ and the fixed structural template (main chain and stationary residues), denoted $E_i(r_i)$, for all positions $1 \le i \le N$
2. Rotamers $r_i \in Rots_i$ and $r_j \in Rots_j$, denoted $E_{ij}(r_i, r_j)$, for neighboring (interacting) positions $i, j$

$$E(\mathbf{r}) = \sum_i E_i(r_i) + \sum_{i,j} E_{ij}(r_i, r_j) \tag{11.1}$$

Computationally, side chain placement attempts to find, for the input structure and sequence, the rotamer configuration of minimal energy [776, 781]:

$$\mathbf{r}^* = \arg\min_{\mathbf{r}} E(\mathbf{r}) \tag{11.2}$$

For the generalization to protein design, denote by $T(k)$ the amino acid type of rotamer $k$ and let $T(\mathbf{r}) \equiv T(r_1, \ldots, r_N) = (T(r_1), \ldots, T(r_N))$, i.e., the amino acid sequence corresponding to rotamer assignment $\mathbf{r}$. Let $\mathbf{S} = (S_1, \ldots, S_N)$ denote an assignment of amino acids for all positions, i.e., an amino acid sequence. Computational protein design attempts to find the sequence $\mathbf{S}^*$ of minimal energy [207, 780]. Specifically:

$$\mathbf{S}^* = \arg\min_{\mathbf{S}} E(\mathbf{S}) \tag{11.3}$$

where:

$$E(\mathbf{S}) = \min_{\mathbf{r}:T(\mathbf{r})=\mathbf{S}} E(\mathbf{r}) \tag{11.4}$$

is the minimal rotamer assignment energy for sequence $\mathbf{S}$ (as in Eq. 11.2). This double minimization problem, over the sequence space (Eq. 11.3) and over the per-sequence rotamer space (Eq. 11.4), is combined as:

$$\mathbf{S}^* = T(\arg\min_{\mathbf{r}} E(\mathbf{r})) \tag{11.5}$$

Thus, both for fixed main chain side chain placement and for fixed main chain protein design, the goal is to find the minimal energy rotamer configuration from

11 Prediction of Low Energy Protein Side Chain Configurations

among an exponential number of possible combinations. This state of lowest energy rotamers is also sometimes referred to as the *global minimum energy configuration* (GMEC) of the modeled protein [612].

## 11.3.1 Computational Hardness

Intuitively, since the number of possible rotamer choices increases exponentially with the sequence length, side chain placement (and design) seem to be difficult computational tasks. This is so, since a naive enumeration and evaluation of each rotameric assignment is clearly not tractable. Moreover, it has been formally shown that side chain placement belongs to the computational class of NP-hard combinatorial optimization problems for which no guaranteed polynomial time algorithms are known [193, 577]. Therefore, computational structural researchers cannot expect to find optimal solutions in a short amount of time for arbitrarily defined energy functions and proteins. Nevertheless, recent experience has demonstrated that very low energy side chain configurations can be obtained for real-world problems; see Sect. 11.8 for an example. Furthermore, there are certain state-of-the-art algorithms that have been devised to find optimal low energy solutions (Sect. 11.6.3), although they can theoretically require extremely long run times or exorbitant amounts of computer memory, in the worst case scenario. In summary, most algorithms will typically be required to compromise either on their provable exactness or on reasonable run times, but there nevertheless do exist algorithms that have fared extremely well in practice.

## 11.4 Recent Work: Algorithms for Predicting Low Energy Side Chain Configurations

Notwithstanding the inherent challenges described above, computational side chain placement and protein design procedures have been applied to numerous and wide-ranging areas of biology. Below we outline some of the more well-known of these algorithms that have resulted from many years of extensive research.

## 11.4.1 Heuristic Algorithms

This class of algorithms is the most natural approach to solving difficult discrete combinatorial optimization problems and were thus the first to be applied for the prediction of low energy rotamer configurations. These algorithms are usually intuitive, simple to implement, and provide quick and reasonable results. However, they typically do not come with a guarantee regarding the quality of the predictions, theoretical insight as to how the results are affected by the multiple parameters of the respective algorithms, or knowledge of the run times required to provide "good" results in practice.

**Monte Carlo simulated annealing (MCSA):** Monte Carlo (MC) methods are a category of computational search techniques that iteratively employ randomization in performing calculations. In the case of searching for the minimal energy side chain configuration [405, 800], each protein position is initially assigned a randomly chosen rotamer. Then, at each iteration, a position is randomly chosen and a "mutation" to a random rotamer side chain is made at that position. This mutation is then accepted or rejected in a probabilistic manner, dependent on the "temperature" of the simulation and the change in energy due to the mutation, where smaller increases (or a decrease) in side chain energy or a higher temperature will increase the chance that the mutation is accepted and kept for the next iteration. The concept behind simulated annealing is to start the Monte Carlo iterations at a high temperature and slowly "cool" the system to a stable equilibrium (low energy solution), analogously to the physical process of annealing in the field of statistical mechanics. In either Monte Carlo variant, the purpose of randomly allowing for rotamer configurations with higher energy than currently observed is to escape local minima in the rugged configuration energy landscape and attempt to find the globally optimal low energy side chain configuration.

**Genetic algorithms (GAs):** In a fashion similar to MCSA, genetic algorithms also utilize randomized sampling in an iterative manner in an attempt to find the lowest energy rotamer configuration [349]. However, as opposed to MCSA, which deals with only a single configuration at a time, GAs maintain a population of hetero-geneous, "mutant" rotamer configurations. And, at each iteration, the "individuals" of the population are subjected to randomized mutations and recombinations, and then to subsequent selection. This process rudimentarily mimics the evolutionary process, where the survival of the "fittest" rotamer configuration is expected under certain conditions on the population dynamics and mutation rates and with sufficient time. As with MC methods, there is typically also an annealing component to the algorithm (selection strength), so as to overcome energetic barriers in the energy landscape.

**Self-consistent mean field theory (SCMF):** Self-consistent mean field theory provides a method to calculate probabilities for each possible rotamer at each protein position [95, 381, 389, 800]. Since high probabilities correspond to low energies, the optimal rotamer configuration can then be predicted by choosing the highest probability rotamer at each position. The per-position rotamer probabilities are calculated in an iterative fashion. At each iteration, the multiple interactions with a particular position are averaged into a "mean field" that is being exerted on this position, based on the current rotamer probabilities at all other positions and their

11 Prediction of Low Energy Protein Side Chain Configurations

interactions with this position. This mean field is then utilized in order to recalculate the rotamer probabilities at the position of interest. At convergence, rotamer probabilities are obtained and the predicted lowest energy rotamer configuration is output.

**Fast and accurate side-chain topology and energy refinement (FASTER):** The FASTER method [148] is a heuristic iterative optimization method loosely based on the framework of the DEE criteria (see below). An initial rotamer configuration is chosen (possibly randomly). Then, in the simplest stage of FASTER, the current rotamer assignment is held fixed at all but one protein position, and this remaining position is exhaustively optimized, i.e., by considering all possible rotamers at that position in a first-order "quenching" procedure. This process continues iteratively for all positions until no further changes can be made, or until a predetermined number of steps has been reached. Next, such a quenching procedure is performed on all positions after fixing a particular position to a given rotamer choice; this is repeated for all rotamer choices at the particular position, after which a different position is chosen to be fixed. Despite the largely heuristic nature of FASTER, it was demonstrated to find the optimal rotamer configuration in many real-world problems [3].

**Side-chains with a rotamer library (SCWRL):** The SCWRL program [98] is one of the longest-standing algorithms for rapid protein side chain placement. In fact, due to its relatively high accuracy and speed, it has been incorporated into various world-wide web servers, including 3D-PSSM [364], which uses SCWRL to generate models of proteins from structure-derived profile alignments. SCWRL 3.0 is considerably faster than the original algorithm, since it uses simple but elegant graph theory to make its predictions. In detail, it models each protein position as a node in an undirected graph. Subsequently, it breaks up clusters of interacting side chains into the biconnected components of this graph, where a biconnected component is one that cannot be broken apart by the removal of a single vertex. The minimal energy rotamer choices are then recursively calculated for each component, and the approximate lowest energy configuration is recovered through this "divide and conquer" framework. Recently, SCWRL 4.0 was released, which achieves higher accuracy at comparable speed by using a new main chain-dependent rotamer library, a more elaborate energy function, and a junction tree related search algorithm (see Sect. 11.6.2) [398].

## 11.4.2 Computationally Exact Algorithms

As opposed to the algorithms detailed above, the algorithms presented below are accompanied with formal guarantees of optimality (i.e., they will find the lowest energy configuration). Nonetheless, they come with no assurance that the run time will be reasonably short, since, after all, the pertinent computational problem is highly difficult (see Sect. 11.3.1).

**Dead-end elimination (DEE):** Dead-end elimination is a rotamer pruning technique that guarantees to only remove those rotamers that do not participate in the lowest energy rotamer configuration [147, 237]. Thus, if a sufficient number of rotamers are removed, then DEE is guaranteed to find the optimal configuration. Conceptually, the basic DEE criterion compares two rotamers at a given protein position and determines if the first one is "better" than the second, in that any low energy configuration using the second can always be made to have lower energy by using the first rotamer in its place. In such a case, the second rotamer can be eliminated without detriment, thus simplifying the computational problem. This procedure is iteratively repeated, reducing the conformational space at each step. In order to account for more difficult problems, more sophisticated DEE criteria have been developed, which include the elimination of pairs of rotamers [247], conditioning on neighboring rotamers [448, 578], and unification of pairs of positions [247]. Although DEE is often successful in many practical side chain placement and design problems, it can require extremely long run times, and it is sometimes not at all feasible.

**Integer linear programming (LP/ILP):** Linear programming (LP) is a general mathematical method for the global optimization of a linear function, subject to linear constraints. Integer linear programming (ILP) problems require that the solution consists of integer values, but makes the computational problem more difficult. The ILP approach to side chain placement and protein design [374] is based on the observation that Eq. 11.1 decomposes the energy for a complete rotamer configuration into a sum of energies for positions and pairs of positions. Thus, by defining a variable for the rotamer choice at each position and for each pair of positions, the energy minimization task can be written as a linear function of these variables. Also, linear equalities are added to ensure that a unique rotamer is consistently chosen at each position. In the ILP formulation, the rotamer variables are constrained to be binary integers (0 or 1), where a value of 1 indicates that the rotamer was chosen. In the LP relaxation, the variables are allowed to assume continuous values between 0 and 1. This LP relaxation is solved with computational efficiency, possibly yielding a solution in which all variables are integers. Otherwise, a more exhaustive ILP solver is used, although with no guarantee of a non-exponential run time. In practice, an LP solver provided quick solutions for many side chain placement problems that were tested, whereas long runs of the ILP solver were often required for the protein design cases assessed. See Sect. 11.6.3 for related algorithms that use a Markov random field-based approach.

For a review of the MCSA, GA, SCMF, and DEE algorithms (and computational benchmarking results), see [742]; see also [653] for a general review of search algorithms for protein design. Table 11.1 provides a short, non-comprehensive list of some of the more notable success stories of computational protein design, with emphasis that the methods outlined above be represented.

11 Prediction of Low Energy Protein Side Chain Configurations

**Table 11.1** A short summary of notable cases of experimentally validated protein design research

| Method | Design, redesign of |
| --- | --- |
| MCSA | • A protein with a novel $\alpha/\beta$ fold [406]<br>• Endonuclease DNA binding and cleavage specificity [16]<br>• Biologically active retro-aldol enzyme [347] and kemp elimination enzyme [615] |
| GA | • Core residues of the phage 434 cro protein [146]<br>• Ubiquitin [422]<br>• Native-like three-helix bundle [393] |
| SCMF | • 88 residues of a monomeric helical dinuclear metalloprotein, in both the apo and the holo forms [95]<br>• Four-helix bundle protein that selectively binds a non-biological cofactor [117]<br>• Ultrafast folding Trp-cage mutant [89] |
| FASTER | • A monoclonal antibody directed against the von Willebrand factor that inhibits its interaction with fibrillar collagen [685]<br>• Engrailed homeodomain [643] |
| DEE | • Zinc finger domain [132]<br>• Enzyme-like protein catalysts [66]<br>• Novel sensor protein [449] |
| ILP | • 16-member library of *E. Coli/B. Subtilis* dihydrofolate reductase hybrids [623] |

### 11.4.3 Generalizations for the Prediction of Low Energy Ensembles

Adapting the heuristic algorithms (Sect. 11.4.1) to predict an ensemble of $M$ low energy side chain configurations tends to be straightforward: rather than retaining the single, lowest energy configuration, the algorithms keep track of the $M$ best configurations observed throughout their "sampling" of the rotamer space.

A DEE criterion (Sect. 11.4.2) can be generalized to predict a set of lowest energy configurations by only eliminating rotamers that do not participate in a rotamer assignment with energy within a threshold $\epsilon > 0$ of the minimal rotamer assignment; note that when using $\epsilon = 0$, this generalized DEE reduces to the original DEE. Eliminating in this way is guaranteed to preserve all configurations whose energy is less than the minimal energy plus $\epsilon$. Unfortunately, generalized DEE reduces the search space far less than the original DEE criterion. Furthermore, it is usually unknown what value of $\epsilon$ corresponds to the given number of desired configurations $M$. The authors of [425] applied generalized DEE to reduce the state space and then applied $\mathbf{A}^*$ to find (in increasing order) all side chain configurations whose energy was less than the minimal energy plus $\epsilon$. In addition, the search space-partitioning approach of X-DEE has been recently proposed to predict gap-free lists of low energy configurations [380].

The LP/ILP framework for predicting a single low energy configuration can also be extended to provide multiple solutions. This is performed by incorporating additional linear inequality constraints into the previously defined system of linear inequalities, such that all previously predicted configurations will be disallowed by these inequalities [374, 622]. One of the drawbacks with this approach is that, in practice, very often these inequalities require that the computationally expensive ILP solver be used. Nonetheless, there does exist a state-of-the-art method (STRIPES: Spanning Tree Inequalities and Partitioning for Enumerating Solutions), which adds spanning tree-based inequalities, that has been empirically shown to perform well without requiring an ILP solver [204]. Finally, for cases of protein design, direct application of these inequalities for the prediction of low energy rotamer configurations will not necessarily preclude the iteration over multiple low energy configurations corresponding to the same amino acid sequence; note that this same issue arises with the DEE-based methods as well. On the other hand, see Sect. 11.6.4.1 for an example of how Markov random field-based approaches have been readily generalized for this task.

## 11.5 Side Chain Placement and Protein Design as a Probabilistic Inference Problem

Since each of the side chain placement and protein design tasks pose a discrete optimization problem (Sect. 11.3) and the energy function consists of a sum of pairwise interactions, the problem can be transformed into a *probabilistic graphical model* (*Markov random field*, MRF) with pairwise potentials [776]; see Fig. 11.2. A random variable is defined for each position, whose values represent the rotameric choices (including amino acid) at that position. Clearly, an assignment for all variables is equivalent to rotamer choices for all positions (where, for the case of protein design, the rotamer choices uniquely define an amino acid sequence).

The pre-calculated rotamer energies taken as input to the problem (see Sect. 11.2.3) are utilized to define *probabilistic potential functions*, or *probabilistic factors*, in the following manner. The singleton energies specify probabilistic factors describing the self-interactions of the positions in their permitted rotamer states:

$$\psi_i(r_i) = e^{\frac{-E_i(r_i)}{T}} \tag{11.6}$$

And, the pairwise energies define probabilistic factors describing the direct interactions between pairs of rotamers in neighboring positions:

$$\psi_{ij}(r_i, r_j) = e^{\frac{-E_{ij}(r_i, r_j)}{T}} \tag{11.7}$$

where $T$ is the system temperature.

In the next step, a graph is constructed, wherein each node corresponds to a variable and the node's values correspond to the variable's values (Fig. 11.2). For a pair of variables $i, j$, the matrix of pairwise probabilistic factors ($\psi_{ij}$) corresponds to an edge between them in the graph. Since the energy functions typically used for

11 Prediction of Low Energy Protein Side Chain Configurations 267

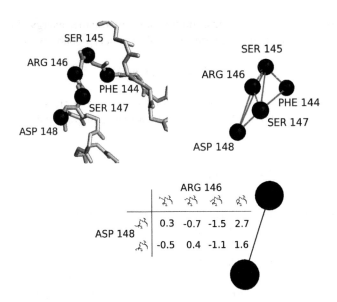

**Fig. 11.2** Side chain placement formulated as a structure-based graphical model. *Top*: Short segment of the protein main chain for PDB 1AAY (see Fig. 11.1) and its corresponding graph, where an edge between two positions describes the pairwise energies between them (Eq. 11.7). *Bottom*: An example pairwise potential matrix for the energetic interactions between the rotamer side chains of ARG 146 and those of ASP 148

side chain prediction essentially ignore interactions occurring between atoms more distant than a certain threshold, this implies that the corresponding graph will often have a large number of missing edges (positions too distant to directly interact). Thus, the locality of spatial interactions in the protein structure induces path separation in the graph and conditional independence in the probability distribution of the variables. Formally, it can be shown that, if node subset $Y$ separates nodes $X$ and $Z$, then $X$ and $Z$ are independent for any fixed values of the variables in $Y$: $p(X, Z|Y) = p(X|Y) \cdot p(Z|Y)$, or $X \perp\!\!\!\perp Z|Y$.

Mathematically, the probability distribution for the rotamer assignment $\mathbf{r} = (r_1, \ldots, r_N)$ decomposes into a product of the singleton and pair probabilistic factors:

$$p(\mathbf{r}) = \frac{1}{Z} \prod_i \psi_i(r_i) \prod_{i,j} \psi_{ij}(r_i, r_j) \qquad (11.8)$$

$$= \frac{1}{Z} e^{\frac{-E(\mathbf{r})}{T}} \qquad (11.9)$$

where $Z$ is the probability normalization factor (*partition function*), and Eq. 11.9 derives from substitution of Eqs. 11.6 and 11.7 into Eq. 11.8 and the energy decomposition of Eq. 11.1. Thus, minimization of rotamer energy (Eqs. 11.2 and 11.5) is equivalent to the maximization of rotamer probability (a probabilistic *inference* task):

$$\mathbf{r}^* = \arg\max_{\mathbf{r}} p(\mathbf{r}) \tag{11.10}$$

Now, we proceed to discuss how Eq. 11.10 is solved in practice using general-purpose algorithms devised for probabilistic inference in graphical models.

## 11.6 Message-Passing Algorithms

Message-passing algorithms are a general class of algorithms that attempt to solve global problems by performing local calculations and updates; see, e.g., [199, 300]. Intuitively, a useful message-passing algorithm will model a large, complex system without requiring any one element (e.g., protein position) to directly consider the interaction states of all other elements. Nevertheless, it is desired that this system have the "emergent" property that the input received by a particular element from its neighboring elements consistently encodes global properties of the system (e.g., the lowest energy conformation of the protein). Thus, by passing local messages between neighboring elements, researchers aspire to perform difficult calculations with relative computational ease.

### 11.6.1 The Belief Propagation Algorithm

Max-product *belief propagation* (BP) [568] is a message-passing algorithm that efficiently utilizes the inherent locality in the graphical model representation (see Sect. 11.5). Messages are passed between neighboring (directly interacting) variables, where the message vector describes one variable's "belief" about its neighbor – that is, the relative likelihood of each allowed state for the neighbor. A message vector to be passed from one position to its neighbor is calculated using their pairwise interaction probabilistic factor and the current input of other messages regarding the likelihood of the rotamer states for the position (Fig. 11.3). Formally, at a given iteration, the message passed from variable $i$ to variable $j$ regarding $j$'s rotameric state ($r_j$) will be:

$$m_{i \to j}(r_j) = \max_{r_i} \left( e^{\frac{-E_i(r_i) - E_{ij}(r_i, r_j)}{T}} \prod_{k \in N(i) \setminus j} m_{k \to i}(r_i) \right) \tag{11.11}$$

where $N(i)$ is the set of nodes neighboring variable $i$. Note that $m_{i \to j}$ is, in essence, a message vector of relative probabilities for all possible rotamers $r_j$, as determined at a specific iteration of the algorithm.

# 11 Prediction of Low Energy Protein Side Chain Configurations

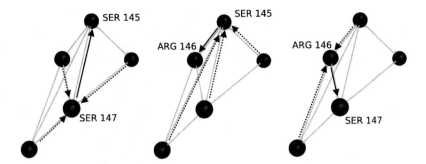

**Fig. 11.3** Message-passing in the max-product belief propagation (BP) algorithm. From *left* to *right*, messages are passed (*solid, bold arrows*) along the following cycle: from SER 147 to SER 145, from SER 145 to ARG 146, and from ARG 146 to SER 147. Note that, for each message passed, all incoming messages (*dotted arrows*) are integrated into the calculation, with the exception of that from the target node; for example, SER 147 ignores the message sent by SER 145 when calculating the message to send it

In detail, messages are typically initialized uniformly. Next, messages are calculated using Eq. 11.11. Now, for each position for which the input message vectors have changed, its output messages are recalculated (Eq. 11.11) and "passed" on to its neighbors. This procedure continues in an iterative manner until numeric convergence of all messages, or a predetermined number of messages has been passed. Finally, max-marginal belief vectors (max-beliefs) are calculated as the product of all incoming message vectors and the singleton probabilistic factor:

$$b_i(r_i) = e^{\frac{-E_i(r_i)}{T}} \prod_{k \in N(i)} m_{k \to i}(r_i) \qquad (11.12)$$

where $b_i(r_i)$ is the max-belief of a particular rotamer $r_i \in Rots_i$ at position $i$.

Thus, max-product *loopy* belief propagation (BP) can be utilized to find the maximal probability rotamer (and sequence) assignments (Eq. 11.10). Specifically, the max-beliefs obtained by BP (Eq. 11.12) are employed as approximations of the exact max-marginal probability values:

$$MM_i(r_i) = \max_{\mathbf{r}' \,:\, r'_i = r_i} p(\mathbf{r}') \qquad (11.13)$$

for which it can be shown [568] that assignment of:

$$r_i^* = \arg\max_{r_i \in Rots_i} MM_i(r_i) \qquad (11.14)$$

yields the most probable rotamer assignment $\mathbf{r}^*$ (as defined in Eq. 11.10).

The belief propagation algorithm was originally formulated for the case where the graphical model is a tree graph, i.e., no "loops" exist [568]. However, since typical side chain placement and protein design problems will have numerous cycles (Fig. 11.2), inexact max-beliefs are thus obtained, and the predicted rotamer configurations are not guaranteed to be optimal. Nonetheless, loopy BP has been shown to be empirically successful in converging to optimal solutions when run on non-tree graphs (e.g., [776]). Furthermore, loopy BP has conceptual advantages over related (statistical) inference techniques, since it does not assume independence between protein positions and yet largely prevents self-reinforcing feedback cycles that may lead to illogical or trivial fixed points. Equation 11.11 attempts to prevent the latter by exclusion of the content of what variable $j$ most recently sent to variable $i$. On the other hand, for example, self-consistent mean field algorithms are forced to make certain positional independence assumptions and thus may fail to perform as well [206, 742].

## 11.6.2 Generalized Belief Propagation and the Junction Tree Algorithm

Whereas the standard belief propagation algorithm for side chain placement involves the passing of messages between neighboring variable nodes (Fig. 11.3), in the *generalized belief propagation* (GBP) algorithm [786], messages are passed between sets of variables (known as *clusters* or *regions*). These regions in the graph are often chosen based on some problem-specific intuition, typically along with methods already developed in the field of physics (e.g., the Kikuchi cluster variation method). GBP is also an approximate inference algorithm, but it incorporates greater complexity in capturing dependencies between a large number of nodes in the original graph (Fig. 11.2). This is done with the intention of achieving higher accuracy and more frequent convergence [776, 786], while still maintaining computational feasibility. Additional research related to side chain placement using GBP, with regions of sequential triplets of protein positions, is detailed in [358].

The *junction tree* (JT) algorithm [420] can be considered to be a special case of GBP, where the regions are chosen in such a way so that message-passing between these nodes is guaranteed to converge to the optimal solution. It is thus considered the "gold standard" in probabilistic inference systems. However, the reason that the JT algorithm is not actually relevant in most real-world scenarios (specifically for side chain placement and protein design) is that the sizes of the resulting regions are so great as to make it infeasible to store them in any computer, yet alone perform computations such as message-passing between them. Nonetheless, it has proven useful in showing that, for small enough problems where the JT algorithm can be applied and the exact GMEC is found, BP usually finds this low energy solution as well [776], validating the use of BP on larger problems where the JT algorithm cannot be utilized.

## 11.6.3 Message-Passing Algorithms with Certificates of Optimality

As noted above, the loopy belief propagation (BP) algorithm has no theoretical guarantees of convergence to the correct results, although BP does, in practice, often converge to optimal results. Nevertheless, for cases for which BP does not converge, or when it is desired that the side chain predictions are guaranteed to be of minimal energy, there do exist algorithms related to BP, and message-passing in general, that are mathematically proven to yield the optimal side chain configuration upon convergence and under certain conditions.

In general, these message passing algorithms essentially aim at solving *dual* linear programming (LP) relaxations to the global minimum energy configuration (GMEC) integer program (as defined in Sect. 11.4.2 and [374]). In particular, they compute a lower bound to the energy functional in Eq. 11.1 and suggest a candidate minimal energy configuration. If the energy associated with this configuration is equal to the energy bound, then it is provably the GMEC and the bound is said to be tight. Otherwise, either the candidate configuration is not optimal but there exists a configuration which does attain the (tight) bound; or, the bound is not tight and hence unattainable by proper (integral) configurations. If the latter is the case, a tighter lower bound can be sought, e.g., by calculating additional, higher order messages [499, 680], albeit with increasing computational complexity. Below we list some of the more notable algorithms that follow this scheme and provide a short description of each.

The **tree-reweighted max-product (TRMP)** message-passing algorithm provides an optimal dual solution to the LP relaxation of the GMEC integer program [747]. As pointed out in Sect. 11.6.1, the belief propagation message-passing algorithm is guaranteed to converge to the GMEC if the underlying graph is singly connected, i.e., it is an acyclic, or tree, graph. The idea behind the tree-reweighted max-product algorithm [748] is to utilize the efficiency and correctness of message-passing on a tree graph to attempt to obtain optimal results on an arbitrary graph with cycles. This is implemented by decomposing the original probability distribution into a convex combination of tree-structured distributions, yielding a lower bound on the minimal energy rotamer configuration in terms of the combined minimal energies obtainable for each of the tree graphs considered. A set of corresponding message update rules are defined within this scheme, where the minimal energy configuration can be exactly determined if TRMP converges and the output "pseudo"-max-marginal beliefs are uniquely optimized by a single rotamer configuration. It was also shown that even in cases where these max-beliefs are not uniquely maximized by a particular configuration, it is sometimes still possible to determine the lowest energy configuration with certainty [763]. A comparison of simple LP relaxations (Sect. 11.4.2) and TRMP was performed in [779].

Similarly, the **max-product linear programming (MPLP)** algorithm [229] is derived by considering the dual linear program of a relaxed linear program (LP) for finding the minimal energy rotamer configuration (similar to that used in [374]).

It is also a message-passing algorithm, but the calculation of max-beliefs is always guaranteed to converge, since it performs block coordinate descent steps in the dual space. As in the TRMP algorithm, the minimal energy rotamer configuration can be determined only if the resulting max-beliefs are not tied; i.e., at each position, they are maximized by a unique rotamer.

**Tightened linear programming relaxations:** The original dual LP relaxation discussed above can be extended by (progressively) augmenting it with equalities that enforce higher-order variable consistency, e.g., a triplet of positions is required to have max-beliefs that are consistent with each of the three pairwise edges it involves [681], hence providing tighter LP relaxations. A similar approach was taken by the authors of [387], where problematic cycles are "repaired" to obtain dual problems with optimal values closer to the primal problem, yielding tighter relaxations so that the minimal energy configuration can be unambiguously determined.

### 11.6.4   The Best Max-Marginal First (BMMF) Algorithm

The BMMF (**B**est **M**ax-**M**arginal **F**irst) algorithm of [777] provides a general framework for using belief propagation (BP) to yield *multiple* optimal solutions to a particular inference problem. Conceptually, it partitions the search space while systematically excluding all previously determined minimal energy rotamer assignments (Fig. 11.4). In detail, the lowest energy rotamer configuration is found by running BP and applying Eq. 11.14. To find the second best rotamer configuration, the max-marginal beliefs calculated for each allowed rotamer at each position (Eq. 11.12) are utilized to identify a minimal discrepancy in rotamer choices between the lowest energy configuration and the next lowest configuration. This is possible since these two configurations must differ in their choices of rotamers for at least a single position, and this position can be identified as that with the second highest rotamer belief value, analogously to the identification of the lowest energy rotamer in Eq. 11.14. The relevant position is then constrained to possess this differential rotamer and BP is run again, yielding the full identity of the second best rotamer configuration.

Mathematically, let $\mathbf{r}^1$ and $\mathbf{r}^2$ denote the first and second lowest energy rotamer configurations, respectively; note that $\mathbf{r}^1$ was also designated $\mathbf{r}^*$ above. Then, BMMF can find $\mathbf{r}^2$ by running BP while enforcing the constraint that the rotamer choice at position $i$ is in fact the correct rotamer present in $\mathbf{r}^2$:

$$\text{Positive constraint} : r_i = r_i^2 \tag{11.15}$$

where $r_i^1 \neq r_i^2$, i.e., position $i$ differs between these two configurations. Note that the identities of position $i$ and rotamer $r_i^2$ were determined using the maximal sub-optimal rotamer beliefs (Eq. 11.12). After this, an additional run of BP is

# 11 Prediction of Low Energy Protein Side Chain Configurations

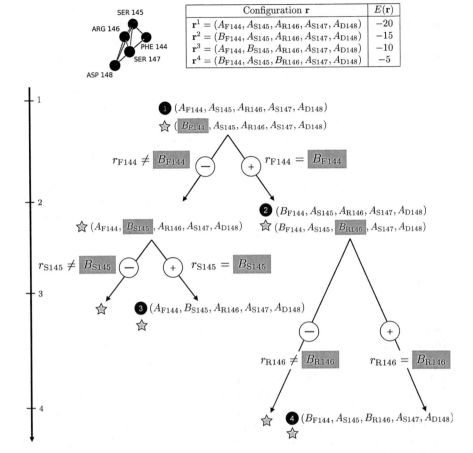

**Fig. 11.4** Calculation of the four lowest energy configurations by the BMMF sub-space partitioning algorithm, for the side chain placement of the five positions in the protein structure modeled in Fig. 11.2. In this toy example, each of the five positions has two possible rotamers, *A* and *B*, and the four lowest energy conformations (and their energies) are as listed at the top. At each iteration (marked by the vertical axis), the next rotamer configuration (denoted by a *circle*) is determined using BP, and the max-marginal beliefs resulting from BP are utilized to determine which rotamer choice at which position (marked by a *gray box*) should be used in subsequent partition steps to find an additional low energy configuration (denoted by a *star*). At each iteration, positive and negative constraints are as marked (Adapted from [207])

required in order to calculate the next lowest energy configuration within the current configurational sub-space, where this second run is performed while requiring that:

$$\textbf{Negative constraint}: r_i \neq r_i^2 \quad (11.16)$$

Again, the next lowest energy configuration within this sub-space (stars in Fig. 11.4) is distinguished by having the best rotamer max-belief at a particular position (besides those of $\mathbf{r}^1$). Now, subsequent low energy configurations can be found in either the positively constrained rotamer sub-space (Eq. 11.15) or the negatively constrained sub-space (Eq. 11.16), where the search space has effectively been partitioned in a manner that will facilitate efficient enumeration of the low energy rotamer choices. Thus, this successive partitioning and constraining procedure continues in an iterative manner until the desired number of lowest energy configurations is output; see [207, 777] for full details. In cases where loopy belief propagation (BP) yields *exact* max-marginal (MM) probabilities (i.e., Eq. 11.12 yields results identical to the computationally intractable Eq. 11.13), the BMMF algorithm is guaranteed to find the top configurations for side chain placement or protein design.

For the case of protein design, there exists a generalization of the BMMF algorithm that directly yields successive low energy sequences, without considering sub-optimal rotamer configurations for any particular sequence during intermediate steps of the algorithm. We now discuss this algorithm.

### 11.6.4.1 Generalization of BMMF for Protein Design

As opposed to side chain placement, for the case of protein design, it is highly undesirable to simply apply the BMMF algorithm in a straightforward manner. This derives from the fact that such a run would often entail the prediction of *multiple* low energy rotamer configurations that are available to each low energy amino acid sequence, without providing additional sequence design predictions. To overcome this obstacle, the tBMMF (type-specific BMMF) algorithm [207] exploits the formulation for protein design described in Sect. 11.5 and generalizes the BMMF algorithm. It does so by requiring that the successive low energy rotamer configurations sought out by the BMMF algorithm correspond to distinct amino acid sequences (see Fig. 11.6).

Conceptually, tBMMF operates by partitioning the rotamer search space by the amino acid identities of the low energy rotamers, in order to prevent the same amino acid sequence from being repeated twice (with various low energy rotamer configurations for that sequence). Formally, tBMMF replaces the positive (Eq. 11.15) and negative (Eq. 11.16) constraint operations of BMMF with:

$$\textbf{Positive constraint}: r_i \text{ such that } T(r_i) = T(r_i^2) \qquad (11.17)$$

$$\textbf{Negative constraint}: r_i \text{ such that } T(r_i) \neq T(r_i^2) \qquad (11.18)$$

where $T(r_i)$ denotes the amino acid corresponding to rotamer $r_i$, and, analogously to the BMMF algorithm, position $i$ differs in amino acid identity between the $\mathbf{r}^1$ and $\mathbf{r}^2$ configurations. Additionally, the identities of position $i$ and rotamer $r_i^2$ were determined using the maximal sub-optimal rotamer beliefs (Eq. 11.12). Thus, tBMMF proceeds through successive *amino acid-based* partitionings of the rotamer

11 Prediction of Low Energy Protein Side Chain Configurations

space, and it outputs an ensemble of predicted low energy rotamer configurations, each with a unique amino acid sequence; see [207] for more details.

## 11.7 Obtaining the Lowest Energy Amino Acid Side Chain Configuration (GMEC)

As described above, the primary goal of both protein side chain placement and protein design is the prediction of the lowest energy rotamer configuration, either for a given sequence (side chain placement) or among the rotamers for multiple sequences (protein design). Recall that this configuration is also known as the global minimum energy configuration (GMEC) [612] .

Alas, this minimum energy side chain configuration may be considerably different from the "correct" configuration, as captured in the static protein crystal structure. This incompatibility is, in part, a consequence of the widely used modeling paradigm, which, for the sake of computational feasibility, limits the conformational search space to discrete rotamers [574]. But, in fact, even within this restricted conformational space, there often exist more "native-like" configurations that are assigned (much) higher energy than the GMEC. Thus, the biological quality of a side chain placement search algorithm is measured primarily using structural attributes, such as the root mean square deviation (RMSD) between the minimal energy configuration and the native structure, or the fraction of side chain dihedral angles predicted correctly (up to some threshold), and only rarely using the energy itself. However, since we focus here on the search techniques themselves, we also consider the comparison of the actual minimal energies, as obtained by the various algorithms.

### 11.7.1 Belief Propagation for the GMEC of Side Chain Placement

To calculate the rotameric GMEC when the protein is represented as a probabilistic graphical model (Sect. 11.5), belief propagation is applied (Eq. 11.11) and the highest probability (lowest energy) rotamer configuration is determined using the max-marginal rotamer belief probabilities (Eqs. 11.13 and 11.14) [207, 776, 778]. Note also that other related message-passing algorithms that calculate the GMEC (directly, or using max-marginal beliefs) have also been computationally applied [229, 374, 681, 763, 779, 781].

In general, the belief propagation algorithm, when applied to cyclic graphs, is not guaranteed to (numerically) converge. Practically, BP converges for many side chain placement and protein design problems, but the convergence rate decreases when the problems become "harder", e.g., for larger models or using energy functions which account for longer distance interactions [778]. Even better convergence rates might be obtained by using a different message update schedule, e.g., as suggested by [171]. Importantly, it is always known when BP has failed to converge; in these cases, one can use the best intermediate results (that is, the one with lowest energy), or run another algorithm (e.g., GBP or Monte Carlo simulated annealing).

Notwithstanding these convergence problems, BP has been shown to obtain the *global* minimum side chain configuration, for the vast majority of models where this configuration is known (using DEE, the junction tree exact algorithm, or one of the message passing algorithms with an optimality certificate) [776,778]. Moreover, when BP converged, it almost always obtained a lower energy solution than that of the other state-of-the-art algorithms. This finding agrees with the observation that the max-product BP solution is a "neighborhood optimum" and therefore guaranteed to be better than all other assignments in a large conformational region around it [762].

The run time of BP is typically longer than that devoted, in practice, to exploring the conformation space by heuristic methods (Sect. 11.4.1). Message passing algorithms with a certificate of optimality are, in general, even slower, but they usually obtain the GMEC solution faster than other computationally exact methods (see Sect. 11.4.2).

For cases of computational side chain placement, reasonably accurate results have been observed using either BP [778] or the tree-reweighted max-product message-passing variant (TRMP, see Sect. 11.6.3) [781] on a standard test set of 276 single chain proteins: approximately 72% accuracy for correct prediction of both $\chi_1$ and $\chi_2$ angles. Furthermore, it was found that optimizing the energy function employed (Sect. 11.2.3), also utilizing TRMP, was capable of significantly improving the predictive accuracy up to 83%, when using an extended rotamer library [781].

## 11.7.2 Message-Passing for the Protein Design GMEC

In contrast to the placement of side chains of a single amino acid sequence on a main chain structure, defining a biologically correct solution for a given protein design problem is a much more convoluted issue. This solution is, by definition, the amino acid sequence that most stably (thermodynamically) folds to the given main chain structure, but experimentally finding that sequence is currently an infeasible task. As a matter of fact, in most cases, the only sequence definitively known to fold to the given main chain structure is the "native" (wild-type) sequence of the input structure. Therefore, this sequence is typically used as the "ground truth" for benchmarking comparisons [316]. Alternatively, predicted ensembles of low energy protein sequences (see Sect. 11.8) can be compared to evolutionary profiles, as derived from sequences of homologous proteins [207, 417, 624].

11 Prediction of Low Energy Protein Side Chain Configurations

For the rotamer minimization scenario of protein design, no experimental work directly validating the predictions of BP or other message-passing variants has been performed. Nonetheless, it is expected to be highly successful in this arena since it has been shown to outperform other state-of-the-art computational search methods in providing low energy sequence predictions [207, 776, 778], and even (provably) obtaining the globally optimal sequence [681]. Also, a recent method combining both dead-end elimination (DEE) and the TRMP message-passing variant was shown to find optimal energetic results within a reasonable time frame [312]. In the experimental realm, BP was used to accurately predict residue-residue clashes in hybrid protein sequences of *Escherichia coli* and human glycinamide ribonucleotide transformylases [521], suggesting that BP and related message-passing approaches will prove to be highly successful in directly designing novel protein sequences.

## 11.8 Predicting an Ensemble of Lowest Energy Side Chain Configurations

For the case of side chain placement, the BMMF algorithm (Sect. 11.6.4) [776] utilizes the speed and observed accuracy of belief propagation (BP) to predict the collection of lowest energy rotamer configurations for a single amino acid sequence and structure (Fig. 11.5). In [778], computational benchmarking for this task of finding multiple low energy rotamer configurations for side chain placement was performed. It was found that BMMF performed as well as, and typically better than, other state-of-the-art algorithms, including Gibbs sampling, greedy search, generalized DEE/A\*, and Monte Carlo simulated annealing (MCSA). Specifically, BMMF was unique in its ability to quickly and feasibly find the lowest energy rotamer configuration (GMEC) and to output successive low energy configurations, without large increases in energy. The other algorithms often did not succeed since they were either not computationally practical (and thus did not converge at all), or their collection of predicted rotamer configurations contained many rotamer choices with overly high energies. On the other hand, whenever an exact algorithm was tractable and capable of providing the optimal ensemble of configurations, BMMF provided these optimal predictions as well.

For protein design, the message-passing-based tBMMF algorithm was computationally benchmarked against state-of-the-art algorithms on a dataset of design cases of various biological qualities, sizes, and difficulties [207]. It was found that tBMMF was always capable of recovering the optimal low energy sequence ensemble (e.g., see Fig. 11.6) when this set could be feasibly calculated (by DEE). As in the BMMF benchmark above, tBMMF was the fastest, most adept, and most accurate algorithm, since it almost always managed to converge to extremely low energy sequences, whereas other algorithms might have converged quickly but to high energy solutions, or not have converged at all. Furthermore, tBMMF did not suffer from the sharp increases in predicted sequence energy that the MCSA algorithms

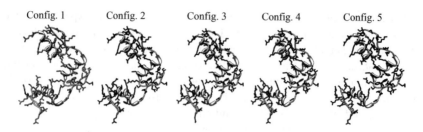

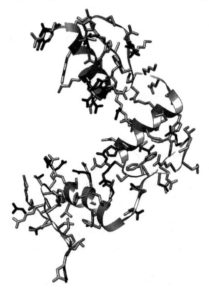

**Fig. 11.5** Side chain placement using BMMF. *Top*: Three-dimensional structures of low energy solutions for PDB 1AAY, using the SCWRL energy function [98]. Backbones are colored *light gray*, as well as the side chains for the lowest energy configuration (Config. 1). For comparison, the side chains of the 4 next lowest conformations are colored *black*. *Bottom*: Superposition of the 1,000 lowest energy conformations, where the side chains of all but the lowest energy structure are colored *black*; note the large amount of side chain variation within this structural ensemble, as depicted by *black* side chains (Structures were drawn using Chimera [575])

encountered. Finally, the authors of [207] report that the tBMMF predictions made using the Rosetta energy function [405], popularly used for computational protein design, typically included many extremely similar low energy sequences. To overcome this phenomenon, they suggested a modified tBMMF algorithm, which successfully bypasses amino acid sequences of biochemical nature similar to those already output. The BMMF and tBMMF methods described here are implemented in the freely available Side-chain PRediction INference Toolbox (SPRINT) software package [208].

# 11 Prediction of Low Energy Protein Side Chain Configurations

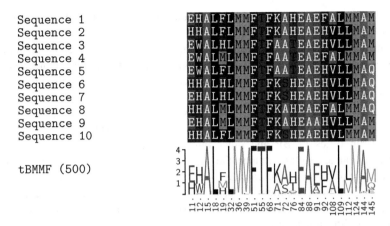

**Fig. 11.6** tBMMF design output: multiple sequence alignment of predictions for 24 core calmodulin positions (marked on the *bottom*) designed [207] for binding smooth muscle myosin light chain kinase (PDB 1CDL) using the Rosetta energy function [405]. tBMMF was used to predict low energy amino acid sequences; amino acid identities for the 10 lowest energy solutions are shown, and the sequence logo summarizes 500 best sequences. Graphics were generated using TeXshade [30]

**Table 11.2** Summary of message-passing methods for the calculation of low energy protein side chain configurations.

| Task | Application | Configuration(s) obtained | Algorithm [Benchmark refs] |
|------|-------------|--------------------------|---------------------------|
| Single low energy side chain configuration | Side chain placement, protein design | Low energy (empirical) | Belief propagation (BP) [171, 240, 776, 778] |
|  |  | Globally optimal (provable) | TRMP [779, 781] |
|  |  |  | MPLP [229] |
|  |  |  | LP relaxations [681] |
| Ensemble of low energy side chain configurations | Side chain placement | Low energy rotamer ensembles (empirical) | BMMF [777, 778] |
|  | Protein design | Low energy sequence ensembles (empirical) | tBMMF [207] |

## 11.9 Discussion and Future Directions

Belief propagation, BMMF, and their variants have been benchmarked on large and diverse sets of real world side chain placement and protein design problems and shown to predict lower energy side chain configurations, compared to state-of-the-art methods; for a summary, see Table 11.2. In fact, some of the algorithms outlined in Sect. 11.6.3, including TRMP [781], MPLP [229], and tightened LP relaxations [681], have provably obtained the *global* minimum side chain configurations for many of the instances in the side chain placement and protein design benchmark sets of [779].

The ability to predict lower energy configurations is, by all means, encouraging. However, computational structural researchers are usually more concerned with the "biological correctness" of the predicted models than with their corresponding energies. Unfortunately, using contemporary energy functions, the lower energy configurations obtained by BP and its variants are only negligibly (if at all) more native-like than those obtained by sub-optimal algorithms [781]. The usually slower run times of BP and other inference algorithms and the possibility of non-convergence further reduce the attractiveness of these algorithms from the perspective of biological researchers. Consequently, an important direction in future research is the development of faster, better converging, and more accurate BP-related algorithms; structural biology provides a challenging, real world test bed for such algorithms and ideas.

The following subsections describe a few variants of the "classical" side chain placement and protein design tasks and discuss the potential utility of message-passing algorithms in addressing these related applications.

### 11.9.1 The Partition Function and Protein Side Chain Free Energy

In defining the probability of a particular rotamer configuration in Sect. 11.5 (Eq. 11.9), we came across the *partition function* ($Z$) as the normalization factor of the configuration probabilities. This partition function is in fact closely related to the side chain conformational free energy of the molecular ensembles accessible to a protein (sought out, for example, in [359, 442]).

In the physics literature, the partition function corresponds to the statistical physical properties of a system with a large number of microstates [318]. Likewise, the partition function has a natural interpretation in the field of probabilistic graphical models, and Markov random fields in particular. Thus, many probabilistic inference algorithms have directly focused their attention on this task. In fact, there exist corresponding versions of the belief propagation (BP), generalized belief propagation (GBP), junction tree, tree-reweighted max-product (TRMP), max-product linear programming (MPLP, see Sect. 11.6), and self-consistent mean field theory (SCMF, see Sect. 11.4.1) algorithms that try to bound or otherwise approximate this quantity.

Importantly, the per-sequence partition function accounts for the ensemble of *all* possible side chain configurations, where each one is weighted by an appropriate Boltzmann probability (so that lower energy conformations have exponentially higher probabilities). The reasoning goes that only by accounting for multiple low energy rotamer configurations can the favorability of a particular interaction, or the stability of a particular amino acid sequence, be correctly evaluated [141]. The side chain free energy (partition function) was efficiently estimated for a dataset of fixed main chain protein structures and sequences in [358] by applying the GBP algorithm. BP was also enlisted to predict the free energies of association for protein-protein interactions, with the additional advantage that both the side chains and the protein main chain were permitted to be somewhat flexible [359].

In the context of protein design, the use of side chain conformational free energy plays an essential role in what has been termed *probabilistic protein design* [560]. As opposed to the protein design framework described in this chapter (designated *directed protein design* by [560]), probabilistic design starts off by computationally predicting the amino acids at each design position with a high probability of being compatible with the structure. Next, these predicted probabilities are used to bias the experimental synthesis of random libraries of billions of gene sequences, which are subsequently selected for relevant biological function in a high-throughput screen. These probabilities have been successfully predicted using SCMF-related techniques [389,560]. Furthermore, in [206], it was found that the corresponding version of BP outperforms other state-of-the-art methods at this probability prediction task. In a different direction, BP has been used within a protein design protocol in order to calculate the per-sequence partition function within the larger framework of a stochastic Monte Carlo simulated annealing (MCSA) search through the sequence space [639].

## 11.9.2 Relaxation of the Protein Main Chain and Rotamer Side Chains

Throughout this chapter, we have formulated the procedures of side chain placement and protein design as operating on a fixed main chain structure that is given as input to the computational problem. In addition, the amino acid side chains are modeled as being one of a number of discrete possibilities (rotamers) sampled from a large library of protein structures. Both of these simplifying assumptions essentially define the classical paradigm for these problems, within which researchers have made major inroads into fundamentally understanding proteins at the atomic level, e.g., the design of a sequence that assumes the structure of a zinc finger domain without the need for zinc [132]. Also, see [612] for an overall review on the successes of protein design in the past decades. Nevertheless, this paradigm of a fixed main chain and rigid side chains artificially limits the natural degrees of freedom available to a protein sequence folding into its stable conformation(s). For example, in protein design, we may energetically disallow a certain amino acid sequence due to steric clashes between side chains, whereas small main chain perturbations would have completely removed the predicted atomic overlaps. We now outline some of the methods that bypass the limitations inherent in making these modeling assumptions.

The simplest technique that incorporates main chain flexibility into computational side chain placement and protein design procedures consists of the use of a pre-calculated ensemble of fixed main chain structures similar to the target structure [201, 418]. A more sophisticated approach entails cycles of fixed main chain design and main chain improvements on an as-needed basis, e.g., when random Monte Carlo steps are rejected at a high rate due to the main chain strains induced by sequence mutation [787]. Some protein design methods take this a step further, where the main chain improvements are regularly performed every number of iterations by optimizing the main chain for the current amino acid sequence using atomic-resolution structure prediction algorithms [406, 624]. The most elegant, yet computationally burdensome, method involves simultaneous energetic optimization within the main chain and side chain spaces [213, 214, 674]. Such optimization should most closely resemble the simultaneous latitude available to proteins that have naturally evolved within the structure-sequence space.

Similarly, permitting side chain flexibility within the framework of side chain placement and design has been implemented in a number of ways. The most straightforward of these approaches is to still model the side chains with a discrete number of rotamer states, but to use more rotamers by "super-sampling" from the empirical rotamer library. This technique has been employed in many works, e.g., [207, 325, 624]. More refined methods for permitting side chain flexibility include the stochastic sampling of sub-rotamer space [151], flexible side chain redesign (iterative application of sub-rotameric minimization within an MCSA search) [151], and exact optimization within the sub-rotamer space using computationally costlier DEE-based methods [215]. Another possibility for the sampling of sub-rotamer space is to perform energy calculations for a particular rotamer from the library based on stochastically generated versions of that rotamer [500]. Alternatively, probabilistic modeling of side chain conformational space could allow sampling in continuous space [290].

Most importantly, recent protein design studies have shown that the relaxation of the rigid main chain and rotamers can produce more "realistic" side chain and amino acid variability [151, 201, 624, 674], i.e., boosting the accuracy of protein modeling procedures in describing natural phenomena.

The preliminary use of message-passing algorithms that somewhat model protein main chain flexibility (in predicting the free energies of association for protein-protein interactions) was demonstrated in [359], where a probabilistic ensemble of protein main chains was employed. Future work that utilizes belief propagation to model molecular flexibility will strive to directly incorporate simultaneous optimization within the main chain and side chain spaces (e.g., as noted above for [213, 214, 674]). An intriguing possibility for this approach would be to use continuous-valued (Gaussian) Markov random fields [682] and their corresponding belief propagation algorithms [465, 571, 761].

## 11.9.3 Multistate Protein Design

In the classical paradigm of computational protein design detailed above, the stability of a sequence in a single three-dimensional structure is sought. A natural generalization involves the incorporation of multiple structural states into the design process. These structural states can range from alternative conformations of the target structure to protein-protein interactions of the target with various binding partners.

One of the earlier works in this direction integrated the explicit requirement of higher free energy for the denatured state, thus essentially designing the global shape of the protein folding funnel in order to better produce sequences with well-defined three-dimensional structures [348]. This process was aptly termed *negative design*. A conceptually related approach is to single out a small number of states considered to be undesirable, in order to find a sequence that will fold to the target structure and not to these negative states [4]. For example, the computational design of stable and soluble homodimeric-specific coiled-coil proteins required negative design states modeling each of the unfolded, aggregated, and heterodimeric states [297], leading the researchers to conclude that multiple design goals require explicit computational consideration of multiple structural states. Be that as it may, there may be cases where performing single-state design and simply ignoring these negative states suffices to design against compatibility with them, e.g., for the calmodulin interface designs of [205, 654].

A complementary design strategy is derived from the observation that, in nature, many proteins assume multiple structural configurations (e.g., [4, 729, 799]) and partake in numerous protein-protein interactions. Thus, in a fashion inversely similar to the negative design procedure, the process of *multispecific protein design* was considered, where the goal is to construct a sequence compatible with multiple structural states [10, 209, 325]. Computational protein design in which both negative and multispecific design may be included is generally known as *multistate design*.

A promising proof of the multispecific design concept was the fully computational optimization of a single sequence for two distinct target structures, where the resulting peptide can switch conformations from a 2Cys-2His zinc finger-like fold to a trimeric coiled-coil fold depending upon pH or metal ion concentration [9]. Another success story involves the experimental design of selective peptide partners for basic-region leucine zipper (bZIP) transcription factors, which demonstrated that human bZIPs have only sparsely sampled the possible interaction space accessible to them [258]. Finally, large-scale computational analyses have shown that the character of natural protein sequences can be better explained when accounting for the multiple in vivo binding states that confer the totality of the biological function of a protein with multiple interactions [205, 209, 325].

In the context of message-passing algorithms, the works in [205, 209] employ Markov random field approaches for multispecific protein design, using both the tBMMF and BP algorithms (see Sect. 11.6). These studies were performed by combining the Markov random fields corresponding to each individual protein structure and enforcing sequence consistency between them on a position-by-position basis. Subsequent work will still need to address the general goal of multistate design (i.e., negative design) within this framework.

**Acknowledgements** We would like to thank Amir Globerson and Talya Meltzer for their discussions on message-passing algorithms with certificates of optimality.

# Part VI
# Inferring Structure from Experimental Data

# Chapter 12
# Inferential Structure Determination from NMR Data

**Michael Habeck**

## 12.1 Introduction

### 12.1.1 Overview

The standard approach to biomolecular structure calculation from nuclear magnetic resonance (NMR) data is to solve a non-linear optimization problem by simulated annealing or some other optimization method. Despite practical success, fundamental issues such as the definition of a meaningful coordinate error, the assessment of the goodness of fit to the data, and the estimation of missing parameter values remain unsolved. Inferential structure determination (ISD) is a principled alternative to optimization approaches. ISD applies Bayesian reasoning to represent the unknown molecular structure and its uncertainty through a posterior probability distribution. The posterior distribution also determines missing parameter values such as NMR alignment tensors and error parameters quantifying the quality of the data. The atomic coordinates and additional unknowns are estimated from their joint posterior probability distribution with a parallel Markov chain Monte Carlo algorithm.

---

M. Habeck (✉)
Department of Protein Evolution, Max-Planck-Institute for Developmental Biology, Spemannstr. 35, Tübingen, Germany
e-mail: michael.habeck@tuebingen.mpg.de

Department of Empirical Inference, Max-Planck-Institute for Intelligent Systems, Spemannstr. 38, Tübingen, Germany

T. Hamelryck et al. (eds.), *Bayesian Methods in Structural Bioinformatics*, Statistics for Biology and Health, DOI 10.1007/978-3-642-27225-7_12,
© Springer-Verlag Berlin Heidelberg 2012

## 12.1.2 Background

The aim of high-resolution protein structure determination is to infer the spatial position of each individual atom making up a protein from experimental data such as X-ray crystallographic diffraction patterns, nuclear magnetic resonance (NMR) spectra or micrographs measured with electron microscopy. Biomolecular structure determination is a complicated data analysis problem that requires objective, reliable and automated methods. The first protein structures were built physically using wireframes. Such tedious procedures were replaced by more and more sophisticated computer-assisted methods. Today, most of the work that previously had to be done manually is automatized and replaced by computer programs.

The final step in a protein structure determination involves the fitting of atom coordinates against the experimental data. The generally accepted attitude towards protein structure calculation is to solve a (constrained) optimization problem. An objective function is minimized that assesses the fit between the observed data and data back-calculated from the protein structure [81]. The total cost function comprises additional terms that alleviate problems with unfavorable data-to-parameter ratios and missing data. In structure determination by solution NMR [772], for example, only distances smaller than 5 Å between the magnetically active nuclei (typically protons) can be measured. To position the non-hydrogen atoms, we have to rely on our knowledge about the chemical structure of amino acids. Therefore the total cost function, often called *hybrid energy*, includes a molecular force field. The hybrid energy is given by:

$$E_{\text{hybrid}}(\mathbf{x}) = w_{\text{data}} \, E_{\text{data}}(\mathbf{x}) + E_{\text{phys}}(\mathbf{x}) \tag{12.1}$$

where $E_{\text{data}}$ evaluates the goodness of fit for a specific structure $\mathbf{x}$ (in case of crystallographic data this could be the R-factor), and $E_{\text{phys}}$ is a molecular force field which guarantees the integrity of the structures in terms of covalent parameters and van der Waals contacts. The weighting factor $w_{\text{data}}$ balances the two terms governing the hybrid energy function.

Jack and Levitt [330] introduced this way of refining protein structures against crystallographic data. In the original publication they already remark that the best choice of the weighting factor $w_{\text{data}}$ "is something of a problem". This hints at a more general problem with minimization approaches, namely the question of how to set *nuisance parameters*. Nuisance parameters are parameters that are not of primary interest but need to be introduced in order to model the data and their errors. In the above example, the weighting factor is such a nuisance parameter. Other nuisance parameters are related to the theories that we use to calculate mock data from a given protein structure. These could be crystallographic temperature factors, NMR calibration scales, peak assignments and alignment tensors, etc. In optimization approaches, nuisance parameters are chosen manually or set to some default value. The problem with manual intervention in protein structure determination is that one might impose personal beliefs and biases on the structure.

Typically the data measured by X-ray crystallography, NMR spectroscopy and electron microscopy are not complete enough to determine the entire molecular structure. Moreover, often not the raw data are used but data that have passed several pre-processing and pre-analysis steps which themselves can be subject to personal biases. For X-ray crystallography there is a remedy: Although the observed structure factors might not be used in the initial model building but only in the refinement, they will be used finally to judge the reliability of the structure in terms of an objective free R-factor [79]. In NMR structure determination, this is not the case. The quality of a structure is reported in terms of restraint violations and ensemble root-mean-square deviations (RMSDs) [487]. But the restraints themselves are the result of preceding analysis steps including sequential and cross-peak assignment as well as calibration and classification of cross-peaks. Moreover, the spread of a structure ensemble can be influenced heavily by choosing weighting factors or distance bounds in a biased fashion [105, 547, 684].

Therefore no generally accepted measure for the quality of an NMR structure exists. NMR structures are typically presented as ensembles [487, 695]. But there is some controversy over the meaning of NMR ensembles. In the beginning of NMR structure determination, structure ensembles were calculated by distance geometry [296]. The ensembles obtained by distance geometry reflect the conformational space that is compatible with an incomplete set of distance bounds. This view is close to a Bayesian perspective according to which the ensemble is a statistical sample from the posterior distribution. However, issues with sampling bias [295, 502] show that distance geometry lacks a probabilistic concept that could unambiguously determine the correct multiplicity.

According to a more pragmatic view, NMR ensembles reflect the robustness of the structure calculation procedure [262]. This interpretation arose with the advent of non-linear optimization approaches in NMR structure calculation, with molecular dynamics-based simulated annealing (MDSA) [263, 546] being the most prominent. Structure calculation by MDSA typically generates an ensemble by starting at random positions and velocities; each calculation will converge to a different solution. But an ensemble obtained in such a multi-start fashion is not a proper statistical sample and does not necessarily reflect the precision of the structure. A correlation between the spread of the ensemble and the error of the structure could be rather due to the fact that a hybrid energy derived from complete data is easier to optimize and will therefore yield sharper ensembles than a hybrid energy derived from sparse data.

Often it is argued that NMR ensembles represent the "solution structure", that is the ensemble is viewed as a sample from the thermodynamic ensemble of the solvated protein. But this argument is fallacious. Although NMR parameters are indeed time- and ensemble-averaged quantities they are seldom modeled as such (notwithstanding the recent renaissance [116, 445] of ensemble calculations [629, 722]). In standard structure calculations, NMR parameters are treated as instantaneous measurements made on a single structure. Consequently, the standard NMR ensemble is not an experimentally determined thermodynamic ensemble.

Sometimes it is also stated that NMR ensembles reflect the dynamics of the protein in solution. Also this statement has to be judged with some caution. Indeed there may exist a correlation between the structural heterogeneity in NMR ensembles and the conformational fluctuations predicted, for example, with molecular dynamics [1]. But again this is not because the data are interpreted correctly but more or less accidentally: nuclear spins that are subject to stronger dynamics produce broader peaks that eventually disappear in the noise. That is, one tends to observe and assign fewer NMR signals in mobile than in less flexible regions. Variations in the completeness of data along the protein chain result in ensembles that are in some parts better defined than in others. But it could also well be that a lack of restraints is caused simply by missing chemical shift or cross-peak assignments rather than dynamics.

The source common to the above-mentioned issues is that we use the wrong tools to tackle protein structure determination. An optimization approach can only determine a single, at best globally optimal structure but not indicate its quality. Missing parameters have to be set manually or determined by techniques such as cross-validation. Inferential structure determination (ISD) [267, 601] is a principled alternative to the optimization approach. The reasoning behind ISD is that protein structure determination is nothing but a complex data analysis or inference problem. In this view, the main obstacle is in the incompleteness of the information that is provided by experiments and in the inadequacy of optimization methods to cope with this incompleteness. Instead of jumping the gun by formulating protein structure calculation as an optimization problem, let us step back and contemplate on the question: What are the adequate mathematical tools to make quantitative inferences from incomplete information?

The basic problem with inferences from incomplete information is that, contrary to deductive logic, no unique conclusions can be drawn. Rather, one has to allow propositions to attain truth values different from just True or False and introduce a whole spectrum spanning these two extrema. It was Cox [127] who showed that, rather surprisingly, the rules that govern the algebra of such a scale of truth values are prescribed by the simple demand that they comply with the rules of standard logic. Cox showed that the calculus of probability theory is isomorphic to the algebra of probable inference. Cox's work was the culmination of a long-lasting effort to establish the principles and foundations of Bayesian inference beginning with the work of Bayes [27] and Laplace [416]. In recent years, the view of probability theory as an extended logic has been promoted most fiercely by Jaynes [342].

These developments established Bayesian inference as a unique and consistent framework to make quantitative inferences from limited information. Today, past quarrels between Bayesians and Frequentists (see for example Jaynes' polemics on significance tests [338]) appear as records of some ideological stone age. Also the advent of powerful computers helped to develop a more constructive attitude towards Bayesianism. Advanced numerical methods and computer power make it possible to solve problems that resisted a Bayesian treatment because they were too complex to be tackled with pencil and paper.

12 Inferential Structure Determination from NMR Data    291

Bayesian inference is particularly well-suited to solve complex data analysis problems. Modern Bayesian studies easily involve several thousand parameters that are estimated with Markov chain Monte Carlo methods (cf. for example Neal's treatment of Bayesian neural nets as an early account [536]). Markov chain Monte Carlo methods are discussed extensively in Chap. 2. Yet only few attempts have been made to apply Bayesian principles to protein structure calculation from experimental data.

ISD is the first entirely probabilistic effort to calculate protein structures from experimental data. To do this, ISD employs modern Markov chain Monte Carlo sampling algorithms. One big advantage of a Bayesian over an optimization approach is that Bayesians do not distinguish between parameters of primary or secondary interest and do not fall back on the notion of a "random variable". Bayesian probabilities do not express frequencies of occurrence but states of knowledge. A probability quantifies our ignorance about a parameter, it does not relate to a physical propensity or fluctuation.

Applied in the context of macromolecular structure determination, the use of Bayesian concepts allows the determination of all nuisance parameters that we need to model the data. The nuisance parameters will be estimated along with the atomic coordinates. This approach also solves the problem of generating statistically meaningful structure ensembles because there is no space left to choose the weighting factor $w_{data}$ or other parameters and thereby tweak the spread of the ensemble as it pleases. Still ISD works with restraints and not the raw data; one of the future developments will be to include the raw measurements rather than restraints.

In this chapter, I introduce the theoretical background and algorithms employed in inferential structure determination. I will outline the generic probabilistic formalism and then apply it to typical structural data measured in biomolecular NMR spectroscopy. My goal is to provide details on how we do it and why we do it. The formalism will be illustrated using a very simple example, the determination of a single dihedral angle from scalar coupling constants. I will then proceed by outlining the inference algorithm used in ISD. The algorithm is a Gibbs sampler embedded in a generalized two-temperature replica Monte Carlo scheme. The major problem is posed by the conformational degrees of freedom, and I will describe how we deal with these parameters which are notoriously difficult to sample. I will then briefly illustrate that the "sampling" provided by multi-start simulated annealing gives conformations with incorrect multiplicities and is therefore inappropriate for statistical sampling. In the final section, I will discuss practical issues. These include probabilistic models for the most important structural parameters determined by NMR spectroscopy. I will discuss the estimation of specific NMR parameters such as alignment tensors or Karplus coefficients. I will address the question of how experimental data should be weighted relative to each other and the molecular force field. I will illustrate that the estimated weights are useful figures of merit. Finally, I present an outlook on future developments.

## 12.2 Inferential Structure Determination

ISD starts from the observation that protein structure determination is nothing but a complex inference problem. Given experimental data (from X-ray crystallography, NMR spectroscopy, electron microscopy) we want to determine a protein's three-dimensional structure. The Bayesian way to solve this problem is straight-forward. First, we recall and formalize what we already know about our particular protein and protein structures in general. We know the amino acid sequence of the protein, we know that amino acids have a rather rigid covalent structure, we know that each amino acid is composed of atoms that occupy some volume in space and should not overlap. The second ingredient is a probabilistic model for the observations that we want to use for structure determination.

Here an advantage of a Bayesian approach comes into play. Data analysis problems in general are *inverse problems*: we measure data that are related to our parameters of interest and hope to invert this relation to gain insight into the parameters themselves. Bayesians do not try to solve inverse problems by direct inversion but delegate the inversion task to the Bayesian inference machinery [341]. In our context this means that we do not derive from the NMR spectra distance bounds which will then be used to generate conformations. Rather we aim at modeling the data as realistically as possible and apply Bayes' theorem to do the inversion for us. Modeling the data probabilistically comprises two steps. First, we calculate idealized "mock" data using a *forward model*, i.e. a theory or physical law. Second, we need to take account of the fact that the observations will deviate from our predictions. There are many possible sources for discrepancies (experimental noise, theoretical short-comings, systematic errors because some effects such as protein dynamics or averaging were neglected); they all need to be quantified using an *error model*.

### 12.2.1 Formalism

Let us introduce some mathematical notation [267]. We consider a set of $n$ measurements $\mathbf{d} = \{y_1, \ldots, y_n\}$. The parameters that we want to determine from the data are the atom positions $\mathbf{x}$ and additional parameters $\boldsymbol{\theta}$ which we will treat as nuisance parameters. Bayes' theorem allows us to estimate the parameters from the data and general background knowledge "$I$":

$$p(\mathbf{x}, \boldsymbol{\theta} | \mathbf{d}, I) = \frac{p(\mathbf{d} | \mathbf{x}, \boldsymbol{\theta}, I) \, p(\mathbf{x}, \boldsymbol{\theta} | I)}{p(\mathbf{d} | I)} \tag{12.2}$$

Bayes' theorem converts the probability of the data into a probability over the joint space of protein conformations and nuisance parameters. This posterior distribution $p(\mathbf{x}, \boldsymbol{\theta} | \mathbf{d}, I)$ is everything we need to solve a structure determination problem.

# 12 Inferential Structure Determination from NMR Data

It determines the most likely protein conformations and specifies their precision in terms of the posterior probability mass that is associated with each of them. To arrive at the posterior distribution we simply consider the likelihood a function of the parameters $\mathbf{x}$ and $\boldsymbol{\theta}$ (instead of inverting the data directly). Furthermore, we express our data-independent knowledge through a prior probability:

$$p(\mathbf{x}, \boldsymbol{\theta}|I) = p(\boldsymbol{\theta}|\mathbf{x}, I)\, p(\mathbf{x}|I) \tag{12.3}$$

where $p(\mathbf{x}|I)$ is the prior distribution of protein conformations and $p(\boldsymbol{\theta}|\mathbf{x}, I)$ the prior probability of the nuisance parameters given the structure. Here we will assume that, a priori, the nuisance parameters are independent of the structure: $p(\boldsymbol{\theta}|\mathbf{x}, I) = p(\boldsymbol{\theta}|I)$. Moreover, it is not necessary to calculate the normalization constant $p(\mathbf{d}|I)$, the "evidence" [457], for the questions addressed in this chapter (further details on the use of evidence calculations in ISD can be found in [266]).

Because the probability of the data plays such an important role we will give it its own name, *likelihood*, and denote it by $L$. The likelihood is the probability of the data conditioned on the molecular structure $\mathbf{x}$ and nuisance parameters $\boldsymbol{\theta}$ viewed as a function of these parameters. That is,

$$L(\mathbf{x}, \boldsymbol{\theta}) = p(\mathbf{d}|\mathbf{x}, \boldsymbol{\theta}, I). \tag{12.4}$$

In NMR structure calculation, it is appropriate to model the probability of an entire data set (NOEs, scalar coupling constants, etc.) as independent measurements. The likelihood is then a product over the individual probabilities of the single measurements: $L(\mathbf{x}, \boldsymbol{\theta}) = \prod_i p(y_i|\mathbf{x}, \boldsymbol{\theta})$.

In case we do not have any prior knowledge, the posterior distribution $p(\mathbf{x}, \boldsymbol{\theta}) = p(\mathbf{x}, \boldsymbol{\theta}|I)$ is proportional to the likelihood. What are we doing when maximizing the posterior/likelihood? We can rewrite the likelihood in a different form:

$$L(\mathbf{x}, \boldsymbol{\theta}) = \exp\left\{\int dy \sum_i \delta(y - y_i) \log p(y|\mathbf{x})\right\} = \exp\left\{-n\, D[Q\|p](\mathbf{x}, \boldsymbol{\theta})\right\} \tag{12.5}$$

Let $Q(y) = \sum_i \delta(y - y_i)/n$ denote the empirical distribution of the data which is a rather solipsistic estimate because it only assumes data where they were actually observed. Then the exponent in the previous equation involves the cross-entropy between the empirical distribution and our model $p(y|\mathbf{x}, \boldsymbol{\theta})$:

$$D[Q\|p](\mathbf{x}, \boldsymbol{\theta}) = -\int dy\, Q(y) \log p(y|\mathbf{x}, \boldsymbol{\theta}) \tag{12.6}$$

viewed as a function of the inference parameters $\mathbf{x}$ and $\theta$. Maximum likelihood adapts the parameters $\mathbf{x}$ and $\boldsymbol{\theta}$ so as to minimize the cross-entropy (or equivalently the Kullback-Leibler divergence) between the empirical distribution of the data and our model. This is achieved if the overlap between the empirical distribution $q$ and the model $p$ is maximal.

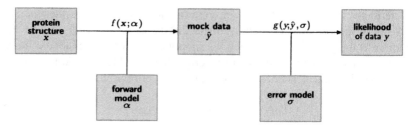

**Fig. 12.1** Bayesian modeling of structural data. To set up the likelihood function for some observable $y$, we first use a forward model $f(\mathbf{x}; \alpha)$, a deterministic function to predict mock data $\hat{y}$ from the structure. The forward model is parameterized with additional unknowns $\alpha$. The mock data are then forwarded into an *error model* $g(y|\hat{y}, \sigma)$ which assesses the discrepancies between observed and mock data. $g$ is a conditional probability distribution that is normalized for the first argument (i.e. the observed data) and involves additional nuisance parameters $\sigma$

The likelihood can be viewed as a score for the fit between the distribution of the data and a parameterized model. Additional prior information modulates this score by down-weighting regions where parameter values are unlikely to fall and up-weighting those regions that are supported by our prior knowledge. The prior distribution of the molecular structure is the Boltzmann distribution [336]:

$$\pi(\mathbf{x}) \equiv p(\mathbf{x}|I) = \exp\{-\beta E(\mathbf{x})\}/Z(\beta). \tag{12.7}$$

Note, that there is no simple way of calculating the partition function $Z(\beta)$, which is not a problem if we are only interested in parameter estimation.

In the current implementation of ISD, we use a very basic force field $E(\mathbf{x})$, commonly employed in NMR structure determination [446]. The force field maintains the stereochemistry of the amino acids, i.e. it restrains bond lengths and bond angles to ideal values as defined in the Engh-Huber force field [173]. In addition we penalize the overlap of atoms using a purely repulsive potential [301]. Van der Waals attraction and electrostatic interactions are usually not included in NMR structure calculations. These simplifications are valid for data with favorable data-to-parameter ratios. For sparse and/or low quality NMR data it becomes necessary to include additional terms [266] either derived from the data base of known protein structures [410] or a full-blown molecular force field [111]. One of the future developments of ISD will be to incorporate a more elaborate force field such as the full-atom energy of Rosetta [76] and to add new potentials that restrict the main chain dihedral angles to the allowed regions of the Ramachandran plot, or the side chain dihedral angles to the preferred rotamers. The latter goal can be accomplished by the combined use of the probabilistic models presented in Chap. 10 as a prior distribution, as was recently shown [552].

The generic approach for formulating an appropriate likelihood is to first come up with a reasonable *forward model* that allows the calculation of idealized data from a given structure (see Fig. 12.1). The forward model often involves parameters $\alpha$ such that for an idealized measurement $\hat{y} = f(\mathbf{x}; \alpha)$. The second ingredient

## 12 Inferential Structure Determination from NMR Data

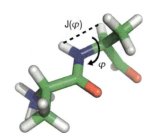

**Fig. 12.2** Example system with one conformational degree of freedom. The dihedral angle $\varphi$ parameterizes the different conformations of an alanine dipeptide and can be determined from scalar coupling measurements

of the likelihood function is a probabilistic model for the discrepancies between observed and predicted data. We will call this probability distribution an *error model* $g(y|\hat{y}, \sigma)$ and denote its parameters by $\sigma$. The error model is a conditional probability distribution for the measurements, i.e. $\int dy \, g(y|\hat{y}, \sigma) = 1$. A simple error model is a Gaussian distribution, in this case $\sigma$ is the standard deviation. The total likelihood is:

$$L(\mathbf{x}, \boldsymbol{\theta}) \equiv L(\mathbf{x}, \boldsymbol{\alpha}, \sigma) = \prod_i g(y_i | f_i(\mathbf{x}; \boldsymbol{\alpha}), \sigma) \qquad (12.8)$$

and the nuisance parameters $\boldsymbol{\theta}$ comprise the theory parameters $\boldsymbol{\alpha}$ and the parameters of the error models $\sigma$. Using the notation introduced in this paragraph the posterior distribution is:

$$p(\mathbf{x}, \boldsymbol{\alpha}, \sigma) \propto L(\mathbf{x}, \boldsymbol{\alpha}, \sigma) \, \pi(\mathbf{x}) \, \pi(\boldsymbol{\alpha}) \, \pi(\sigma). \qquad (12.9)$$

Here we introduced prior distributions for the forward and error model parameters $\boldsymbol{\alpha}$ and $\sigma$. Again, we made the assumption that all parameters are independent a priori. This could be relaxed if needed.

### 12.2.2 An Illustrative Example

Let us illustrate the basic machinery of the ISD approach for a simple example. We consider a structure that can be parameterized by a single conformational degree of freedom, a dihedral angle $\varphi$ (Fig. 12.2). We want to infer the angle from NMR measurements. One way to determine dihedral angles by NMR is to measure three-bond scalar coupling constants (also called $J$ couplings). There is an approximate theory developed by Karplus [361] that relates the strength of a three-bond scalar coupling $J$ to the dihedral angle of the intervening bond $\varphi$:

$$J(\varphi) = A \cos^2 \varphi + B \cos \varphi + C. \qquad (12.10)$$

The Karplus curve (Eq. 12.10) is basically a Fourier expansion involving three expansion coefficients (Karplus coefficients) $A$, $B$, and $C$ that we assume to be known for now. The Karplus curve is our forward model. To analyze $n$ measured couplings $\mathbf{d} = \{J_1, \ldots, J_n\}$, we need to introduce an error model. We choose a Gaussian distribution which is the least biased choice according to the maximum entropy principle. The likelihood function can be written as [269]:

$$L(\varphi, \sigma) = \exp\left\{-\frac{1}{2\sigma^2} \sum_i [J_i - J(\varphi)]^2\right\} \bigg/ Z(\sigma) \qquad (12.11)$$

where $Z(\sigma) = (2\pi\sigma^2)^{n/2}$ is the $n$th power of the normalization constant of a Gaussian distribution. The exponent involves the $\chi^2$ residual:

$$\chi^2(\varphi) = \sum_i [J_i - J(\varphi)]^2 = n\,[\overline{J} - J(\varphi)]^2 + n\,\mathrm{var}(J) \qquad (12.12)$$

where $\overline{J}$ is the sample average and $\mathrm{var}(J)$ the sample variance over all measured scalar couplings.

As we saw in Sect. 12.2.1, the likelihood function can be interpreted in terms of the cross-entropy between the empirical distribution of the data and a parametric model. In this example, the model has two free parameters, the conformational degree of freedom $\varphi$ and the error of the couplings $\sigma$. By varying these parameters we can control how much the model and the distribution of the couplings overlap. Let us assume that we know the error $\sigma$ and focus on the angle $\varphi$. Changes in the angle shift the center of the Gaussian model in a non-linear fashion. The optimal choice for the model maximizes the overlap and is obtained for angles $\varphi$ such that $J(\varphi) = \overline{J}$. Because the Karplus curve is a non-linear forward model the posterior distribution of the angle is not a Gaussian density (see Fig. 12.3). Moreover, the Karplus curve is not one-to-one, we therefore obtain a multi-modal posterior distribution.

Without taking into account prior structural knowledge on the dihedral angle $\varphi$, the posterior distribution is multi-modal. However, we know that in proteins for all amino acids except glycine the right half of the Ramachandran plot is almost not populated. This fact is encoded in the force field $E(\varphi)$ used in ISD. Figure 12.4 shows the prior distribution, the likelihood and the resulting posterior distribution for the alanine dipeptide example. The posterior distribution is the product of the prior and likelihood. Hence, regions in conformational space that are unlikely to be populated are masked out. The resulting posterior distribution has only a single mode, in contrast to the multi-modal likelihood function.

Similar arguments hold for $\sigma$. Assuming that we know the correct angle, the posterior distribution of the inverse variance (precision) $\lambda = \sigma^{-2}$ is the Gamma distribution:

$$p(\lambda | \mathbf{d}, \varphi) = \frac{(\chi^2(\varphi)/2)^{n/2}}{\Gamma(n/2)} \,\lambda^{n/2-1} \exp\left\{-\lambda\,\chi^2(\varphi)/2\right\}. \qquad (12.13)$$

12 Inferential Structure Determination from NMR Data

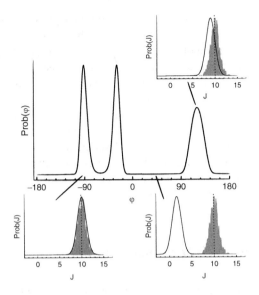

**Fig. 12.3** Posterior distribution of the $\varphi$ angle in the alanine dipeptide example. Regions with high posterior probability $p(\varphi)$ corresponding to $\varphi$ values that maximize the overlap between the empirical distribution of the data (*grey*) and the model (*dashed*) where only the $\varphi$ angle is considered a free parameter and the error $\sigma$ is fixed. Low posterior probability regions correspond to angular values that minimize the fit between the distribution of the data and the model. Because the Karplus curve is a non-linear, many-to-one forward model the posterior density is non-Gaussian and multi-modal

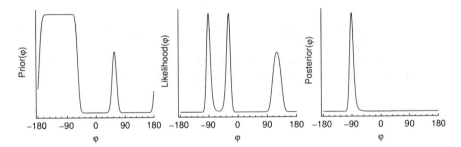

**Fig. 12.4** Prior probability, likelihood, and posterior density of the alanine dipeptide example. A priori the $\varphi$ angle is mostly restricted to negative values (*left panel*). This helps to disambiguate the likelihood function (*middle panel*). Multiplication of the prior with the likelihood results in a posterior distribution (*right panel*) that exhibits a single mode whose width encodes how accurately $\varphi$ is determined by the data

The expected precision and its uncertainty are:

$$\bar{\lambda} \equiv \langle \lambda | \varphi \rangle = \frac{n}{\chi^2(\varphi)}, \quad \langle (\lambda - \bar{\lambda})^2 | \varphi \rangle = \frac{2n}{[\chi^2(\varphi)]^2}. \quad (12.14)$$

This example illustrates that often the conditional posterior distribution of the nuisance parameters are of a standard form, which allows us to estimate them analytically. The estimation of $\sigma$ shows that it is indispensable to use a probabilistic framework to estimate nuisance parameters. As discussed before, the likelihood function evaluates the overlap between the empirical distribution of the data and a parametric model. If the model were not normalized, the cross-entropy score would be meaningless, and the overlap could be maximized by simply increasing the amplitude or width $\sigma$ of the model. However, in normalized models the amplitude and width of the model are directly related via the normalization condition. When working in a hybrid energy framework it becomes very difficult, if not impossible to derive normalization constants and thereby additional terms that allow for a correct extension of the hybrid energy function.

### 12.2.3 Relation to Hybrid Energy Minimization

It is straightforward to derive a hybrid energy function from the posterior distribution. Maximization of the posterior distribution with respect to the conformational degrees of freedom (a so-called MAP estimate, see Sect. 12.3) yields the structure that is most probable given the available data and other information that is incorporated in the likelihood and prior distribution. Instead of maximizing the posterior probability it is numerically more convenient to minimize its negative logarithm (the logarithm is a monotonic one-to-one mapping and does not change the location of optima). The negative logarithm of the posterior distribution (Eq. 12.9) is

$$-\log p(\mathbf{x}, \boldsymbol{\alpha}, \sigma) = -\log L(\mathbf{x}, \boldsymbol{\theta}) + \beta E(\mathbf{x}) - \log \pi(\boldsymbol{\alpha}) - \log \pi(\sigma) \quad (12.15)$$

where constant terms have been dropped on the right hand side. Considering only terms that depend on the conformational degrees of freedom we obtain a pseudo-energy

$$-\log L(\mathbf{x}, \boldsymbol{\theta}) + \beta E(\mathbf{x})$$

that has the same functional form as the hybrid energy (Eq. 12.1) if we identify $-\log L(\mathbf{x}, \boldsymbol{\theta})$ with $w_{\text{data}} E_{\text{data}}(\mathbf{x})$ and $\beta E(\mathbf{x})$ with $E_{\text{phys}}(\mathbf{x})$. Assuming our generic model for NMR data (Eq. 12.8) we obtain the correspondence:

$$w_{\text{data}} E_{\text{data}}(\mathbf{x}) = -\sum_i \log g(y_i \mid f(\mathbf{x}, \boldsymbol{\alpha}), \sigma). \quad (12.16)$$

Therefore, $-\log g(y \mid f(\mathbf{x}; \boldsymbol{\alpha}), \sigma)$, viewed as a function of the structure, implies a restraint potential resulting from measurement $y$.

Bayesian theory can be used to set up the hybrid energy and clarifies some of the shortcomings of the optimization approach to biomolecular structure calculation.

There are two major problems with the hybrid energy approach: First, it fails to treat nuisance parameters as unknowns. The nuisance parameters are treated as fixed quantities that need to be set somehow, additional terms in Eq. 12.15 originating in the prior and normalization constant of the likelihood are neglected. Second, the minimization approach fails to acknowledge the probabilistic origin of the hybrid energy. The hybrid energy is treated merely as a target function for optimization, its interpretation as a conditional posterior distribution over protein conformational space is not seen. Moreover, hybrid energy minimization only locates the maxima of the posterior distribution but does not explore the size of the maxima. Often it is argued minimization by MDSA "samples" the hybrid energy function. We will show later that the meaning of "sampling" by multi-start MDSA from randomized initial conformations and velocities has no statistical meaning.

## 12.3 Applying ISD in Practice

It is relatively straightforward to write down a prior distribution and likelihood function and to obtain a posterior distribution after invoking Bayes' theorem. The difficult part is to derive numerical results. For very simple inference tasks it might be possible to calculate analytical estimates. For more complex problems this becomes virtually impossible. In non-linear systems, the posterior distribution tends be complicated in the sense that it is not of a standard form and exhibits multiple modes such that the Laplace approximation [62] that approximates the posterior with a multivariate Gaussian centered about the MAP estimate will be inappropriate.

Often parameter estimates are derived from the posterior distribution by locating its (global) maximum; in case of completely uninformative priors this is equivalent to maximum likelihood estimation. The Bayesian analog is called a Maximum A Posteriori (MAP) estimate. However, reducing the posterior distribution to a MAP estimate has serious flaws (or as Peter Cheeseman put it [109]: "MAP is crap"). First, MAP only locates a single maximum, which might even not be the global one, and therefore is inappropriate whenever the posterior distribution is multi-modal. Second, a MAP procedure does not take into account the multiplicity or "entropy" of posterior modes. If the posterior distribution has a needle-shaped global maximum and a less pronounced but very broad second maximum, this second mode will carry a much larger portion of the total probability mass. In this case there will be a large discrepancy between the global posterior maximum and the posterior mean. Third, MAP only locates the most probable parameter values but fails to make statements about the reliability or uncertainty of the estimates. Hence, it throws away most of the benefits of a probabilistic approach to data analysis. Finally, a MAP estimate is not invariant under parameter transformation. According to Rosenblatt [613] any probability distribution (also multivariate densities) can be transformed to a uniform distribution over the hypercube. In this parameterization, MAP becomes completely meaningless because all possible estimates are equally probable.

**Fig. 12.5** Gibbs sampler used to estimate the conformational degrees of freedom **x**, the parameters of the forward model $\alpha$ and the parameters of the error model $\sigma$. The number of Gibbs sampling iterations $T$ has to be chosen large enough such that the Markov chain converges. After convergence ("burn-in"), all samples are valid samples from the joint posterior

$$
\begin{aligned}
&t \leftarrow 0 \\
&\textbf{while } t < T \textbf{ do} \\
&\quad \alpha^{(t+1)} \sim p(\alpha | \mathbf{x}^{(t)}, \sigma^{(t)}) \\
&\quad \sigma^{(t+1)} \sim p(\sigma | \mathbf{x}^{(t)}, \alpha^{(t+1)}) \\
&\quad \mathbf{x}^{(t+1)} \sim p(\mathbf{x} | \alpha^{(t+1)}, \sigma^{(t+1)}) \\
&\quad t \leftarrow t + 1
\end{aligned}
$$

The most adequate numerical method of Bayesian inference is statistical sampling from the posterior distribution [457, 535]. Almost all quantities that one is interested in involve the evaluation of a high-dimensional integral over the posterior distribution. Correctly sampled posterior states can be used to approximate this otherwise intractable integration. A correct sampling algorithm generates posterior samples such that in the limit of infinitely many samples a histogram over the samples converges to the posterior distribution. If this is guaranteed, the samples can be used to calculate averages, marginal distributions, and other integrals over the posterior distribution.

### 12.3.1 The Gibbs Sampler

The sampling problem that we encounter in ISD is to generate properly weighted samples from the joint posterior density of the protein conformational degrees of freedom and nuisance parameters. We use a Gibbs sampling scheme [212] to decompose the sampling of all these parameters into sampling sub-tasks. We successively update the theory parameters, the error parameters, and the conformational degrees of freedom. Each of these updates involves the generation of a sample from the conditional posterior distribution in which the parameters that are not updated are set to their current values. The Gibbs sampler then cycles through the various sampling sub-tasks (see Fig. 12.5).

In case of the models for NMR parameters, we can use off-the-shelf random number generators to generate random samples of the theory and error parameters. An update of the error parameters involves the generation of a random sample from the Gamma distribution [270]. Theory parameters such as alignment tensors, Karplus parameters, calibration factors, etc. are normally or log-normally distributed. Thus sampling these parameters does not pose a problem.

To update the conformational degrees of freedom is much more problematic. The conditional posterior distribution $p(\mathbf{x} | \alpha, \sigma)$ is the product of the likelihood functions and the canonical ensemble. To sample protein structures properly from the canonical ensemble is already a formidable, highly non-trivial task [288]. This is so because in a protein structure the conformational degrees of freedom are strongly correlated. Changes in one atom position can influence the allowed spatial region of atoms that are very far away in sequence. A random walk Metropolis scheme will be hopelessly inefficient because even small changes in dihedral angles can add up to cause major structural changes in the following part of the polypeptide chain. It is therefore important to take these correlations into account when proposing a new structure.

We use the Hamiltonian Monte Carlo (HMC) method [161, 536] to sample dihedral angles from their conditional posterior probability. The trick of HMC is to first blow up the problem by the introduction of angular momenta that follow a Gaussian distribution. The negative logarithm of the product of the distributions of the momenta and dihedral angles defines a Hamiltonian. The proposal of a new conformation proceeds as follows. A Molecular dynamics (MD) calculation in dihedral angle space is started from the current angles and from random momenta generated from a Gaussian distribution. The angles and momenta are updated using the leapfrog method [5]. Finally, the move is accepted according to Metropolis' criterion. The acceptance probability is determined by the difference in Hamiltonian before and after running the MD calculation.

### 12.3.2 Replica-Exchange Monte Carlo

The Gibbs sampler has various shortcomings, among which the most serious is non-ergodicity. The Markov chain gets trapped in a single mode and fails to sample the entire posterior distribution. Posterior samples that have been generated with a non-ergodic sampler will yield biased estimates. The remedy is to apply an idea from physics. A system that is trapped in a meta-stable state can reach the equilibrium after heating and subsequent annealing (as opposed to quenching). This idea is used, for example, in simulated annealing (SA). A problem here is that SA can still get trapped if the cooling schedule is not appropriate (e.g. if the system is not cooled slow enough). Moreover SA is an optimization not a sampling method. SA does not aim at generating properly weighted samples from a probability distribution. An inherent problem is that SA generates a single trajectory in state-space and thus can still end up in a suboptimal state.

The replica-exchange Monte Carlo (RMC) or parallel tempering method [704] fixes some of the flaws of simulated annealing. Similar to SA, RMC uses a temperature-like parameter to flatten the probability density from which one seeks to generate samples. However, in contrast to SA, RMC does not keep only a single heat-bath whose temperature is lowered but maintains multiple copies of the system, so-called replicas, at different temperatures. These systems do not interact and are sampled independently but they are allowed to exchange sampled states if the probability for such an exchange satisfies the Metropolis criterion. Hence RMC is a hierarchical sampling algorithm. The individual replicas are simulated using some standard sampling method (referred to as "local sampler") such as random walk MC, molecular dynamics, or Gibbs sampling. The local samplers are combined using a meta-Metropolis sampler that generates "super-transitions" of the entire replica system.

Obviously, the choice of the temperatures influences the sampling properties of the replica-exchange method. Several issues need consideration. First, the temperature of the "hottest" replica has to be chosen such that the local sampler can operate ergodically. Second, the temperature spacing between neighboring replicas has to be such that the exchange rate is high enough (i.e. the energy distributions need to overlap considerably). The advantages of RMC over SA are manifold. First, RMC is a proper sampling method that, after convergence, yields correctly weighted samples, SA does not generate statistically correct samples. Second, RMC maintains heat-baths at all temperatures throughout the entire simulation. Therefore states that are trapped in local modes can still escape the modes when continuing the simulation. Third, several indicators monitor the performance of RMC sampling. Low exchange rates indicate that mixing of the replica chain is probably not achieved within short simulation times. Trace plots of the replica energies and the total energy can be used to check for non-convergence.

In ISD, we use two temperature-like parameters that separately control the influence of the data and force field [268]. The data are weighted by scaling the likelihood function $L^{\lambda}$. The force field is modified by using the Tsallis ensemble instead of the Boltzmann ensemble. In the Tsallis ensemble, a new parameter $q$ is introduced that controls the degree of non-linearity in the mapping of energies. For $q > 1$, energies are transformed using a logarithmic mapping:

$$E_q(\mathbf{x}) = \frac{q}{\beta(q-1)} \log\{1 + \beta(q-1)(E(\mathbf{x}) - E_{min})\} + E_{min} \qquad (12.17)$$

where $E_{min} \leq E(\mathbf{x})$ must hold for all configurations $\mathbf{x}$. In the low energy regime $\beta(q-1)(E(\mathbf{x}) - E_{min}) \ll 1$, the Tsallis ensemble reduces to the Boltzmann ensemble. In particular it holds that $E_{q=1}(\mathbf{x}) = E(\mathbf{x})$. The logarithmic mapping of energies facilitates conformational changes over high-energy barriers.

### 12.3.3 ISD Software Library

ISD is implemented as a software library [603] that can be downloaded from

http://www.isd.bio.cam.ac.uk/

The software is written in Python, but computation-intensive routines are implemented in C. ISD includes some of the following features. Protein structures are parameterized in dihedral angles. Covalent parameters are kept fixed but could, in principle, be estimated as well. Cartesian forces are mapped into dihedral angle space using a recursive algorithm. The stepsize used during HMC is updated using the following heuristic. Acceptance of a new sample increases the stepsize by multiplication with a factor slightly larger than one. Rejection of a new sample reduces the stepsize by multiplication with a factor slightly smaller than one. After a user-defined exploration phase, the stepsizes are fixed to their median values. The library provides support for running RMC posterior simulations on computer clusters.

### 12.3.4 Posterior Sampling Versus Hybrid Energy Minimization

MDSA is the standard method for protein structure calculation. It is sometimes claimed that multi-start MDSA samples protein conformational space. Here the notion of sampling is different from random sampling from a probability density function. To illustrate this point we applied MDSA to the alanine dipeptide example with one conformational degree of freedom. To obtain a multi-modal hybrid energy/posterior distribution the force field is neglected in the calculations. We assume that a coupling of $J = 10$ Hz has been measured for Karplus parameters $A = 10$ Hz, $B = 2$ Hz, and $C = 1$ Hz. Equation $J = J(\varphi)$ has two exact solutions: $\varphi = -91.36°$ and $\varphi = -28.64°$ (torsion degrees with a phase of $60°$). Both angles correspond to the highest probability modes of the posterior density (see Fig. 12.6). The third mode at $\varphi = 120°$ for which the measured angle is not exactly reproduced is locally closest to the observed coupling. In the multi-start MDSA calculations, 1,000 independent minimizations were run using CNS [83]. Figure 12.6 shows the posterior histogram of the $\varphi$ angles calculated from the structures optimized with CNS. The optima are located correctly but the width of the ensemble is completely underestimated. Moreover, the relative population of the modes is incorrect. This shows that MDSA is a means to explore the hybrid energy with the aim of locating

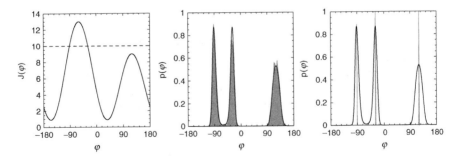

**Fig. 12.6** Comparison of RMC with multi-start MDSA. The *left panel* shows the Karplus curve used to generate noise-free mock data indicated by the dashed *horizontal line*. The *middle* and *right panels* show the posterior distribution of the dihedral angle as *solid black line*. Histograms have been compiled (shown in *gray*) from sampled dihedral angles using the RMC method (*middle panel*) and multi-start MDSA (*right panel*)

global maxima. If the MDSA protocol were perfect the same structure would be found irrespective of the starting positions and velocities, and the ensemble would collapse to a single structure. That is, "sampling" multi-start MDSA produces ensembles that are not statistically meaningful. MDSA ensembles may be too sharp if the protocol is very efficient. But they can also produce ensembles that are too heterogeneous if the protocol is not adapted to the quality of the data. This is often the case for sparse data [601].

## 12.4 Applications

### 12.4.1 Probabilistic Models for Structural Parameters Measured by NMR

#### 12.4.1.1 Modeling Nuclear Overhauser Effects

Nuclear Overhauser effect (NOE) data are by far the most important and informative measurements for protein structure determination [772]. The NOE is a second order relaxation effect. Excited nuclear spins relax back to thermal equilibrium, the speed of the relaxation process is modulated by the environment of the spins. The isolated spin-pair approximation (ISPA) [677] neglects that the relaxation involves the entire network of nuclear spins and is governed by spin diffusion. Still the ISPA is the most widespread theory to model NOE intensities and volumes. The ISPA simplifies the exchange of magnetization to a pair of isolated, spatially close spins of distance $r$. A more formal derivation builds on relaxation matrix theory [460] and considers short mixing times (initial-rate regime). According to the ISPA the intensity $I$ of an NOE is proportional to the inverse sixth power of the distance such that our forward model for NOE data is

$$I(\mathbf{x}) = \gamma \left[ r(\mathbf{x}) \right]^{-6} \qquad (12.18)$$

and involves the calibration factor $\gamma > 0$ which will be treated as a nuisance parameter.

What is an appropriate error model for NOEs? The absolute scale of a set of NOEs has no physical meaning. Therefore deviations between scaled and unscaled intensities should have the same likelihood. Mathematically this means that the error model has to show a scale invariance:

$$g(\gamma I | \gamma I(\mathbf{x}))\gamma = g(I | I(\mathbf{x})).$$

This relation must hold for any choice of the scale. For $\gamma = 1/I(\mathbf{x})$ we have:

$$g(I | I(\mathbf{x})) = g(I/I(\mathbf{x}) | 1)/I(\mathbf{x}) = h(I/I(\mathbf{x}))/I(\mathbf{x})$$

12 Inferential Structure Determination from NMR Data 305

where $h(\cdot) = g(\cdot|1)$ is a univariate density defined on the positive axis. We are free in the choice of $h$. A maximum entropy argument yields the lognormal distribution as the least biasing distribution if we assume that the average log-error of the ratio $I/I(\mathbf{x})$ is zero and its variance to be $\sigma^2$. Our error model for NOE intensities is

$$g(I|I(\mathbf{x}),\sigma) = \frac{1}{\sqrt{2\pi\sigma^2}\,I} \exp\left\{-\frac{1}{2\sigma^2}\log^2(I/I(\mathbf{x}))\right\}.$$

If we combine this error model with the forward model (ISPA, Eq. 12.18) we obtain:

$$p(I|\mathbf{x},\sigma,\gamma) = \frac{1}{\sqrt{2\pi\sigma^2}\,I} \exp\left\{-\frac{1}{2\sigma^2}\log^2(I/\gamma\,[r(\mathbf{x})]^{-6})\right\} \quad (12.19)$$

One neat feature about the lognormal model (Eq. 12.19) is that it does not distinguish between restraints involving intensities or distances. A problem with classical relaxation matrix calculations that try to fit NOE intensities in a more quantitative way than the ISPA is that a harmonic restraint potential has the effect that it over-emphasizes large intensities corresponding to short, non-informative distances between spins of the same residue or sequentially adjacent residues. On the other hand, a harmonic potential defined on intensities under-emphasizes small intensities that correspond to tertiary contacts defining the protein topology. The functional form of the log-normal model is invariant under scaling transformations of the intensities. Therefore, calculations based on intensities or distances lead to the same results, only the range of the error $\sigma$ is different.

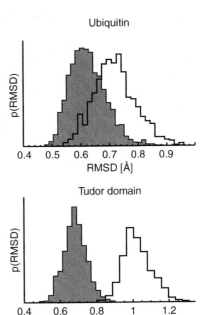

**Fig. 12.7** Improvement in accuracy for the lognormal model. The *upper panel* shows results for a structure calculation using NOE data measured on Ubiquitin (PDB code 1D3Z). The *lower panel* shows results for a structure calculation of the Tudor domain [642]. Displayed are histograms of $C\alpha$-RMSD values calculated between the X-ray structure and ensemble members calculated with a lognormal model (*gray fill*) and the standard *lower-upper*-bounds potential (*no fill*) as implemented in CNS

We could show that use of the lognormal model results in more accurate structures than the standard approach involving distance bounds [602]. Figure 12.7 shows the distribution of $C\alpha$-RMSD values between X-ray and NMR structures for two structure calculations from NOE data. Both data sets were analyzed using a lognormal model and a lower-upper-bound potential (FBHW, flat-bottom harmonic-wall) as implemented in CNS. The RMSD histograms are systematically shifted towards smaller values when using a lognormal model indicating that this model is able to extract more information from the data which results in more accurate structures. The lognormal model is also useful in conventional structure calculation where it implies a log-harmonic restraint potential, which was proven to be superior to distance bounds [547].

### 12.4.1.2 Modeling Three-Bond Scalar Coupling Constants

The basic model for scalar coupling constants was already introduced in the illustrative example (Sect. 12.2.2). In a fully Bayesian approach, also the Karplus coefficients need to be considered as free nuisance parameters [269]. The posterior distribution of the three Karplus coefficients is a three-dimensional Gaussian because they enter linearly into the forward model and because the error model is also Gaussian. Figure 12.8 shows the estimated Karplus curves from data measured on Ubiquitin. The sampled Karplus curves correspond well to those fitted to the crystal structure and reported in the restraint file 1D3Z.

### 12.4.1.3 Modeling Residual Dipolar Couplings

It is possible to measure dipolar couplings which usually average to zero for an isotropically tumbling molecule in solution. The measurement of dipolar couplings requires that the molecule is partially aligned either by an appropriate alignment medium [718] or an external electric or magnetic field [721]. The measured residual dipolar couplings (RCDs) provide orientational information on single- or two-bond vectors [26]. The strength of an RDC $d$ depends on the degree of alignment which is encoded in the alignment tensor $\mathbf{S}$ (also called Saupe order matrix) [625]. The alignment tensor is symmetric and has zero trace: $\mathbf{S}^T = \mathbf{S}$ and $\mathrm{tr}\,\mathbf{S} = 0$. We have

$$d = \mu\,\mathbf{r}^T\mathbf{S}\,\mathbf{r}/r^5 \tag{12.20}$$

where $\mathbf{r}$ is the three-dimensional bond vector that is involved in the dipolar coupling and $r$ is the length of the bond vector. Because of the constraints on $\mathbf{S}$, we use a parameterization involving five independent tensor elements $s_1, \ldots, s_5$:

$$\mathbf{S} = \begin{pmatrix} s_1 - s_2 & s_3 & s_4 \\ s_3 & -s_1 - s_2 & s_5 \\ s_4 & s_5 & 2s_2 \end{pmatrix}. \tag{12.21}$$

# 12 Inferential Structure Determination from NMR Data

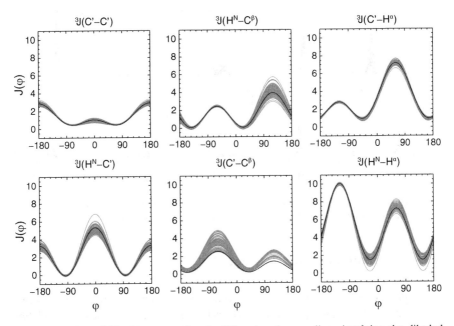

**Fig. 12.8** Estimated Karplus curves for six different scalar couplings involving the dihedral angle $\varphi$. The couplings were measured for ubiquitin. All parameters (structures, Karplus coefficients, and errors) were estimated simultaneously using the replica algorithm. The *gray curves* are sampled Karplus curves, the *black line* indicates the Karplus curve given in the literature (used in the determination of 1D3Z). The coupling type is indicated in the title of the panels

Using the vector representation $\mathbf{s} = (s_1, \ldots, s_5)^T$ we write an RDC as a vector product:

$$d = \mu \, \mathbf{s}^T \mathbf{a}(\mathbf{r}) \tag{12.22}$$

where

$$\mathbf{a}(\mathbf{r}) = (x^2 - y^2, 3z^2 - r^2, 2xy, 2xz, 2yz)^T / r^5 \tag{12.23}$$

and $\mathbf{r} = (x, y, z)^T$. The elements of the alignment tensor thus enter the forward model linearly. If we use a Gaussian error model the conditional posterior distribution of the alignment tensor will be a five dimensional Gaussian. We can therefore estimate the alignment tensor simultaneously with the structure [271] in very much the same way as we estimate the parameters of the Karplus curve. This has advantages over standard approaches that estimate the eigenvalues of the alignment tensor before the structure calculation and optimize only the relative orientation. Figure 12.9 shows the posterior histograms of the five tensor elements for two different alignments of Ubiquitin (data sets from PDB entry 1D3Z).

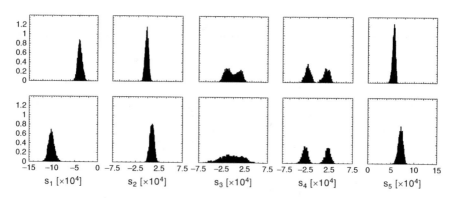

**Fig. 12.9** Estimated alignment tensors for two different alignments of ubiquitin. The *top* row shows the five tensor elements for the first alignment, the *bottom* row for the second alignment

## 12.4.2 Weighting Experimental Data

The hybrid energy function used in conventional optimization approaches involves a weighting constant $w_{data}$ that needs to be set before the actual structure calculation. Often the weighting constant is set to some default value which might not be harmful when working with complete and high-quality data. However, in less favorable situations or when dealing with multiple heterogeneous data sets it becomes vital to weight the data adaptively. In the hybrid energy minimization framework, the most objective method of weighting the experimental data is cross-validation [79]. The basic idea behind cross-validation is to not use all data in the actual structure calculation but to put a certain fraction of the data aside and use them only to assess if the calculated structures suffer from under- or over-fitting. In X-ray crystallography, this recipe results in the free R-value that is the de facto standard quality measure and also reported in the PDB [79]. In NMR, the analog of the free R-factor are cross-validated root mean square (RMS) differences between observed and back-calculated NMR observables such as NOE intensities [82]. The Bayesian approach estimates the weighting factor simultaneously with the structure. To do so, it uses the fact that $w_{data} = 1/\sigma^2$ [270] as is obvious from the connection between likelihood and restraint potential $E_{data}$ (cf. Eq. 12.8).

To compare Bayesian against cross-validated weighting of experimental data, tenfold complete cross-validation [82] was applied to a data set measured on the HRDC domain [447]. The R-value of the working data set was evaluated. Its evolution reflected in the monotonically decreasing blue dashed line in Fig. 12.10 illustrates that large weights bear the risk of over-fitting the data. The black line is the free R-value evaluated on the test data that were not used in the structure calculation. The minimum of the free R-value curve indicates weighting factors that neither lead to under- or over-fitting of the experimental data. It is in this region where also the posterior probability of the weights peak in a Bayesian analysis (gray histogram). Also the accuracy as monitored by the $C\alpha$-RMSD to

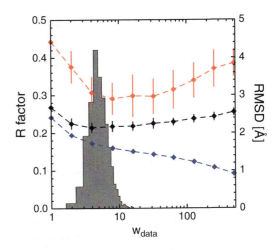

**Fig. 12.10** Bayesian vs. cross-validated weighting of experimental data. The *blue* (*black*) *dashed line* indicate the R-value of the work (test) data. The *gray histogram* is the posterior distribution of the weights in a Bayesian analysis. The *red curve* shows the accuracy of the structures as monitored by the $C\alpha$-RMSD to the crystal structure

the crystal structure becomes minimal for these weights. That is, cross-validation and Bayesian weighting provide very similar results and yield the most accurate structures. The advantages of the Bayesian approach over cross-validation is that it uses all data, whereas cross-validation relies on some test data. If the data are intrinsically coupled the question arises whether to choose the test data randomly or in a more sophisticated manner that takes the correlations into account [82]. Moreover, in case of sparse data the removal of test data may significantly affect and eventually spoil the convergence of the optimization protocol. A Bayesian analysis is extensible to multiple data sets without increasing the computational burden significantly, whereas multi-dimensional cross-validation would become very time-consuming.

### 12.4.3 A Probabilistic Figure of Merit

The estimated error not only balances the various sources of structural information but also provides a useful figure of merit. This is intuitively clear because the error evaluates the quality of a data set, and high quality data provide more reliable structures than low quality data. In the same way cross-validated "free" R-values [79] or RMS values [82] are used to validate crystal or NMR structures. To illustrate the use of $\sigma$ for validation let us discuss the structure determination of the Josephin domain. At the time of analysis, two NMR structures [466, 542] of the Josephin domain were available. The two structures agree in the structural core but differ significantly in the conformation of a long helical hairpin that is potentially involved in functional interactions. Basically, one structure showed an "open" hairpin (PDB code 1YZB) whereas the alternative structure has a "closed" hairpin (PDB code 2AGA). A recalculation of the two controversial structures using ISD showed a clear difference in their estimated errors [543]. Figure 12.11 shows the distribution

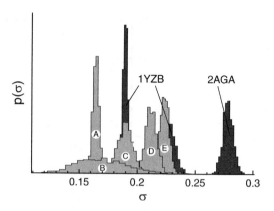

**Fig. 12.11** Estimated error as figure of merit. Shown are posterior histograms of the estimated error $\sigma$. The *light gray* histograms are results from ISD calculations on reference data sets (A: Ubiquitin, B: Fyn-SH3 domain [601], C: Bpti [53], D: Tudor domain, E: HRDC domain). The *dark gray* histograms are the $\sigma$ distributions for the two NMR structures of the Josephin domain

of $\sigma$ values for 1YZB and 2AGA in comparison with other NMR data sets. Several conclusions can be drawn from this figure. First, the 2AGA structure with the closed hairpin has a larger error than the reference data sets. This indicates that the 2AGA data are of a lower quality than the other data sets. Moreover, the errors of the 1YZB data sets lie well within the region that is expected from the reference data. The left peak of the 1YZB data set corresponds to unambiguously assigned NOE data, whereas the right peak is the error distribution of the ambiguous distance data. This is reasonable because the unambiguous data are more precise and have a higher information content than the ambiguous data. These findings suggest that the 1YZB structure ("open") is more reliable than the 2AGA structure ("closed"). This is confirmed by complementary data that were not used in the structure calculation [543]: Additional RDC measurements can be better fitted with 1YZB and also small-angle scattering curves are more compatible with an open structure.

## 12.5 Conclusion and Outlook

Bayesian inference has some distinct advantages over conventional structure determination approaches based on non-linear optimization. A probabilistic model can be used to motivate and interpret the hybrid energy function. A probabilistic hybrid energy (the negative log-posterior probability) comprises additional terms that determine nuisance parameters that otherwise need to be set heuristically or by cross-validation. A Bayesian approach requires that we generate random samples from the joint posterior distribution of all unknown parameters. This can take significantly more time (one or two orders of magnitude) than a structure calculation by minimization, which is the major drawback of a Bayesian approach. Table 12.1

# 12 Inferential Structure Determination from NMR Data

**Table 12.1** Comparison of conventional NMR structure determination by hybrid energy minimization and inferential structure determination

| Hybrid energy minimization | Inferential structure determination |
|---|---|
| Hybrid energy | Posterior probability |
| Force field | Conformational prior distribution |
| Restraint potential | Likelihood |
| Non-linear optimization | Posterior sampling |
| Heuristics for nuisance parameters | Estimation of nuisance parameters |
| Ensemble from multi-start MDSA | Ensemble by statistical sampling |

provides a compact comparison between the conventional approach based on hybrid energy minimization and inferential structure determination.

Originally, ISD has been applied in the context of NMR structure determination with the intention to provide a solid basis for validating NMR structures. Other applications have emerged in the meantime. ISD has been used to calculate the first three-dimensional structure of a mitochondrial porin from hybrid crystallographic and NMR data [28]. The ISD software has also been used to explore the conformational variability of the N domains of an archaeal proteasomal ATPase [152]. The formalism presented in Sect. 12.2 is completely general and readily applies to other structural data. In the future, several developments are envisioned including structure determination from new NMR parameters such as chemical shifts, sparse NMR data and low-resolution data such as electron density maps reconstructed from single-particle cryo-electron microscopy. The incorporation of concepts and ideas from protein structure prediction methods will turn ISD into a tool that allows probabilistic structure calculation also from low-quality data.

**Acknowledgements** ISD has been developed in close collaboration with Wolfgang Rieping and Michael Nilges (Institut Pasteur, Paris). This work has been supported by Deutsche Forschungsgemeinschaft (DFG) grant HA 5918/1-1 and the Max Planck Society.

# Chapter 13
# Bayesian Methods in SAXS and SANS Structure Determination

**Steen Hansen**

## 13.1 Introduction

### 13.1.1 Small-Angle Scattering

Small-angle scattering (SAS) is an experimental technique which may be used to derive size and shape information about large molecules in solution (e.g. [177,228]). Also molecular interaction and aggregation behavior may be studied by small-angle scattering. Solution scattering experiments using X-rays (SAXS) were initiated in the 1950s and scattering experiments using neutrons (SANS) followed about a decade later. Static light scattering (SLS) which probes larger sizes than SAXS and SANS is frequently used as a complementary technique e.g. for detection of aggregation.

Small-angle scattering is especially well suited for studying many types of biological structures as shown in Fig. 13.1. For biomolecules it is useful that this technique allows them to be studied in solution and thus in a state which is more likely to preserve the biologically active form. Furthermore, the scattering takes place in a volume which contains a large number of scatterers and this provides information about *an average* of the structures in the solution.

Due to differences in contrast between SAXS and SANS as shown in Fig. 13.2, for some biological materials it may be possible to obtain extra information by using both techniques. Using SANS the scattering length of the solvent can be varied by substituting $H_2O$ with $D_2O$ to enhance different parts of a complex scatterer. Another advantage of SANS is that radiation damage to biological molecules can be avoided completely. However due to easier access to facilities and to the higher

---

S. Hansen
Department of Basic Sciences and Environment, University of Copenhagen, Faculty of Life
Sciences, Thorvaldsensvej 40, DK-1871 FRB C, Copenhagen, Denmark
e-mail: slh@life.ku.dk

T. Hamelryck et al. (eds.), *Bayesian Methods in Structural Bioinformatics*,
Statistics for Biology and Health, DOI 10.1007/978-3-642-27225-7_13,
© Springer-Verlag Berlin Heidelberg 2012

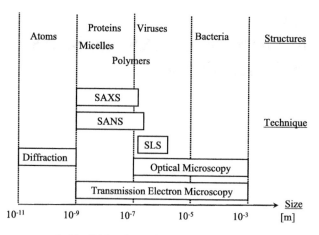

**Fig. 13.1** Size ranges probed by SAS and complementary techniques

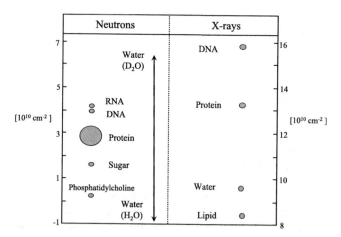

**Fig. 13.2** SAXS and SANS scattering lengths relevant for biological samples

flux available for X-rays SAXS is more frequently used. Using high-flux SAXS from e.g. a synchrotron it may be possible to follow processes involving structural changes in real time.

The information content of experimental SAS data is usually relatively low. For SAXS it is not unusual to be able to determine less than ten parameters from the data, while SANS suffering from experimental smearing and lower counting statistics may only offer half this number. Therefore small angle scattering is frequently used in combination with other experimental techniques for solving structural problems.

## 13.1.2 Data Analysis in Small-Angle Scattering

The data analysis applied to small-angle scattering data has undergone significant changes since the introduction of the technique. In the early days of small-angle scattering only a few simple parameters characterizing the scatterer were deduced from the experimental data, like e.g. the scattering-length-weighted radius of gyration or the volume of the molecule. The computational advances of the 1970s and 1980s made it possible to make simple trial and error models of the scatterer using a few basic elements (spheres, cylinders etc.). Also, it allowed the estimation of a one dimensional real space representation (a "pair distance distribution function" [225]) of the scattering pattern by indirect Fourier transformation. The real space representation of the scattering data facilitated the interpretation of the measurements. In combination these direct and indirect procedures made it possible to extract much more information about the scatterer than previously.

During the last decade estimation of three-dimensional structures from the one-dimensional scattering pattern [699, 701] has proven to be a new powerful tool for analysis of small-angle scattering data – especially for biological macromolecules. However the difficulties associated with the assessment of the reliability of the three dimensional estimates still complicate this approach. Much prior information (or assumptions) has to be included in the estimation to obtain real space structures which do not differ too much when applying the method repeatedly. Furthermore the three dimensional estimation is frequently rather time consuming compared to the more traditional methods of analysis.

## 13.1.3 Indirect Fourier Transformation

A direct Fourier transform of the scattering data to obtain a real space representation of the scattering data would be of limited use due to noise, smearing and truncation of the data. An *indirect Fourier transformation* (IFT) also preserves the full information content of the experimental data, but an IFT is an underdetermined problem where several (and in this case often quite different) solutions may fit the data adequately. Consequently some (regularization-)principle for choosing one of the many possible solutions must be used if a single representation is wanted. Various approaches to IFT in SAS have been suggested [278, 282, 520, 698, 700], but the most frequently used principle is that of Glatter [225–227, 524], who imposed a smoothness criterion upon the distribution to be estimated giving higher preference to smoother solutions. This is in good agreement with the prior knowledge that most SAS experiments have very low resolution. Consequently this method has demonstrated its usefulness for analysis of SAS data for more than three decades.

For application of the original method of Glatter it was necessary to choose (1) a number of basis functions (usually on the order of 20–40), (2) the overall noise level as well as (3) the maximum dimension of the scatterer. Various ad hoc guidelines for how to make these choices were provided in the original articles

### 13.1.4  Bayesian Methods

Due to improved computing facilities with regard to both hardware and software, the restriction in the number of basis functions mentioned above must now be considered redundant. However the choice of the overall noise level as well as the maximum dimension of the scatterer still hamper the applicability of Glatter's method – following the guidelines on how to choose these *hyperparameters* one is often left with a relatively wide range of choices. Other methods for IFT in SAS face similar problems with the determination of hyperparameters. As the method of Glatter is by far the most frequently used for IFT in SAS the applicability of the Bayesian method for selection of hyperparameters is demonstrated in the following using Glatter's method as an example.

Estimation of the noise level of the experiment may not be of great interest on its own and consequently this parameter may often be integrated out using Bayesian methods. However the maximum diameter of the scatterer is clearly an important structural parameter, which has been estimated by various ad hoc methods (for example, [526]).

Using a Bayesian approach a two-dimensional probability distribution for the hyperparameter associated with the noise level and for the maximum diameter may be calculated by probability theory. From this distribution it is possible to make unique choices for these two hyperparameters. Also reliable error estimates for the real space distribution to be estimated as well as for the hyperparameters may be calculated from the probability distribution.

Using the Bayesian method the parameters and hyperparameters are all determined uniquely by probability theory and in principle the only choice left is which method of regularization should be used. However the most likely regularization method may also be found by integration over all the hyperparameters yielding the posterior probability for each method.

### 13.1.5  Non-dilute Solutions

The original method of Glatter treated measurements which were done at low concentrations where inter particle effects could be neglected and where the measured data only referred to intra (single) particle contributions to the scattering intensity. This is useful as it is frequently possible to make measurements using very diluted samples of just a few mg/ml. However this may not always be the case. For many problems the assumption of a dilute solution does not hold. In some cases the structure of interest is only to be found in non-dilute solutions. This means

# 13 Bayesian Methods in SAXS and SANS Structure Determination

that experiments have to be done with relatively concentrated samples and that the subsequent analysis of the measured data has to take inter particle effects into account.

For such non-dilute systems the "Generalized Indirect Fourier Transformation" (GIFT) extension of IFT was introduced by Brunner-Popela and Glatter [84]. In GIFT inter particle effects are taken into account by including a structure factor in the calculations. The inclusion of the structure factor leads to a non linear set of equations which has to be solved either iteratively or by Monte Carlo methods.

Using GIFT the interaction between the scatterers has to be specified by the user, who has to choose a specific structure factor. On one hand this requires some extra a priori information about the scattering system, but on the other hand the choice of a structure factor allows estimation of relevant parameters describing the interaction such as the charge of scatterers and their interaction radius. Further input parameters may be needed such as temperature and dielectric constant of the solvent. The estimation of parameters from the model may be useful (provided of course that the chosen model is correct), but correlations between the parameters may also reduce the advantages of the approach [203].

Using a Bayesian approach and expressing the function describing the (real space) structure of the scatterer as a combination of an intra particle contribution and an inter particle contribution with appropriate constraints, it is possible to separate the contributions in real space leading to the form factor and the structure factor in reciprocal space (assuming that the scatterers are not very elongated). In this situation it is not necessary to specify a structure factor to be used for the indirect transformation. Only a rough estimate of the shape of the scatterer is necessary and this estimate may be made from the scattering profile itself. The downside of this approach is that less detailed information may be deduced from the data. However this is also the case for the original IFT-method which nonetheless has proved to be a most useful supplement to direct model fitting for analysis of small-angle scattering data.

## 13.2 Small-Angle Scattering

### 13.2.1 Overview

In small-angle scattering the intensity $I$ is measured as a function of the length of the scattering vector $q = 4\pi \sin(\theta)/\lambda$, where $\lambda$ is the wavelength of the radiation and $\theta$ is half the scattering angle (Fig. 13.3). For scattering from a dilute solution of randomly oriented monodisperse molecules of maximum dimension $d$, the intensity can be written in terms of the distance distribution function $p(r)$ [228]:

$$I(q) = 4\pi n V \int_0^d p(r)\frac{\sin(qr)}{qr}\,dr. \qquad (13.1)$$

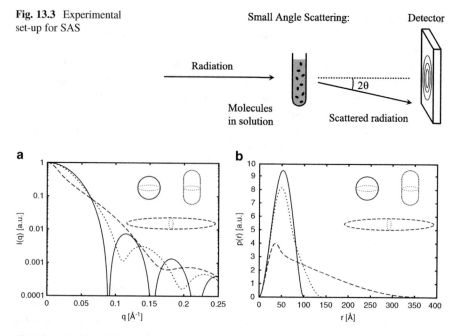

**Fig. 13.3** Experimental set-up for SAS

**Fig. 13.4** (a) Scattering intensities for homogeneous scatterers of different shapes. (b) Corresponding distance distribution functions

where $n$ is the (average) number density of the particles and $V$ is the volume of one particle.

The distance distribution function is related to the density-density correlation $\gamma(r)$ of the scattering length density $\rho(\mathbf{r})$ by

$$p(r) = r^2\gamma(r) = r^2 < \int_V \rho(\mathbf{r}')\rho(\mathbf{r}+\mathbf{r}')\,d\mathbf{r}' >, \qquad (13.2)$$

where $\rho(\mathbf{r})$ is the scattering contrast, given by the difference in scattering density between the scatterer $\rho_{sc}(\mathbf{r})$ and the solvent $\rho_{so}$, i.e. $\rho(\mathbf{r}) = \rho_{sc}(\mathbf{r}) - \rho_{so}$, $< \cdot >$ means averaging over all orientations of the molecule.

Examples of scattering intensities for a few simple geometrical shapes and their corresponding distance distribution functions are shown in Fig. 13.4

For uniform scattering density of the molecule the distance distribution function is proportional to the probability distribution for the distance between two arbitrary scattering points within the molecule.

If the distance distribution is known, *the Guinier radius $R_g$* may be calculated from $p(r)$ according to the formula [228]:

$$R_g^2 = \frac{\int p(r)r^2\,dr}{2\int p(r)\,dr}, \qquad (13.3)$$

13 Bayesian Methods in SAXS and SANS Structure Determination

It is seen that the Guinier radius is the scattering length density weighted radius of gyration for the scatterer.

Also from Eq. 13.1 *the forward scattering* $I(0)$ is related to $p(r)$ through

$$I(0) = 4\pi n V \int_0^d p(r)\,dr \qquad (13.4)$$

from which the volume $V$ of the scatterer may be calculated when the contrast and the concentration of the sample is known.

If $p(r)$ is not known, it may be possible to estimate $R_g$ and $I(0)$ from a plot of $\ln(I(q))$ against $q^2$ as

$$\ln(I(q)) \approx \ln(I(0)) - R_g^2 q^2/3 \qquad (13.5)$$

This approximation is generally valid for $q \ll 1/R_g$, which is often a very small part of the scattering data (measurements close to zero angle are prevented by the beam stop, the purpose of which is to shield the sensitive detector from the direct beam).

For non-uniform scattering density the distance distribution may have negative regions (if the scattering density of some region of the scatterer is less than the scattering density of the solvent).

## 13.2.2 Non-dilute Solutions

The simple interpretation of the distance distribution function $p(r)$ has to be modified for high concentrations to take the inter particle effects into account. The most obvious effect of an increase in concentration is usually that the calculated $p(r)$ exhibits a negative part around the maximum diameter of the scatterer. This is mainly caused by the excluded volume effect and the consequent reduction in the effective scattering length density near the scatterer.

In this case the total distance distribution function may be divided into three parts from the intra- and inter-particle contributions according to Kruglov [399]

$$p(r) = p_1(r) + \eta p_{excl}(r) + \eta p_{struct}(r) \qquad (13.6)$$

where $\eta$ is the volume fraction ($\eta = nV$), $p_1(r)$ is the distance distribution function of a single particle, $p_{excl}(r)$ is the distance distribution function of the excluded volume and $p_{struct}(r)$ is the remaining part of the total distance distribution function which depends on the mutual arrangement of the scatterers outside the excluded volume.

The different contributions to $p(r)$ for a sphere of diameter $100\,\text{Å}$ and a volume fraction of $\eta = 0.1$ can be seen in Fig. 13.5.

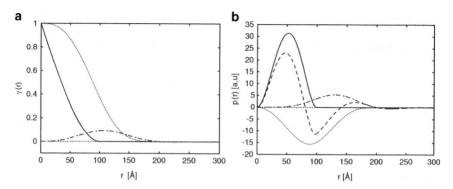

**Fig. 13.5** (a) *Full line*: $\gamma_1(r)$ for sphere of diameter 100 Å. *Dotted line*: Corresponding $-\gamma_{excl}(r)$. *Dashed-dotted line*: $\gamma_{struct}(r)$. (b) *Full line*: $p_1(r)$ for sphere of diameter 100 Å. *Dotted line*: $p_{excl}(r)$ for spheres of diameter 100 Å and volume fraction $\eta = 0.1$. *Dashed-dotted line* $p_{struct}(r)$ for spheres of diameter 100 Å $\eta = 0.1$. *Dashed line*: Total distance distribution function $p(r)$ according to Eq. 13.6 [280]

For a monodisperse solution $p_{excl}(r)$ is due to the perturbation of the distribution of distances caused by the fact that the centers of two molecules cannot come closer than the minimum dimension of the molecules. At distances larger than twice the maximum dimension $p_{excl}(r) = 0$. The introduction of inter particle effects increases the integration limit of Eq. 13.1 from the maximum dimension $d$ of the single molecule to that of the maximum length of the interaction (which may in principle be infinite). The first term on the right hand side of Eq. 13.6 determines the form factor $P(q)$ when Fourier transformed according to Eq. 13.1 and the last two terms determine the structure factor $S(q)$ specified below. Correspondingly the intensity in Eq. 13.1 can be divided into a part which is due to intra particle effects – the form factor $P(q)$ – and a part which is due to the remaining inter particle effects – the structure factor $S(q)$.

$$I(q) \propto S(q)P(q), \tag{13.7}$$

For dilute solutions $S(q) = 1$ and the measured intensity is given by the form factor $P(q)$. Equation 13.7 is valid for spherical monodisperse particles, but frequently it is assumed to hold true also for slightly elongated particles with some degree of polydispersity [202].

The structure factor can be written

$$S(q) = 1 + 4\pi n \int_0^\infty h(r)r^2 \frac{\sin(qr)}{qr} dr \tag{13.8}$$

where $h(r)$ is the total correlation function [554], which is related to the radial distribution (or pair correlation) function $g(r)$ [790] for the particles by

$$h(r) = g(r) - 1 \tag{13.9}$$

13   Bayesian Methods in SAXS and SANS Structure Determination

For hard spheres $g(r) = 0$ for $r < d$, where $d$ is the diameter of the sphere. For a system of monodisperse hard spheres the functional form for $S(q)$ is known [572]. Hayter and Penfold [299] have given the analytical solution for $S(q)$ for a system of particles interacting through a screened Coulomb potential.

From Fig. 13.5b it is also seen that for spheres $p_1(r)$ and $p_{\text{struct}}(r)$ have their support mainly in different regions of space. This means that if $p_{\text{excl}}(r)$ is given, it may be possible to estimate $p_1(r)$ and $p_{\text{struct}}(r)$ separately from experimental data. As the form of $p_{\text{excl}}(r)$ is only dependent upon the geometry of the particle, it requires less information for an IFT than a complete determination of a structure factor $S(q)$ which requires specification of the interaction between the particles.

## 13.3   Indirect Fourier Transformation

From Eq. 13.1 for dilute scattering the distance distribution function $p(r)$ may be approximated by $\mathbf{p} = (p_1, \ldots, p_N)$ and the measured intensity at a given $q_i$ written as

$$I(q_i) = \sum_{j=1}^{N} \mathbf{T}_{ij} \, p_j + e_i \tag{13.10}$$

where $e_i$ is the noise at data point $i$ and and matrix $\mathbf{T}$ is given by

$$\mathbf{T}_{ij} = 4\pi \Delta r \sin(q_i r_j)/(q_i r_j), \tag{13.11}$$

where $\Delta r = r_j - r_{j-1}$. The aim of the indirect Fourier transformation is to restore $\mathbf{p}$ which contains the full information present in the scattering profile.

### 13.3.1   Regularization

The estimation of $\mathbf{p}$ from the noisy scattering data is an underdetermined and ill-posed problem. To select a unique solution among the many which may fit the data adequately, regularization by the method of Tikhonov and Arsenin [717] may be used. Tikhonov and Arsenin estimated a distribution $\mathbf{p} = (p_1, \ldots, p_N)$ by minimizing a new functional written as a weighted sum of the chi-square $\chi^2$ and a regularization functional $K$:

$$\alpha K(\mathbf{p}, \mathbf{m}, \rho) + \chi^2 \tag{13.12}$$

where $\alpha$ is a Lagrange multiplier, which may be found by allowing the $\chi^2$ to reach a predetermined value (assuming the overall noise level to be known). The regularizing functional is given by the general expression

$$K(\mathbf{p}, \mathbf{m}, \rho) = \|\mathbf{p} - \mathbf{m}\|^2 + \rho \|\mathbf{p}'\|^2 \tag{13.13}$$

where the prime indicates a derivative (first and/or higher) of $\mathbf{p}$. The first term minimizes the deviation of $\mathbf{p}$ from the prior estimate $\mathbf{m} = (m_1, \ldots, m_N)$ with respect to a given norm and the second term imposes a smoothness constraint on the distribution to be estimated.

The $\chi^2$ is defined in the conventional manner i.e.

$$\chi^2 = \sum_{i=1}^{M} \frac{(I_{\mathrm{m}}(q_i) - I(q_i))^2}{\sigma_i^2} \tag{13.14}$$

where $I_{\mathrm{m}}(q_i)$ is the measured intensity and $\sigma_i$ is the standard deviation of the noise at data point $i$.

For choice of regularization functional the expression $K = \int p''(x)^2 \mathrm{d}x$ is frequently used giving preference to smooth functions $p(r)$ (double prime here indicating the second derivative). Assuming $p(0) = p(d) = 0$, this regularization expression takes the discrete form

$$K = \sum_{j=2}^{N-1} \left( p_j - \frac{(p_{j-1} + p_{j+1})}{2} \right)^2 + \frac{1}{2} p_1^2 + \frac{1}{2} p_N^2 \tag{13.15}$$

## 13.3.2 Smoothness Constraint

The method of Glatter [225] is an implementation of the method of Tikhonov and Arsenin and uses only the last term in Eq. 13.13 to impose a smoothness constraint upon $\mathbf{p}$. In Glatter's method the distance distribution function was written as a sum of cubic B-splines: $p(r) = \sum_{j=1}^{N} a_j B_j(r)$. The smoothness constraint (similar to Eq. 13.15) was given by the sum $K = \sum_{j=1}^{N}(a_{j+1} - a_j)^2$ which was minimized subject to the constraint that the $\chi^2$ took some sensible value [225]. This problem left two parameters to be determined: the maximum diameter used $d$ and the noise level $\alpha$ determining the relative weighting of the constraints from the data and the smoothness respectively. The number of basis functions $N$ should be chosen sufficiently large as to accommodate the structure in the data.

The estimation of $\mathbf{p}$ is very sensitive to the choice of $\alpha$ and $d$. The noise level was found by the so called 'point of inflexion' method by plotting $K$ and the $\chi^2$ as a function of the Lagrange multiplier $\alpha$. Using this method a plateau in $K$ was to be found when $\chi^2$ had reached a low value and this region determined the correct noise level. The maximum diameter $d$ was found in a similar manner by plotting the forward scattering $I(0)$ against $d$. A problem with the point of inflexion method is that the plateaus which should be used may not exist. Furthermore when a

13 Bayesian Methods in SAXS and SANS Structure Determination

plateau does exist, the point on the plateau which should be selected is not uniquely determined as the plateaus may be relatively wide.

### 13.3.3 The Maximum Entropy Method

Using only *first* term in Eq. 13.12 will give a regularization similar to that of *the maximum entropy method*. The norm is then to be replaced by the Shannon entropy which measures the distance between two distributions **f** and **m** [407, 646].

For regularization by *the maximum entropy method* [668] a constraint

$$K = \int [p(r)\ln(p(r)/m(r)) - p(r) + m(r)]\,dr \tag{13.16}$$

is used, which takes the discrete form

$$K = \sum_{j=1}^{N} p_j \ln(p_j/m_j) - p_j + m_j \tag{13.17}$$

where $(m_1, \ldots, m_N)$ is a prior estimate of $(p_1, \ldots, p_N)$. Using this method will bias the estimate towards the prior (i.e. for the case of no constraints from the experimental data, minimizing Eq. 13.12 will lead to $\mathbf{p} = \mathbf{m}$).

A second order Taylor approximation for $\mathbf{p} \approx \mathbf{m}$ of Eq. 13.17 will lead to [687]

$$K \approx \sum_{j=1}^{N} [(p_j - m_j)^2/2m_j] \tag{13.18}$$

From this equation it can be seen that using a prior $m_j = (p_{j+1} + p_{j-1})/2$ the maximum entropy constraint corresponds to the smoothness constraint Eq. 13.15 in a new metric defined by the denominator $2m_j$ in Eq. 13.18. Using this metric will combine the positivity constraint of Eq. 13.17 with the smoothness constraint of Eq. 13.15.

### 13.3.4 Non-dilute Solutions

For non-dilute solutions $p$ in Eq. 13.1 may be replaced by the sum of $p_1$, $\eta p_{\text{excl}}$ and $\eta p_{\text{struct}}$ as given by Eq. 13.6. In the examples shown in Sect. 13.4.4 $p_1$ was regularized using Eq. 13.18 with $m_j = (p_{j+1} + p_{j-1})/2$ as mentioned above, while the conventional constraint Eq. 13.15 was used for $p_{\text{struct}}$.

For the shape $p_{\text{excl}}$ of an ellipsoid of revolution may be used. This only requires one extra parameter – the axial ratio for the ellipsoid – because the maximum diameter is known from $p_1$. The axial ratio may be estimated from the shape of $p_1$ or alternatively it can enter the estimation as an additional hyperparameter.

Furthermore for non-dilute solutions it can assumed that $p_{\text{struct}} \approx 0$ for $r < 0.5d$ (Fig. 13.5) which holds true for scatterers which do not have large axial rations.

Also as $S(q) \to 1$ for $q \to \infty$ Eq. 13.7 gives:

$$I(q) \propto S(q)P(q) \to P(q) \quad \text{for} \quad q \to \infty \tag{13.19}$$

which can be written

$$\text{FT}[p_1(r) + p_{\text{excl}}(r) + p_{\text{struct}}(r)] \to \text{FT}[p_1(r)] \quad \text{for} \quad q \to \infty \tag{13.20}$$

where FT denotes the Fourier transform of Eq. 13.1. Consequently it must hold that

$$\text{FT}[p_{\text{excl}}(r) + p_{\text{struct}}(r)] \to 0 \quad \text{for} \quad q \to \infty \tag{13.21}$$

which may also used for constraining $p_{\text{struct}}$.

## 13.4 Bayesian Analysis of SAS Data

### 13.4.1 General Case

To incorporate IFT in a Bayesian framework, the functional form of the regularization constraints ("smoothness", "maxent" etc.) are considered to be "models" and the parameters of the models determine the distribution of interest. The hyperparameters, such as the maximum diameter of the scatterer or the Lagrange multiplier associated with the noise level of the experiment, are considered to be part of the models. Using Gaussian approximations around the maximum probability for the parameters in each model, the total probability of each model (including the hyperparameters) can be calculated by integration over the parameters as shown below.

When applied to the problem of inferring which model or hypothesis $H_i$ is most plausible after data $D$ have been measured, Bayes' theorem gives for this posterior probability $p(H_i|D)$ that

$$p(H_i|D) = p(D|H_i) \cdot p(H_i)/p(D). \tag{13.22}$$

where $p(D|H_i)$ is the probability of the data $D$ assuming that the model $H_i$ is correct, $p(H_i)$ denotes the prior probability for the model $H_i$ which is assumed constant for "reasonable" hypotheses (i.e. different hypotheses should not be ascribed different prior probabilities) and $p(D)$ is the probability for measuring the data which amounts to a renormalization constant after the data have been measured.

# 13 Bayesian Methods in SAXS and SANS Structure Determination

The *evidence* $p(D|H_i)$ for a hypothesis can be found by integrating over parameters $\mathbf{p} = (p_1, \ldots, p_N)$ in the model $H_i$ which can be written:

$$p(H_i|D) \propto p(D|H_i) = \int p(D, \mathbf{p}|H_i) \, \mathrm{d}^N \mathbf{p}, \qquad (13.23)$$

where $N$ is the number of parameters in the model, and $d$ is the maximum diameter of the sample. Again using Bayes' theorem

$$p(D|H_i) = \int p(D|H_i, \mathbf{p}) p(\mathbf{p}|H_i) \, \mathrm{d}^N \mathbf{p} \qquad (13.24)$$

where the likelihood is written

$$p(D|H_i, \mathbf{p}) = \exp(-L)/Z_L \qquad (13.25)$$

$$Z_L = \int \exp(-L) \, \mathrm{d}^M D \qquad (13.26)$$

$M$ being the number of data points. For the usual case of Gaussian errors $L = \chi^2/2$ and $Z_L = \prod(2\pi\sigma_i^2)^{-1/2}$ where $\sigma_i$ is the standard deviation of the Gaussian noise at data point $i$.

Correspondingly it is now assumed that the prior probability for the distribution $\mathbf{p}$ can be expressed through some functional $K$ (to be chosen) and written

$$p(\mathbf{p}|H_i) = \exp(-\alpha K)/Z_K \qquad (13.27)$$

$$Z_K = \int \exp(-\alpha K) \mathrm{d}^N \mathbf{p} \qquad (13.28)$$

By this expression the model $H_i$ is the hypothesis that the prior probability for the distribution of interest $\mathbf{p}$ can be written as above with some functional $K$ and a parameter $\alpha$ which determines the "strength" of the prior (through $K$) relative to the data (through $\chi^2$). Both the functional form of $K$ as well as the value for the parameter $\alpha$ are then part of the hypothesis and are subsequently to be determined.

Inserting Eqs. 13.25 and 13.27 in Eq. 13.24 and writing $Q = -\alpha K - \chi^2/2$, the evidence is given by

$$p(D|H_i) = \frac{\int \exp(-\alpha K) \exp(-\chi^2/2) \, \mathrm{d}^N p}{Z_K Z_L} = \frac{\int \exp(Q) \, \mathrm{d}^N p}{Z_K Z_L} \qquad (13.29)$$

Using Gaussian approximations for the integrals and expanding $Q$ around the maximum in $\mathbf{p}$ writing $\mathbf{A} = \nabla\nabla K$, $\mathbf{B} = \nabla\nabla\chi^2/2$ evaluated at the maximum value of $Q(\mathbf{p}_0) = Q_0$ where $\nabla Q = 0$:

$$p(D|H_i) = \frac{(2\pi)^{-N/2} \exp(Q_0)\det^{-1/2}(\alpha\mathbf{A} + \mathbf{B})}{(2\pi)^{-N/2} \exp(K_{\max})\det^{-1/2}(\alpha\mathbf{A}) \prod(2\pi\sigma_i^2)^{-1/2}} \qquad (13.30)$$

where $K_{\max}$ is the maximum for the functional $-K$. Usually $K_{\max} = 0$ which will be assumed in the following – otherwise just a renormalizing constant is left. Furthermore the term $\prod(2\pi\sigma_i^2)^{-1/2}$ from the experimental error is redundant for comparison of different hypothesis and is left out in the following. The probability for different hypothesis each being equally probable a priori can then be calculated from the expression

$$p(D|H_i) = \det^{1/2}(\mathbf{A}) \exp(Q_0)\det^{-1/2}(\mathbf{A} + \alpha^{-1}\mathbf{B}) \qquad (13.31)$$

$$\log p(D|H_i) = \frac{1}{2} \log \det(\mathbf{A}) - \alpha K_0 - \chi_0^2/2 - \frac{1}{2} \log(\det(\mathbf{A} + \alpha^{-1}\mathbf{B})) \qquad (13.32)$$

In the previous the hyperparameters were implicitly included in the models or hypothesis. Now writing $H_i$ for the model *without* the hyperparameters $\alpha$ and $d$ again we obtain from Bayes' theorem that the posterior probability is determined by

$$p(D, \alpha, d|H_i) = p(D|\alpha, d, H_i)p(\alpha)p(d), \qquad (13.33)$$

where $p(\alpha)$ is the prior probability for $\alpha$ and $p(d)$ is the prior probability for $d$. Assuming $\alpha$ to be a scale parameter [340] gives $p(\alpha) = \alpha^{-1}$ which should be used for the prior probability of $\alpha$. For a parameter $d$ with relatively narrow a priori limits, the prior probability should be uniform within the allowed interval.

The Lagrange multiplier $\alpha$, the maximum diameter $d$ of the scatterer and – for the case of non-dilute solutions: the volume fraction $\eta$ – are all hyperparameters which can be estimated from their posterior probability $p$ for a set $(\alpha, d, \eta)$ after data have been measured. This probability is calculated using Gaussian approximations around the optimal estimate $\mathbf{p}_{opt}$ for a given set of hyperparameters and integrating over all solutions $\mathbf{p}$ for this particular set of hyperparameters [261, 456]. Using the regularization from Eq. 13.15, writing $\mathbf{A} = \nabla\nabla K$ and $\mathbf{B} = \nabla\nabla\chi^2/2$, the probability of a set of hyperparameters $(\alpha, d, \eta)$ may be written [278]:

$$p(\alpha, d, \eta, a) \propto \frac{\exp(-\alpha K - \chi^2/2)}{\det^{1/2}(\mathbf{A} + \alpha^{-1}\mathbf{B})} \qquad (13.34)$$

In Eq. 13.34 both matrices as well as $(-\alpha K - \chi^2/2)$ have to be evaluated at the point $\mathbf{p}$ where $\exp(-\alpha K - \chi^2/2)$ takes its maximum value.

Using Eq. 13.34 the most likely value for each hyperparameter can be found from the optimum of the probability distribution and an error estimate for the hyperparameters can be provided from the width of the distribution.

As the Bayesian framework ascribes a probability to each calculated solution $\mathbf{p}$, an error estimate for the (average) distribution of interest is provided from the

13 Bayesian Methods in SAXS and SANS Structure Determination 327

individual probabilities of all the solutions (each solution corresponding to a specific choice of noise level, volume fraction and maximum dimension of the scatterer).

## 13.4.2 Dilute Monodisperse Samples

### 13.4.2.1 Estimation of $d$ and $\alpha$: Simulated Data

The first simulated example shown in Fig. 13.6a was taken from May and Nowotny [281, 282, 490]. The original distance distribution function for the simulated scatterer is shown in Fig. 13.6b.

In Fig. 13.6c is shown the conventional plot used to find the "point of inflexion", which is used to determine the noise level of the experiment using Glatter's method.

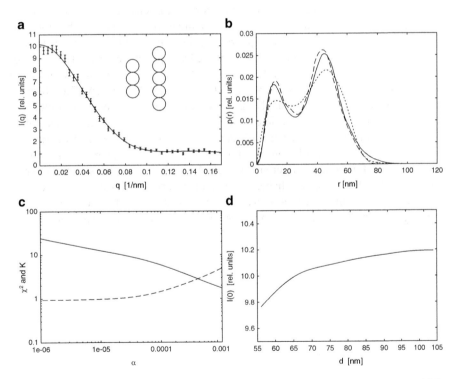

**Fig. 13.6** (a) Simulated data and fit to data. Insert shows the structure of the scatterer (eight spheres). (b) Distance distribution functions corresponding to data in (a). *Full line*: Original distribution. *Long dashes*: Calculation using maximum entropy. *Dotted*: Calculation using smoothness regularization. (c) The $\chi^2$ (*dashed*) and $K$ (*full line*), rescaled for clarity, calculated as a function of the Lagrange multiplier $\alpha$ using the correct maximum diameter $d = 78$ nm. (d) The forward scattering $I(0)$ calculated as a function of the maximum length $d$ of the scatterer using the correct noise level [278]

The curve showing $\chi^2$ as a function of $\alpha$ has a very large plateau for decreasing $\alpha$. The correct value is $\chi^2 = 1.0$. The curve showing $K$ has a negative slope for all $\alpha$'s, leaving a range of possible $\alpha$'s to be chosen for the "best" solution.

In Fig. 13.6d is shown the "forward scattering" $I(0)$ plotted as a function of the maximum diameter calculated assuming the correct noise level to be known a priori. This curve is to be used for estimation of the maximum diameter $d$. Similarly to the choice of over-all noise level it is not clear which diameter should be chosen.

Using the Bayesian method for selection of $\alpha$ and $d$ the posterior probability $p(D, \alpha, d|H_i)$ is shown in Fig. 13.7 displays a clear maximum making it simple to select the most likely set of hyperparameters $(\alpha, d)$.

Associated with each point on the two dimensional surface $p(\alpha, d)$ in Fig. 13.7 is the distribution **p** which has the largest posterior probability for the given $(\alpha, d)$. From these distributions with their associated probabilities an average value for **p** as well as a standard deviation for each $p_i$ can be calculated.

In spite of the relatively large variation in diameters and noise levels used for calculation of the average **p**, the posterior probabilities ensure that unlikely values for the hyperparameters are ascribed relatively small weights, making the "total" or average estimate of the distribution **p** – or representation of $p(r)$ – well behaved.

In connection with the determination of the hyperparameters from the posterior probability distribution is should be noticed that the exact determination of the maximum diameter of the scatterer may be a rather difficult task. The distance distribution function $p(r)$ usually has a smooth transition to zero around the maximum diameter of the particle. This means the the maximum diameter may have to be estimated from a relatively small proportion of the total $p(r)$. E.g. for a prolate ellipsoid of revolution of semi axis $(45, 55)$ Å less than 1% of the total area of $p_1$ is found in the interval $[90; 110]$ Å and less than 0.1% in the interval $[100; 110]$ Å. This makes a reliable estimate of $d$ very difficult in these cases as a truncation of the tail of $p(r)$ around $d$ will give a better value for the regularization constraint without significant implications for the Fourier transformation of $p(r)$

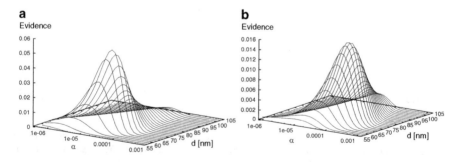

**Fig. 13.7** (a) Evidence for $\alpha$ and $d$ calculated from the simulated data shown in Fig. 13.6 using smoothness regularization. (b) Evidence for $\alpha$ and $d$ calculated from the simulated data shown in Fig. 13.1 using maximum entropy with a spherical prior [278]

and the corresponding quality of the fit of the data. To improve the estimates in such cases an additional smoothness constraint on $p(r)$ could be implemented in the region around the maximum diameter. Alternatively the maximum diameter may be estimated from the chord length distribution for the scatterer, which may also be estimated by IFT [279]. The chord length distribution does not suffer from a large contribution of inner point distances in the scatterer which makes it easier to estimate the maximum dimension.

### 13.4.2.2 Estimation of $d$ and $\alpha$: Experimental Data

Experimental SANS data from measurements on casein micelles and the estimated distance distribution function are shown in Fig. 13.8a. The posterior distribution for $\alpha$ and $d$ in Fig. 13.8b was calculated using a smoothness constraint and has a clear and well defined maximum. The average distance distribution in Fig. 13.8a appears to be free from artifacts although the data are fitted quite closely. The maximum dimension calculated from Fig. 13.8b is in good agreement with that calculated from a model fit. The error bars on $p(r)$ have been calculated from the posterior probability distributions as described above.

By comparison of the different methods for selection of hyperparameters it becomes evident that the Bayesian method provides clearer and correct answers. E.g. using Glatter's original ad hoc method for the determination of the noise level and the maximum diameter of the scatterer will leave the user with the subjective choice within a large interval of values.

Furthermore it is not necessary to restrict the analysis by using only the most likely distance distance distribution, but all solutions may be used with the associated probabilities. The means that deduced parameters of interest like

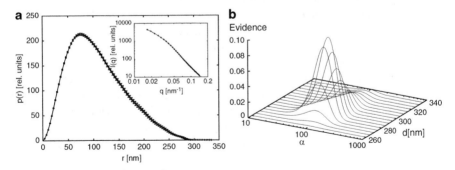

**Fig. 13.8** Bayesian estimation of hyperparameters for experimental data from SANS on casein. (**a**) Estimated distance distribution function. *Insert* shows data and fit to data. (**b**) Evidence for $\alpha$ and $d$. [278]

## 13.4.2.3 Comparison of Models

As mentioned previously, Bayesian methods may also be used for determining which regularizing functional should be used. After integration of the posterior probabilities over all the hyperparameters, a "total" probability for each method or hypothesis is obtained, which allows different methods for IFT to be compared. However in this case it must be considered that various experimental situations can and usually will correspond to different states of prior knowledge, which may consequently ascribe different prior probabilities to different models or regularization methods. E.g. in a situation where information about the scatterer is available from electron micrographs, the information about the shape may be expressed as a distance distribution function which may be used as a prior distribution using the maximum entropy method for the data analysis.

If regularizing by maximum entropy is assumed to have identical prior probability to the regularizing by smoothness (that is, if no prior information about the scatterer can be used), it appears that the smoothness criterion is most likely to perform best. The reason that Bayesian analysis ascribes larger probability to the smoothness criterion is that the smoothness criterion leads to broader probability distributions for the both the distance distribution function as well as the hyperparameters.

This agrees well with the intuitive notion that a more restrictive a model is a worse solution. Having to choose between to equally "complicated" models – in the sense of having equally many parameters and each model fitting the data equally well at the most likely choice of parameters, that is, having the identical likelihoods – the one should be chosen according to the largest *range* of parameters fitting the data adequately (largest accessible volume of hypothesis space). In other words, the model to be preferred is characterized by the property that the measurement of the data leads to the least reduction in the accessible volume of hypothesis space (the least informative model).

A numerical integration over all hyperparameters parameters allowing the total probabilities of each regularization method to be calculated confirms the lower probability for the maximum entropy method using a spherical prior. The maximum entropy method using a prior with more degrees of freedom – such as, for example, an ellipsoid of revolution – will be less restrictive and might perform better compared to a smoothness criterion.

Calculations indicate that *without* any form of prior knowledge about the scatterer a smoothness constraint is more likely to give a reliable result than the (conventional) maximum entropy method as the smoothness constraint represents a more general model thus introducing less prior information in the regularization.

## 13.4.3 Dilute Solutions and Two Size Distributions

### 13.4.3.1 Estimation of $p(d)$: Simulated Data

For testing the sensitivity of the Bayesian estimation of $d$ scattering data from mixtures of two different scatterers were simulated.

For comparison a simulation of scattering data from a monodisperse sample of spheres of diameter 20 nm is shown in Fig. 13.9 with the result of a Bayesian estimation of the distance distribution function and associated hyperparameters.

In Fig. 13.10a is shown the simulated scattering data from various mixtures of two spheres. The diameters of the spheres were 18 and 23 nm respectively and the ratio $I_2(0)/I_1(0)$ – index 2 referring to the larger sphere – was varied in the interval $[0.01; 0.20]$. For $I_2(0)/I_1(0) = 0.20$ the estimated distance distribution function $p(r)$ is shown in Fig. 13.10b. For a lower fraction $I_2(0)/I_1(0) = 0.04$ the evidence $p(\alpha, d)$ is shown in Fig. 13.10c. Finally in Fig. 13.10d the evidence for the maximum diameter $d$ is shown for all six different ratios of $I_2(0)/I_1(0)$ as indicated in the figure.

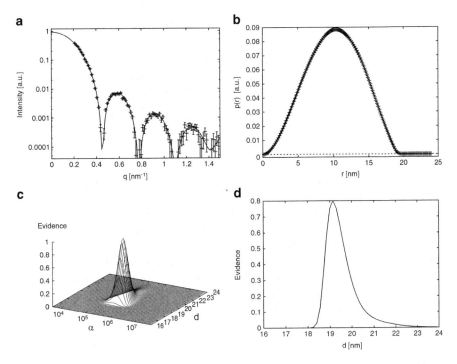

**Fig. 13.9** Estimation of $p(r)$ from simulated scattering data from a sphere of diameter 20 nm. (**a**) *Error bars*: Simulated scattering intensity. *Full line*: Fit. (**b**) Estimated distance distribution function with error bars. (**c**) Evidence $p(\alpha, d)$ scaled to 1. (**d**) $p(d)$ corresponding to (**c**)[740]

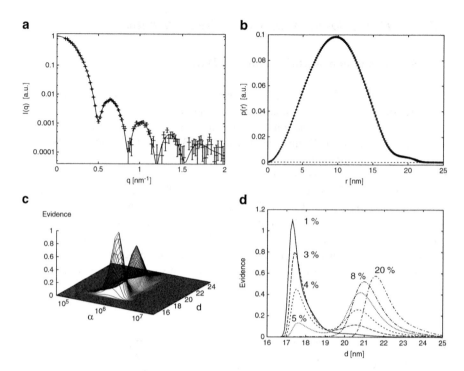

**Fig. 13.10** Estimation of $p(r)$ from simulated scattering data from spheres of diameter 18 and 23 nm in different ratios as described in the text. (**a**) *Error bars*: Simulated scattering intensity. *Full line*: Fit. (**b**) Estimated distance distribution function with error bars from data having 20% large spheres. (**c**) Evidence $p(\alpha, d)$ (scaled to 1) from data having 4% large spheres. (**d**) $p(d)$ from scattering data of spheres in different ratios (percentage of larger sphere indicated near the respective peaks) [740]

In Fig. 13.11 is shown the results of similar analysis of mixtures of ellipsoids of revolution. The diameters of the ellipsoids were (10,10,18) and (13,13,23) nm respectively. For low ratios of $I_2(0)/I_1(0)$ – index 2 referring to the larger ellipsoid – in the interval [1.5; 1.9] the data was truncated at $q = 0.10\,\text{nm}^{-1}$ and the estimate of the evidence for $d$ is shown in Fig. 13.11c. For truncation at a larger $q = 0.15\,\text{nm}^{-1}$ the corresponding results are shown in Fig. 13.11c. The results illustrate that for the higher $q_{\min}$ it is necessary to use higher ratios for the ellipsoids (in the interval [2.0; 3.0]) to obtain similar ratios for the two peaks in $p(d)$.

The influence of the measured $q$-range is seen by comparing Fig. 13.11c, d; as $q_{\min}$ is increased, the relative contribution to the data from the larger ellipsoid is decreased and consequently the ratio of the ellipsoids has to be increased to make both peaks appear in the evidence $p(d)$.

13 Bayesian Methods in SAXS and SANS Structure Determination 333

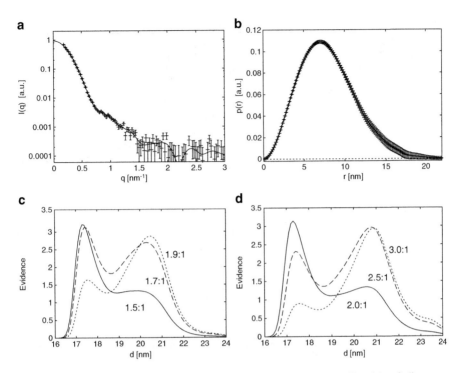

**Fig. 13.11** Estimation of $p(r)$ from simulated scattering data from ellipsoids of diameters (10,10,18) and (13,13,23) nm in different ratios as described in the text. (**a**) *Error bars*: simulated scattering intensity from mixture 2:1. *Full line*: Fit. (**b**) Estimated distance distribution function from data shown in (**a**). (**c**) Evidence $p(d)$ from data having $q_{min} = 0.10$ nm$^{-1}$. *Full line*: Ratio 1.5:1 *Dashed line*: Ratio 1.7:1. *Dotted line*: Ratio 1.9:1. (**d**) Evidence $p(d)$ from data having $q_{min} = 0.15$ nm$^{-1}$. *Full line*: Ratio 2:1. *Dashed line*: Ratio 2.5:1. *Dotted line*: Ratio 3:1 [740]

### 13.4.3.2 Estimation of $p(d)$: Experimental Data

Figure 13.12 shows experimental SAXS data [740] from protein concentrations between 2 and 8 mg/ml. One protein sample consists of molecules with an expected maximum diameter $d$ of 18 nm, the other consists of a mixture of molecules with a $d$ of 18 and 23 nm respectively.

The influence of the measured $q$-range is similar to the case for the simulated data. For the experimental data the two species present in the solution have different overall shapes, thus representing a more complex mixture than two different bodies of similar geometry but with varying radii (Fig. 13.12c). Again it is apparent from Fig. 13.12d that the relative heights of the two main peaks in $p(d)$ is influenced by the truncation at $q_{min}$. Therefore unless the geometry of the scatterers is known well a priori the presence of two peaks in the evidence $p(d)$ should only be used as an indication for polydispersity/degradation of the sample.

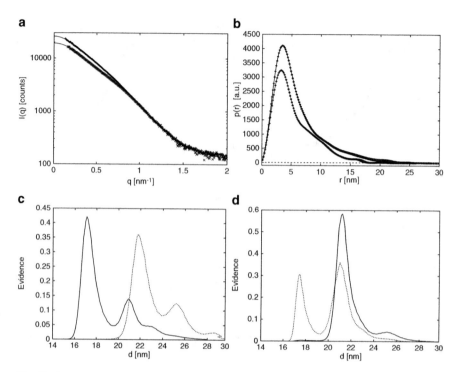

**Fig. 13.12** Estimation of $p(r)$ for experimental scattering data. (**a**) *Points*: Merged scattering data resulting from measurements at 2, 4 and 6 mg/ml. *Upper curve*: Mixture of large and small scatterer respectively. *Lower curve*: Sample of small scatterer. *Full lines*: Corresponding fits. (**b**) Estimated distance distribution functions from data shown in (**a**). *Upper* and *lower curve* as in (**a**). (**c**) Evidence $p(d)$ from data having $q_{min} = 0.15$ nm$^{-1}$. *Full line*: Corresponding to lower curve in (**a**). *Dotted line*: Corresponding to *upper curve* in (**a**). (**d**) Evidence $p(d)$ from mixture 1:1 of data sets shown in (**a**). *Full line*: Estimate using $q$-interval [0.15:2] nm$^{-1}$ *Dotted line*: Estimate using $q$-interval [0.2:0.3] nm$^{-1}$ [740]

From the simulated and experimental examples given here it is evident that the probability distribution for the maximum diameter $d$ of the scatterer which is provided by the Bayesian approach enables additional information to be extracted from the scattering data compared to the conventional ad hoc procedures for selecting $d$.

### 13.4.4 Non Dilute Solutions

#### 13.4.4.1 Estimation of the Excluded Volume Distribution $\eta$ and the Structure Factor $S(q)$

For non dilute solutions all three functions of Eq. 13.6

$$p(r) = p_1(r) + \eta p_{excl}(r) + \eta p_{struct}(r) \qquad (13.35)$$

contributing to the total distance distribution $p(r)$ have to be determined in addition to an extra hyperparameter; the volume fraction $\eta$. The correlation function $\gamma_1(r) = p_1(r)/r^2$ for a homogeneous particle [228] is proportional to the intersection volume of the particle and its ghost, shifted a distance $r$ (index 1 referring to intra particle contributions). To describe interactions between particles a "cross correlation function" $\gamma(r)_{\text{cross}}$ may be introduced [399] as the intersection volume of a particle with the ghost of second particle. Again the ghost is shifted a distance $r$ from the original position of the second particle. The excluded volume correlation function $\gamma_{\text{excl}} = p_{\text{excl}}(r)/r^2$ can then be found by integration of $\gamma(r)_{\text{cross}}$ over all positions of the first particle and the ghost of the second particle, where the two "real" particles (not the ghost) overlap. From the positions where the particles do not overlap the cross correlation function becomes $\gamma_{\text{struct}} = p_{\text{struct}}(r)/r^2$.

Excluded volume distributions for ellipsoids of revolution have been estimated using an approximative method similar to the method of Kruglov [399]. Monte Carlo simulations of two ellipsoids of arbitrary orientation and separation were used to estimate the fraction of ellipsoids which overlapped as a function of their center to center separation. This was done to distinguish between the contributions to $\gamma_{\text{excl}}$ and the contributions to $\gamma_{\text{struct}}$ (which is trivial for the case of spheres). For calculation of the average intersection volume of two ellipsoids the expressions for the correlation functions of prolate and oblate ellipsoids were used [525]. Using $p(r) = r^2 \gamma(r)$ some examples of the corresponding excluded volume distance distribution functions are shown in Fig. 13.13 for ellipsoids of revolution of various axial ratios.

For axial ratios $a$ between 0.1 and 10 the calculated excluded volume distributions were parameterized, which allowed the axial ratio of $p_{\text{excl}}(r)$ to be used as a free parameter. Having estimated $p_1(r)$ and $p_{\text{excl}}(r)$ *the volume fraction $\eta$ may then be calculated from*

$$\eta \approx \frac{-\int p_{\text{excl}}(r)\,dr}{8\int p_1(r)\,dr} \qquad (13.36)$$

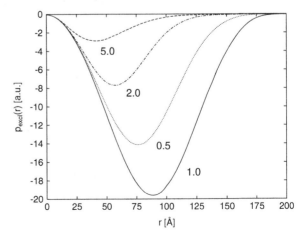

**Fig. 13.13** *Full line*: $p_{\text{excl}}(r)$ for a sphere of diameter 100 Å. *Dotted line*: $p_{\text{excl}}(r)$ for an oblate ellipsoid of maximum dimension 100 Å and axial ratio 0.5. *Dashed-dotted line*: $p_{\text{excl}}(r)$ for a prolate ellipsoid of maximum dimension 100 Å and axial ratio 2.0. *Dashed line*: $p_{\text{excl}}(r)$ for a prolate ellipsoid of maximum dimension 100 Å and axial ratio 5.0 [280]

deduced from the corresponding ratio for homogeneous spheres (where the excluded volume is $2^3 = 8$ times that of the sphere).

### 13.4.4.2 Estimation of $\eta$ and $S(q)$: Simulated Data

Simulated scattering data from monodisperse spheres of radius 50 Å is shown in Fig. 13.14a. From the reconstruction of $p_1$, $p_{excl}$ and $p_{struct}$ shown in Fig. 13.14b compared to the original distributions in Fig. 13.5b it is evident that the shape of the distributions are reproduced well from the simulated data. Calculations using ellipsoids of varying axial ratios give similar results estimating the axial ratio from the evidence or the shape of $p_1$. As expected the best results are obtained for low volume fractions and low noise levels [280].

In Fig. 13.15 are shown the probability distributions for the hyperparameters indicating good agreement with the simulations.

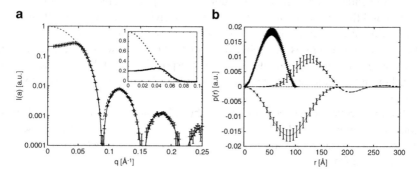

**Fig. 13.14** (a) *Error bars*: Simulated data points from spheres of radius 50 Å, volume fraction 0.2 and noise 1%. *Full line*: Fit of data. *Dotted line*: Estimated $P(q)$. *Insert* shows the intensity on a linear scale for $q < 0.1$ Å$^{-1}$. (b) *Error bars*: $p_1(r)$. Original $p_1(r)$ not discernible from the estimate. *Dotted line*: $p_{struct}(r)$. *Dashed-dotted line*: $p_{struct}(r)$ [280]

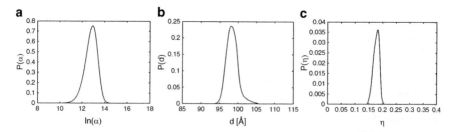

**Fig. 13.15** Posterior probabilities for hyperparameters from Fig. 13.14. (a) Probability for the Lagrange multiplier $\alpha$. (b) Probability for the maximum diameter $d$. (c) Probability for the volume fraction $\eta$ [280]

### 13.4.4.3 Estimation of $\eta$ and $S(q)$: Experimental Data

The results of SANS experiments using SDS at three different ionic strengths are shown in Fig. 13.16 [14]. The corresponding estimates of $p_1$, $p_{\text{excl}}$, $p_{\text{struct}}$ using the resolution function for the specific experimental setup [570] are shown in Fig. 13.16.

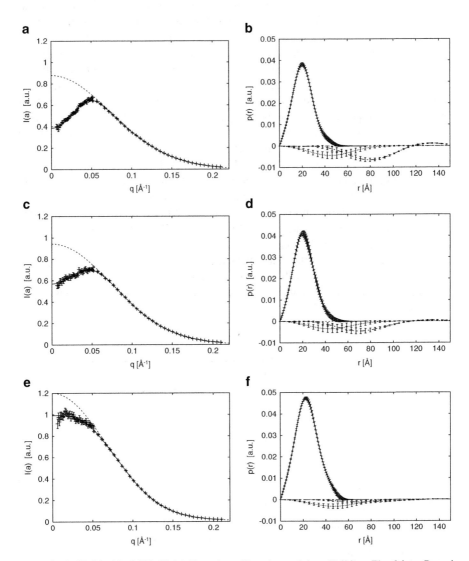

**Fig. 13.16** SDS in 20 mM NaCl. (**a**) *Error bars*: Experimental data. *Full line*: Fit of data. *Dotted line*: FT[$p_1(r)$]. (**b**) *Full line with error bars*: $p_1(r)$. *Dotted line with error bars*: $p_{\text{excl}}(r)$. *Dashed-dotted line with error bars*: $p_{\text{struct}}(r)$. SDS in 50 mM NaCl. (**c**) *Error bars*: Experimental data. *Full line*: Fit of data. *Dotted line*: FT[$p_1(r)$]. (**d**) *Full line with error bars*: $p_1(r)$. *Dotted line with error bars*: $p_{\text{excl}}(r)$. *Dashed-dotted line with error bars*: $p_{\text{struct}}(r)$. SDS in 250 mM NaCl. (**e**) *Error bars*: Experimental data. *Full line*: Fit of data. *Dotted line*: FT[$p_1(r)$]. (**f**) *Full line with error bars*: $p_1(r)$. *Dotted line with error bars*: $p_{\text{excl}}(r)$. *Dashed-dotted line with error bars*: $p_{\text{struct}}(r)$ [280]

The structure factors calculated from the estimates of $p_1$, $p_{excl}$, $p_{struct}$ are shown in Fig. 13.17 for all three examples.

The distance distribution functions $p_1$ in Fig. 13.16 all show spherical structures with a diameter about 5 nm. This is to be expected as the experiments were all done well above the critical micelle concentration for SDS. The small tails around the maximum diameter may indicate a small amount of polydispersity in the solutions which is also expected. The estimated volume fractions are consistent with the initial concentration of SDS and the presence of water molecules in the micelles (more water molecules are expected to be associated with SDS at low ionic strengths). Due to the low volume fractions the corresponding error estimates become relatively large. At decreasing ionic strength a negative region in $p_{struct}$ increases, which indicates the reduced density of micelles at this distance. The reduced density is caused by the repulsion between the charged head groups of the SDS-molecules at the surface of the micelles. The structure factors calculated from the estimates of $p_1$, $p_{excl}$ and $p_{struct}$ are shown in Fig. 13.17.

Figure 13.16 indicates an additional advantage of the free form estimation. Interpretation of data in reciprocal space is usually more difficult than the corresponding representation in real space which is one of the reasons that $p(r)$ is usually preferred to $I(q)$. In the approach suggested here $S(q)$ is represented by the real space distributions $p_{excl}(r)$ and $p_{struct}(r)$ which may allow interactions effects to be interpreted directly from the shape of $p_{struct}(r)$.

The analysis of the simulated and the experimental data shows that by applying Bayesian methods it is possible to obtain simultaneous estimates of $p_1$, $p_{excl}$ and $p_{struct}$, as well as estimates for the posterior probability distributions of the hyperparameters $\alpha$, $d$ and $\eta$.

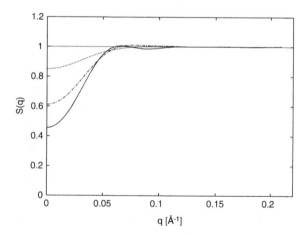

**Fig. 13.17** Calculated structure factors $S(q)$ for SDS experiments shown in Fig. 13.16. *Full line*: 20 mM NaCl. *Dashed-dotted line*: 50 mM NaCl. *Dotted line*: 250 mM NaCl [280]

## 13.5 Information in SAXS and SANS Data

For many purposes it may be useful to be able to quantify the information content of a given set of experimental data.

### 13.5.1 Shannon Channels

The sampling theorem [646] states that a continuous scattering curve $I(q)$ from an object of maximum diameter $d$ is fully represented by its values in a set of points (Shannon channels) at $q_n = n\pi/d$, where $n = (1,\ldots,\infty)$

The number of Shannon channels $N_s$ necessary to represent the intensity $I(q)$ in the interval $[q_{min}; q_{max}]$ is given by

$$N_s = d(q_{max} - q_{min})/\pi \qquad (13.37)$$

Consequently $N_s$ corresponds the maximum number of free (independent) parameters it is possible to determine from the experimental data.

It has been shown that the information in reciprocal space can be related to real space by expressing the scattering data and the corresponding distribution in real space as two series having identical coefficients, but different basis functions [520]. The first Shannon channel $I(q_1)$ at $q_1 = \pi/d$ in reciprocal space then determines the overall dimension of the corresponding Fourier representation of the data in real space. Hence for estimation of the maximum diameter of the scatterer, the first data point $q_{min}$ should be measured at $q_{min} \leq \pi/d$. Adding the higher channels improves the resolution of the data by adding finer details to $p(r)$.

Oversampling the data (i.e. using $\Delta q < \pi/d$) corresponds to measuring the extra Shannon channels within $[q_{min}; q_{max}]$ from the scattering of an object larger than $d$.

The actual number of parameters possible to determine from measured data in a given $q$-range is dependent upon the noise level of the experiment and the instrumental smearing. Furthermore the value of $q_{max}$ which enters into Eq. 13.37 is rarely uniquely defined due to large noise levels at large $q$-values.

### 13.5.2 The Number of Good Parameters

As a more realistic measure of the information content in the data the "number of good parameters" $N_g$ has been suggested using regularization by the *maximum entropy method* [261,525] :

$$N_g = \sum_{j=1}^{N} \frac{\lambda_j}{\alpha + \lambda_j} \qquad (13.38)$$

Here $\lambda_j$ are the eigenvalues of $\mathbf{B}$ and $\alpha$ is the Lagrange multiplier of Eq. 13.12. By this equation $N_g$ "counts" the number of eigenvalues which are large compared to the Lagrange multiplier $\alpha$, balancing the information in the data (eigenvalues for $\mathbf{B}$) against the weight of the regularizing functional or prior (eigenvalues for $\alpha\mathbf{A}$). For entropy regularization $\mathbf{A} = \mathbf{I}$, where $\mathbf{I}$ is the unity matrix. Hence Eq. 13.38 gives the number of directions in parameter space, which are determined well for the given noise level. Expressing the information content of the experimental data through Eq. 13.38 removes ambiguities due to the choice of $q_{max}$ as very noisy data do not contribute to $N_g$.

Overfitting the data will reduce the Lagrange multiplier $\alpha$ and reduce the eigenvalues of $\alpha\mathbf{A} + \mathbf{B}$ towards those of $\mathbf{B}$, increasing $N_g$ above the number of Shannon channels $N_s$.

Underfitting the data to a higher chi-square will increase the Lagrange multiplier $\alpha$. This leads to a lower value for $N_g$ calculated from Eq. 13.38 (consistent with a stronger correlation of the Shannon channels and reduction of information in the experimental data).

For the general case the denominator of Eq. 13.38 has to be replaced by the eigenvalues of the matrix $\alpha\mathbf{A} + \mathbf{B}$ (see e.g. [456]).

Using the eigenvalues of $\mathbf{B}$ for experimental design and analysis of reflectivity data has been suggested by Sivia and Webster [667].

### 13.5.3 Estimation of $N_g$

Deducing $N_g$ good parameters from the data corresponds to reducing the number of degrees of freedom for the $\chi^2$ by $N_g$, which is similar to the conventional reduction of the number of degrees of freedom for the $\chi^2$ by the number of fitting parameters. Fitting the "true" information in the data invariably leads to fitting of some of the noise as well [456]. Writing the reduced chi-square $\chi_r^2$ for $M$ data points leads to

$$\chi_r^2 = \frac{M}{M + N_g} \tag{13.39}$$

In addition to fitting some of the noise the estimate of $p(r)$ may extend beyond the true maximum dimension $d$ for the scatterer (providing information about a larger region of direct space) which may also make $N_g$ exceed $N_s$. For the simulated examples below these effects are seen for the lowest noise levels.

For comparing $N_s$ and $N_g$ calculated by Eqs. 13.37 and 13.38 respectively, simulated scattering data from a sphere of diameter 20 nm was used (Fig. 13.9). The data was simulated in the $q$-interval $[0;1.5]\,\mathrm{nm}^{-1}$ using $M = 100$ points. An absolute noise of $\sigma_i = a I(0) + b I(q_i)$ was added, using various values for $a$ and $b$ (absolute and relative noise respectively). Furthermore the data was truncated at various values for $q_{min}$ and $q_{max}$.

# 13 Bayesian Methods in SAXS and SANS Structure Determination

In Table 13.1 is shown the corresponding values of $N_g$ and $N_s$ calculated for various levels of relative noise $b$ and truncations.

In Fig. 13.18 is shown the variation of $N_g$ also for simulated data from a sphere of diameter 20 nm for $[q_{min}; q_{max}] = [0.1; 1.0]$ nm$^{-1}$ as a function of the absolute noise level $a$ using a fixed relative noise of $b = 0.005$. The horizontal lines in Fig. 13.18 show the number of Shannon channels $N_s = 5.73$ corresponding to $d = 20$ nm and $N_2 = 6.99$ corresponding to $d = 24.4$ nm, the latter being the maximum dimension used for the estimation of $p(r)$.

Table 13.1 and Fig. 13.18 indicate that the number of good parameters $N_g$ calculated for various conditions is sensible compared to the Shannon estimates $N_s$.

The values for $N_g$ – also for the experimental data – are in good agreement with the observation that the number of free parameters rarely exceed 10–15 for SAS experiments [699].

For SAXS experiments the parameter $N_g$ may be used to find the optimal exposure time $t$ of a given sample. A sudden increase in $dN_g/dt$ (which may be calculated in real time) may indicate deterioration of a biological sample due to X-ray exposure and a low value for $dN_g/dt$ may indicate suboptimal use of exposure time.

**Table 13.1** Estimation of the number of good parameters $N_g$ [740]

| $q_{min}$ [nm$^{-1}$] | $q_{max}$ [nm$^{-1}$] | $M$ | $b$ | $N_s$ | $N_g$ |
|---|---|---|---|---|---|
| 0.0 | 1.5 | 100 | 0.05 | 9.55 | 8.91 |
| 0.1 | 1.5 | 93 | 0.05 | 8.91 | 8.48 |
| 0.2 | 1.5 | 87 | 0.05 | 8.28 | 8.23 |
| 0.3 | 1.5 | 80 | 0.05 | 7.64 | 8.17 |
| 0.2 | 1.0 | 54 | 0.10 | 5.09 | 4.59 |
| 0.2 | 1.0 | 54 | 0.05 | 5.09 | 5.19 |
| 0.2 | 1.0 | 54 | 0.02 | 5.09 | 5.53 |
| 0.2 | 1.0 | 27 | 0.02 | 5.09 | 5.63 |

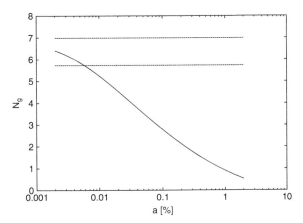

**Fig. 13.18** *Full line*: $N_g$ as a function of absolute noise level $a$ using simulated data for a sphere with $d = 20$ nm as shown in Fig. 13.9. The $q$-range used was $[0.1;1.0]$ nm$^{-1}$. *Dotted lines*: (*top*) $N_g = 6.99$ corresponding to the number of Shannon channels for $d = 24.4$ nm and (*lower*) $N_g = 5.73$ corresponding to the number of Shannon channels for $d = 20$ nm [740]

For SANS experiments it is often relevant to relax the resolution to obtain sufficient flux. Through simulation/calculation of $N_g$ for the experimental set up it may be possible give objective criteria for the optimal experimental settings.

## 13.6 Conclusion

It has been demonstrated that a Bayesian approach to IFT in SAS offers several advantages compared to the conventional methods:

1. It is possible to estimate the hyperparameters (e.g. $\alpha, d, \eta$) relevant for IFT from the basic axioms of probability theory instead of using ad hoc criteria.
2. Error estimates are provided for the hyperparameters.
3. The posterior probability distribution for $d$ may indicate two different sizes when this is relevant.
4. It is possible to separate intra particle and inter particle effects for non dilute solution scattering experiments.
5. The information content of the experimental data may be quantified.

For future applications Bayesian methods may be able to improve upon the recently developed methods for three dimensional structures estimation from SAS data (as mentioned in the introduction) by offering a transparent and consistent way of handling the many new constraints as well as their interrelations.

# References

1. Abseher, R., Horstink, L., Hilbers, C.W., Nilges, M.: Essential spaces defined by NMR structure ensembles and molecular dynamics simulation show significant overlap. Proteins **31**, 370–382 (1998)
2. Akaike, H.: New look at statistical-model identification. IEEE Trans. Automat. Contr. **AC19**, 716–723 (1974)
3. Allen, B.D., Mayo, S.L.: Dramatic performance enhancements for the FASTER optimization algorithm. J. Comp. Chem. **27**, 1071–1075 (2006)
4. Allen, B.D., Mayo, S.L.: An efficient algorithm for multistate protein design based on FASTER. J. Comput. Chem. **31**(5), 904–916 (2009)
5. Allen, M., Tildesley, D.: Computer Simulation of Liquids. Clarendon Press, New York (1999)
6. Amatori, A., Tiana, G., Sutto, L., Ferkinghoff-Borg, J., Trovato, A., Broglia, R.A.: Design of amino acid sequences to fold into $C_\alpha$-model proteins. J. Chem. Phys. **123**, 054904 (2005)
7. Amatori, A., Ferkinghoff-Borg, J., Tiana, G., Broglia, R.A.: Thermodynamic features characterizing good and bad folding sequences obtained using a simplified off-lattice protein model. Phys. Rev. E **73**, 061905 (2006)
8. Amatori, A., Tiana, G., Ferkinghoff-Borg, J., Broglia, R.A.: Denatured state is critical in determining the properties of model proteins designed on different folds. Proteins **70**, 1047–1055 (2007)
9. Ambroggio, X.I., Kuhlman, B.: Computational design of a single amino acid sequence that can switch between two distinct protein folds. J. Am. Chem. Soc. **128**, 1154–1161 (2006)
10. Ambroggio, X.I., Kuhlman, B.: Design of protein conformational switches. Curr. Opin. Struct. Biol. **16**, 525–530 (2006)
11. Anfinsen, C.: Principles that govern the folding of protein chains. Science **181**, 223–230 (1973)
12. Anfinsen, C.B., Haber, E., Sela, M., White, F.H.: The kinetics of formation of native ribonuclease during oxidation of the reduced polypeptide chain. Proc. Natl. Acad. Sci. U.S.A. **47**, 1309–1314 (1961)
13. Appignanesi, G.A., Fernández, A.: Cooperativity along kinetic pathways in RNA folding. J. Phys. A **29**, 6265–6280 (1996)
14. Arleth, L.: Unpublished results. (2004)
15. Artymiuk, P., Poirrette, A., Grindley, H., Rice, D., Willett, P.: A graph-theoretic approach to the identification of three-dimensional patterns of amino acid side-chains in protein structures. J. Mol. Biol. **243**, 327–344 (1994)
16. Ashworth, J., Havranek, J.J., Duarte, C.M., Sussman, D., Monnat, R.J., Stoddard, B.L., Baker, D.: Computational redesign of endonuclease DNA binding and cleavage specificity. Nature **441**, 656–659 (2006)

T. Hamelryck et al. (eds.), *Bayesian Methods in Structural Bioinformatics*,
Statistics for Biology and Health, DOI 10.1007/978-3-642-27225-7,
© Springer-Verlag Berlin Heidelberg 2012

17. Bahar, I., Jernigan, R.L.: Coordination geometry of nonbonded residues in globular proteins. Fold. Des. **1**, 357–370 (1996)
18. Baker, E.N., Hubbard, R.E.: Hydrogen bonding in globular proteins. Prog. Biophys. Mol. Biol. **44**, 97–179 (1984)
19. Baldwin, R.L.: In search of the energetic role of peptide hydrogen bonds. J. Biol. Chem. **278**, 17581–17588 (2003)
20. Baldwin, R.L.: Energetics of protein folding. J. Mol. Biol. **371**, 283–301 (2007)
21. Baldwin, R.L., Rose, G.D.: Is protein folding hierarchic? I. Local structure and peptide folding. Trends Biochem. Sci. **24**, 26–33 (1999)
22. Baldwin, R.L., Rose, G.D.: Is protein folding hierarchic? II. Folding intermediates and transition states. Trends Biochem. Sci. **24**, 77–83 (1999)
23. Bartels, C., Karplus, M.: Multidimensional adaptive umbrella sampling: applications to main chain and side chain peptide conformations. J. Comp. Chem. **18**, 1450–1462 (1997)
24. Bastolla, U., Porto, M., Roman, H., Vendruscolo, M.: Principal eigenvector of contact matrices and hydrophobicity profiles in proteins. Proteins **58**, 22–30 (2005)
25. Bastolla, U., Porto, M., Roman, H., Vendruscolo, M.: Structure, stability and evolution of proteins: principal eigenvectors of contact matrices and hydrophobicity profiles. Gene **347**, 219–230 (2005)
26. Bax, A., Kontaxis, G., Tjandra, N.: Dipolar couplings in macromolecular structure determination. Methods Enzymol. **339**, 127–174 (2001)
27. Bayes, T.: An essay towards solving a problem in the doctrine of chances. Phil. Trans. R. Soc. **53**, 370–418 (1763)
28. Bayrhuber, M., Meins, T., Habeck, M., Becker, S., Giller, K., Villinger, S., Vonrhein, C., Griesinger, C., Zweckstetter, M., Zeth, K.: Structure of the human voltage-dependent anion channel. Proc. Natl. Acad. Sci. U.S.A. **105**, 15370–15375 (2008)
29. Beaumont, M.A., Zhang, W., Balding, D.J.: Approximate Bayesian computation in population genetics. Genetics **162**, 2025–2035 (2002)
30. Beitz, E.: TeXshade: shading and labeling of multiple sequence alignments using LaTeX2e. Bioinformatics **16**, 135–139 (2000)
31. Bellhouse, D.: The Reverend Thomas Bayes, FRS: a biography to celebrate the tercentenary of his birth. Statist. Sci. **19**, 3–43 (2004)
32. Ben-Naim, A.: Statistical potentials extracted from protein structures: are these meaningful potentials? J. Chem. Phys. **107**, 3698–3706 (1997)
33. Ben-Naim, A.: A Farewell to Entropy. World Scientific Publishing Company, Hackensack (2008)
34. Bennett, C.: Efficient estimation of free energy differences from Monte Carlo data. J. Comp. Phys. **22**, 245–268 (1976)
35. Benros, C., de Brevern, A., Etchebest, C., Hazout, S.: Assessing a novel approach for predicting local 3D protein structures from sequence. Proteins **62**, 865–880 (2006)
36. Berg, B.: Locating global minima in optimization problems by a random-cost approach. Nature **361**, 708–710 (1993)
37. Berg, B.A.: Multicanonical recursions. J. Stat. Phys. **82**, 323–342 (1996)
38. Berg, B.A.: Algorithmic aspects of multicanonical simulations. Nuclear Phys. B **63**, 982–984 (1998)
39. Berg, B.: Workshop on Monte Carlo methods, fields institute communications. In: Introduction to Multicanonical Monte Carlo Simulations, vol. 26, pp. 1–23. American Mathematical Society, Providence (2000)
40. Berg, B., Celik, T.: New approach to spin-glass simulations. Phys. Rev. Lett. **69**, 2292–2295 (1992)
41. Berg, B., Neuhaus, T.: Multicanonical algorithms for first order phase transitions. Phys. Lett. B **267**, 249–253 (1991)
42. Berg, B., Neuhaus, T.: Multicanonical ensemble: a new approach to simulate first-order phase transitions. Phys. Rev. Lett. **68**, 9 (1992)

References 345

43. Berger, J., Bernardo, J., Sun, D.: The formal definition of reference priors. Ann. Stat. **37**, 905–938 (2009)
44. Berman, H.M.: The protein data bank: a historical perspective. Acta Crystallogr. A **64**, 88–95 (2008)
45. Berman, H.M., Westbrook, J., Feng, Z., Gilliland, G., Bhat, T.N., Weissig, H., Shindyalov, I.N., Bourne, P.E.: The protein data bank. Nucleic Acids Res. **28**, 235–42 (2000)
46. Berman, H., Henrick, K., Nakamura, H.: Announcing the worldwide protein data bank. Nat. Struct. Biol. **10**, 980 (2003)
47. Bernard, B., Samudrala, R.: A generalized knowledge-based discriminatory function for biomolecular interactions. Proteins **76**, 115–128 (2009)
48. Bernado, J., Smith, A.: Bayesian Theory. Wiley, New York (1994)
49. Bernardo, J.: The concept of exchangeability and its applications. Far East J. Math. Sci. **4**, 111–122 (1996)
50. Bernardo, J.: Reference analysis. In: Dey, D.K., Rao, C.R. (eds.) Handbook of Statistics, vol. 25, pp. 17–90. Elsevier, Burlington (2005)
51. Bernardo, J.: A Bayesian mathematical statistics primer. In: Proceedings of the Seventh International Conference on Teaching Statistics. International Association for Statistical Education, Salvador (Bahia), Brazil (2006)
52. Bernardo, J.: Modern Bayesian inference: foundations and objective methods. In: Philosophy of Statistics. Elsevier, Amsterdam (2009)
53. Berndt, K., Güntert, P., Orbons, L., Wütrich, K.: Determination of a high-quality nuclear magnetic resonance solution structure of the bovine pancreatic trypsin inhibitor and comparison with thee crystal structures. J. Mol. Biol. **227**, 757–775 (1993)
54. Besag, J.: Spatial interaction and the statistical analysis of lattice systems. J. R. Stat. Soc. B **36**, 192–236 (1974)
55. Besag, J.: Statistical analysis of non-lattice data. Statistician **24**, 179–195 (1975)
56. Best, D.J., Fisher, N.I.: Efficient simulation method of the von Mises distribution. Appl. Statist. **28**, 152–157 (1979)
57. Betancourt, M.: Efficient Monte Carlo trial moves for polypeptide simulations. J. Chem. Phys. **123**, 174905 (2005)
58. Betancourt, M.R.: Another look at the conditions for the extraction of protein knowledge-based potentials. Proteins **76**, 72–85 (2009)
59. Betancourt, M.R., Thirumalai, D.: Pair potentials for protein folding: choice of reference states and sensitivity of predicted native states to variations in the interaction schemes. Protein Sci. **8**, 361–369 (1999)
60. Bethe, H.: Statistical theory of superlattices. Proc. R. Soc. London A **150**, 552–575 (1935)
61. Binder, K.: Monte Carlo and Molecular Dynamics Simulations in Polymer Science. Oxford University Press, New York (1995)
62. Bishop, C.: Pattern Recognition and Machine Learning. Springer, New York (2006)
63. Bjornstad, J.F.: On the generalization of the likelihood function and likelihood principle. J. Am. Stat. Assoc. **91**, 791–806 (1996)
64. Blundell, T.L., Sibanda, B.L., Sternberg, M.J., Thornton, J.M.: Knowledge-based prediction of protein structures and the design of novel molecules. Nature **326**, 347–52 (1987)
65. Boas, F.E., Harbury, P.B.: Potential energy functions for protein design. Curr. Opin. Struct. Biol. **17**, 199–204 (2007)
66. Bolon, D.N., Mayo, S.L.: Enzyme-like proteins by computational design. Proc. Natl. Acad. Sci. U.S.A. **98**, 14274–14279 (2001)
67. Bookstein, F.: Size and shape spaces for landmark data in two dimensions. Statist. Sci. **1**, 181–222 (1986)
68. Boomsma, W., Mardia, K., Taylor, C., Ferkinghoff-Borg, J., Krogh, A., Hamelryck, T.: A generative, probabilistic model of local protein structure. Proc. Natl. Acad. Sci. U.S.A. **105**, 8932–8937 (2008)

69. Boomsma, W., Borg, M., Frellsen, J., Harder, T., Stovgaard, K., Ferkinghoff-Borg, J., Krogh, A., Mardia, K.V., Hamelryck, T.: PHAISTOS: protein structure prediction using a probabilistic model of local structure. In: Proceedings of CASP8, pp. 82–83. Cagliari (2009)
70. Borg, M., Mardia, K.: A probabilistic approach to protein structure prediction: PHAISTOS in CASP9. In: Gusnanto, A., Mardia, K., Fallaize, C. (eds.) LASR2009 – Statistical Tools for Challenges in Bioinformatics, pp. 65–70. Leeds University Press, Leeds, UK (2009)
71. Bortz, A., Kalos, M., Lebowitz, J.: A new algorithm for Monte Carlo simulation of Ising spin systems. J. Comp. Phys. **17**, 10–18 (1975)
72. Bottaro, S., Boomsma, W., Johansson, K.E., Andreetta, C., Hamelryck, T., Ferkinghoff-Borg, J.: Subtle Monte Carlo updates in dense molecular systems. J. Chem. Theory Comput., Accepted and published online (2012)
73. Bourne, P.E., Weissig, H.: Structural bioinformatics. In: Methods of Biochemical Analysis, vol. 44. Wiley-Liss, Hoboken (2003)
74. Bowie, J., Eisenberg, D.: An evolutionary approach to folding small alpha-helical proteins that uses sequence information and an empirical guiding fitness function. Proc. Natl. Acad. Sci. U.S.A. **91**, 4436–4440 (1994)
75. Bowie, J., Luthy, R., Eisenberg, D.: A method to identify protein sequences that fold into a known three-dimensional structure. Science **253**, 164–170 (1991)
76. Bradley, P., Misura, K., Baker, D.: Toward high-resolution de novo structure prediction for small proteins. Science **309**, 1868–1871 (2005)
77. Braémaud, P.: Markov Chains. Springer, New York (1999)
78. Brenner, S.E., Koehl, P., Levitt, M.: The ASTRAL compendium for protein structure and sequence analysis. Nucleic Acids Res. **28**, 254–256 (2000)
79. Brünger, A.T.: The free $R$ value: a novel statistical quantity for assessing the accuracy of crystal structures. Nature **355**, 472–474 (1992)
80. Brünger, A.T.: X-PLOR, Version 3.1: A System for X-ray Crystallography and NMR. Yale University Press, New Haven (1992)
81. Brünger, A.T., Nilges, M.: Computational challenges for macromolecular structure determination by X-ray crystallography and solution NMR spectroscopy. Q. Rev. Biophys. **26**, 49–125 (1993)
82. Brünger, A.T., Clore, G.M., Gronenborn, A.M., Saffrich, R., Nilges, M.: Assessing the quality of solution nuclear magnetic resonance structures by complete cross-validation. Science **261**, 328–331 (1993)
83. Brünger, A.T., Adams, P.D., Clore, G.M., DeLano, W.L., Gros, P., Grosse-Kunstleve, R.W., Jiang, J.S., Kuszewski, J., Nilges, M., Pannu, N.S., Read, R.J., Rice, L.M., Simonson, T., Warren, G.L.: Crystallography and NMR system (CNS): a new software suite for macromolecular structure determination. Acta Crystallogr. D **54**, 905–921 (1998)
84. Brunner-Popela, J., Glatter, O.: Small-angle scattering of interacting particles. I. Basic principles of a global evaluation technique. J. Appl. Cryst. **30**, 431–442 (1997)
85. Bryant, S.H., Lawrence, C.E.: The frequency of ion-pair substructures in proteins is quantitatively related to electrostatic potential: a statistical model for nonbonded interactions. Proteins **9**, 108–119 (1991)
86. Bryngelson, J.D., Wolynes, P.G.: Spin glasses and the statistical mechanics of protein folding. Proc. Natl. Acad. Sci. U.S.A. **84**, 7524–7528 (1987)
87. Buchete, N.V., Straub, J.E., Thirumalai, D.: Continuous anisotropic representation of coarse-grained potentials for proteins by spherical harmonics synthesis. J. Mol. Graph. Model. **22**, 441–450 (2004)
88. Buchete, N.V., Straub, J.E., Thirumalai, D.: Development of novel statistical potentials for protein fold recognition. Curr. Opin. Struct. Biol. **14**, 225–32 (2004)
89. Bunagan, M., Yang, X., Saven, J., Gai, F.: Ultrafast folding of a computationally designed Trp-cage mutant: Trp$^2$-cage. J. Phys. Chem. B **110**, 3759–3763 (2006)
90. Burge, J., Lane, T.: Shrinkage estimator for Bayesian network parameters. Lect. Notes Comput. Sci. **4701**, 67–78, Leeds LS2 9JT (2007)

References 347

91. Burnham, K.P., Anderson, D.R.: Model Selection and Multimodel Inference – A Practical Information-Theoretic Approach, 2nd edn. Springer, New York (2002)
92. Butterfoss, G.L., Hermans, J.: Boltzmann-type distribution of side-chain conformation in proteins. Protein Sci. **12**, 2719–2731 (2003)
93. Bystroff, C., Baker, D.: Prediction of local structure in proteins using a library of sequence-structure motifs. J. Mol. Biol. **281**, 565–577 (1998)
94. Bystroff, C., Thorsson, V., Baker, D.: HMMSTR: a hidden Markov model for local sequence-structure correlations in proteins. J. Mol. Biol. **301**, 173–190 (2000)
95. Calhoun, J.R., Kono, H., Lahr, S., Wang, W., DeGrado, W.F., Saven, J.G.: Computational design and characterization of a monomeric helical dinuclear metalloprotein. J. Mol. Biol. **334**, 1101–1115 (2003)
96. Camproux, A., Tuffery, P., Chevrolat, J., Boisvieux, J., Hazout, S.: Hidden Markov model approach for identifying the modular framework of the protein backbone. Protein Eng. **12**, 1063–1073 (1999)
97. Camproux, A., Gautier, R., Tufféry, P.: A hidden Markov model derived structural alphabet for proteins. J. Mol. Biol. **339**, 591–605 (2004)
98. Canutescu, A.A., Shelenkov, A.A., Dunbrack R. L. Jr.: A graph-theory algorithm for rapid protein side-chain prediction. Protein Sci. **12**, 2001–2014 (2003)
99. Carraghan, R., Pardalos, P.: Exact algorithm for the maximum clique problem. Oper. Res. Lett. **9**, 375–382 (1990)
100. Carreira-Perpinan, M., Hinton, G.: On Contrastive divergence learning. In Proceedings of the 10th International Workshop on Artificial Intelligence and Statistics, Jan 6-8 2005, the Savannah Hotel, Barbados p. 217
101. Caruana, R.: Multitask learning. Mach. Learn. **28**, 41–75 (1997)
102. Cawley, S., Pachter, L.: HMM sampling and applications to gene finding and alternative splicing. Bioinformatics **19**(Suppl 2), II36–II41 (2003)
103. Celeux, G., Diebolt, J.: The SEM algorithm: a probabilistic teacher algorithm derived from the EM algorithm for the mixture problem. Comp. Stat. Quart. **2**, 73–92 (1985)
104. Cerf, N., Martin, O.: Finite population-size effects in projection Monte Carlo methods. Phys. Rev. E **51**, 3679 (1995)
105. Chalaoux, F.R., O'Donoghue, S.I., Nilges, M.: Molecular dynamics and accuracy of NMR structures: effects of error bounds and data removal. Proteins **34**, 453–463 (1999)
106. Chan, H.S., Dill, K.A.: Solvation: how to obtain microscopic energies from partitioning and solvation experiments. Annu. Rev. Biophys. Biomol. Struct. **26**, 425–459 (1997)
107. Chandler, D.: Introduction to Modern Statistical Mechanics. Oxford University Press, Oxford (1987)
108. Chandrasekaran, R., Ramachandran, G.: Studies on the conformation of amino acids. XI. Analysis of the observed side group conformation in proteins. Int. J. Protein Res. **2**, 223–233 (1970)
109. Cheeseman, P.: Oral presentation at MaxEnt 2004, Munich (2004)
110. Chen, W.W., Shakhnovich, E.I.: Lessons from the design of a novel atomic potential for protein folding. Protein Sci. **14**, 1741–1752 (2005)
111. Chen, J., Wonpil, I., Brooks, C.L.: Application of torsion angle molecular dynamics for efficient sampling of protein conformations. J. Comp. Chem. **26**, 1565–1578 (2005)
112. Chikenji, G., Fujitsuka, Y., Takada, S.: A reversible fragment assembly method for de novo protein structure prediction. J. Chem. Phys. **119**, 6895–6903 (2003)
113. Chu, W., Ghahramani, Z., Podtelezhnikov, A., Wild, D.L.: Bayesian segmental models with multiple sequence alignment profiles for protein secondary structure and contact map prediction. IEEE Trans. Comput. Biol. Bioinform. **3**, 98–113 (2006)
114. Cieplak, M., Holter, N.S., Maritan, A., Banavar, J.R.: Amino acid classes and the protein folding problem. J. Chem. Phys. **114**, 1420–1423 (2001)
115. Clark, L.A., van Vlijmen, H.W.: A knowledge-based forcefield for protein-protein interface design. Proteins **70**, 1540–1550 (2008)

116. Clore, G.M., Schwieters, C.D.: Concordance of residual dipolar couplings, backbone order parameters and crystallographic B-factors for a small alpha/beta protein: a unified picture of high probability, fast atomic motions in proteins. J. Mol. Biol. **355**, 879–886 (2006)
117. Cochran, F., Wu, S., Wang, W., Nanda, V., Saven, J., Therien, M., DeGrado, W.: Computational de novo design and characterization of a four-helix bundle protein that selectively binds a nonbiological cofactor. J. Am. Chem. Soc. **127**, 1346–1347 (2005)
118. Cooper, N., Eckhardt, R., Shera, N.: From Cardinals to Chaos: Reflections on the Life and Legacy of Stanislaw Ulam. Cambridge University Press, New York (1989)
119. Cootes, T., Taylor, C., Cooper, D., Graham, J.: Training models of shape from sets of examples. In: Hogg, D., Boyle, R. (eds.) British Machine Vision Conference, pp. 9–18. Springer, Berlin (1992)
120. Cootes, T., Taylor, C., Cooper, D., Graham, J.: Image search using flexible shape models generated from sets of examples. In: Mardia, K. (ed.) Statistics and Images, vol. 2. Carfax, Oxford (1994)
121. Corey, R., Pauling, L.: Fundamental dimensions of polypeptide chains. Proc. R. Soc. London B **141**, 10–20 (1953)
122. Cossio, P., Marinelli, F., Laio, A., Pietrucci, F.: Optimizing the performance of bias-exchange metadynamics: folding a 48-residue LysM domain using a coarse-grained model. J. Phys. Chem. B **114**, 3259–3265 (2010)
123. Cossio, P., Trovato, A., Pietrucci, F., Seno, F., Maritan, A., Laio, A.: Exploring the universe of protein structures beyond the protein data bank. PLoS Comput. Biol. **6**, e1000957 (2010)
124. Costa, M., Lopes, J., Santos, J.: Analytical study of tunneling times in flat histogram Monte Carlo. Europhys. Lett. **72**, 802 (2005)
125. Cowles, M., Carlin, B.: Markov chain Monte Carlo convergence diagnostics: a comparative review. J. Am. Stat. Assoc. **91**, 883–904 (1996)
126. Cowell, R., Dawid, A., Lauritzen, S., Spiegelhalter, D.: Probabilistic Networks and Expert Systems. Springer, New York (1999)
127. Cox, R.T.: The Algebra of Probable Inference. John Hopkins University Press, Baltimore (1961)
128. Cox, D., Isham, V.: Point Processes. Chapman and Hall, New York (1980)
129. Crespo, Y., Laio, A., Santoro, G., Tosatti, E.: Calculating thermodynamics properties of quantum systems by a non-Markovian Monte Carlo procedure. Phys. Rev. E **80**, 015702 (2009)
130. Czogiel, I., Dryden, I., Brignell, C.: Bayesian alignment of continuous molecular shapes using random fields. In: Barber, S., Baxter, P., Gusnanto, A., Mardia, K. (eds.) The Art and Science of Statistical Bioinformatics. Leeds University Press, Leeds (2008)
131. Dagpunar, J.: Principles of Random Variate Generation. Clarendon Press, Oxford (1988)
132. Dahiyat, B.I., Mayo, S.L.: De novo protein design: fully automated sequence selection. Science **278**, 82–87 (1997)
133. Das, R., Baker, D.: Automated de novo prediction of native-like RNA tertiary structures. Proc. Natl. Acad. Sci. U.S.A. **104**, 14664–14669 (2007)
134. Das, R., Baker, D.: Macromolecular modeling with ROSETTA. Annu. Rev. Biochem. **77**, 363–382 (2008)
135. Davies, J., Jackson, R., Mardia, K., Taylor, C.: The Poisson index: a new probabilistic model for protein-ligand binding site similarity. Bioinformatics **23**, 3001–3008 (2007)
136. Dawid, A.P.: Some matrix-variate distribution theory: notational considerations and a Bayesian application. Biometrika **68**, 265–274 (1981)
137. Day, R., Paschek, D., Garcia, A.E.: Microsecond simulations of the folding/unfolding thermodynamics of the Trp-cage miniprotein. Proteins **78**, 1889–1899 (2010)
138. Dayal, P., Trebst, S., Wessel, S., Wuertz, D., Troyer, M., Sabhapandit, S., Coppersmith, S.: Performance limitations of flat-histogram methods. Phys. Rev. Lett. **92**, 97201 (2004)
139. de Brevern, A., Etchebest, C., Hazout, S.: Bayesian probabilistic approach for predicting backbone structures in terms of protein blocks. Proteins **41**, 271–287 (2000)

# References

140. de Oliveira, P., Penna, T., Herrmann, H.: Broad histogram method. Brazil. J. Phys. **26**, 677 (1996)
141. Decanniere, K., Transue, T.R., Desmyter, A., Maes, D., Muyldermans, S., Wyns, L.: Degenerate interfaces in antigen-antibody complexes. J. Mol. Biol. **313**, 473–478 (2001)
142. Dehouck, Y., Gilis, D., Rooman, M.: A new generation of statistical potentials for proteins. Biophys. J. **90**, 4010–4017 (2006)
143. DeLano, W.: The PyMOL molecular graphics system. http://www.pymol.org (2002)
144. Dempster, A.P., Laird, N.M., Rubin, D.B.: Maximum likelihood from incomplete data via the EM algorithm. J. R. Statist. Soc. B. **39**, 1–38 (1977)
145. Derrida, B.: Random-energy model: an exactly solvable model of disordered systems. Phys. Rev. B **24**, 2613–2626 (1981)
146. Desjarlais, J.R., Handel, T.M.: De novo design of the hydrophobic cores of proteins. Protein Sci. **4**, 2006–2018 (1995)
147. Desmet, J., De Maeyer, M., Hazes, B., Lasters, I.: The dead-end elimination theorem and its use in protein side-chain positioning. Nature **356**, 539–542 (1992)
148. Desmet, J., Spriet, J., Lasters, I.: Fast and accurate side-chain topology and energy refinement (FASTER) as a new method for protein structure optimization. Proteins **48**, 31–43 (2002)
149. Diamond, R.: On the multiple simultaneous superposition of molecular-structures by rigid body transformations. Protein Sci. **1**, 1279–1287 (1992)
150. Dill, K., Bromberg, S.: Molecular Driving Forces. Garland Science, New York (2002)
151. Ding, F., Dokholyan, N.V.: Emergence of protein fold families through rational design. PLoS Comput. Biol. **2**, e85 (2006)
152. Djuranovic, S., Hartmann, M.D., Habeck, M., Ursinus, A., Zwickl, P., Martin, J., Lupas, A.N., Zeth, K.: Structure and activity of the N-terminal substrate recognition domains in proteasomal ATPases. Mol. Cell **34**, 1–11 (2009)
153. Dodd, L.R., Boone, T.D., Theodorou, D.N.: A concerted rotation algorithm for atomistic Monte Carlo simulation of polymer melts and glasses. Mol. Phys. **78**, 961–996 (1993)
154. Dose, V.: Bayesian inference in physics: case studies. Rep. Prog. Phys. **66**, 1421–1461 (2003)
155. Doucet, A., Freitas, N., Gordon, N. (eds.): Sequential Monte Carlo Methods in Practice. Springer, New York (2001)
156. Dowe, D., Allison, L., Dix, T., Hunter, L., Wallace, C., Edgoose, T.: Circular clustering of protein dihedral angles by minimum message length. Pac. Symp. Biocomput. Edited by Lawrence Hunter and Teri Klein, World Scientific Publishing Co, Singapore, 242–255 (1996)
157. Downs, T.: Orientation statistics. Biometrika **59**, 665–676 (1972)
158. Dryden, I.: Discussion to schmidler. In: Bernardo, J., Bayarri, M., Berger, J., Dawid, A., Heckerman, D. Smith, A., West, M. (eds.) Bayesian Statistics 8, pp. 471–490. Oxford University Press, Oxford (2007)
159. Dryden, I.L., Mardia, K.V.: Statistical Shape Analysis. Wiley Series in Probability and Statistics, Probability and Statistics. Wiley, Chichester (1998)
160. Dryden, I., Hirst, J., Melville, J.: Statistical analysis of unlabeled point sets: comparing molecules in chemoinformatics. Biometrics **63**, 237–251 (2007)
161. Duane, S., Kennedy, A.D., Pendleton, B., Roweth, D.: Hybrid Monte Carlo. Phys. Lett. B **195**, 216–222 (1987)
162. Dunbrack, R.L.: Rotamer libraries in the 21st century. Curr. Opin. Struct. Biol. **12**, 431–440 (2002)
163. Dunbrack, R.L., Karplus, M.: Backbone-dependent rotamer library for proteins application to side-chain prediction. J. Mol. Biol. **230**, 543–574 (1993)
164. Durbin, R., Eddy, S.R., Krogh, A., Mitchison, G.: Biological Sequence Analysis. Cambridge University Press, New York (1998)
165. Dutilleul, P.: The MLE algorithm for the matrix normal distribution. J. Stat. Comput. Sim. **64**, 105–123 (1999)
166. Earl, D., Deem, M.: Parallel tempering: theory, applications, and new perspectives. Phys. Chem. Chem. Phys. **7**, 3910–3916 (2005)

167. Eckart, C.: Some studies concerning rotating axes and polyatomic molecules. Phys. Rev. **47**, 552–558 (1935)
168. Edgoose, T., Allison, L., Dowe, D.: An MML classification of protein structure that knows about angles and sequence. Pac. Symp. Biocomput. **3**, 585–596 (1998)
169. Efron, B., Morris, C.: Limiting the risk of Bayes and empirical Bayes estimators–part II: the empirical Bayes case. J. Am. Stat. Assoc. **67**, 130–139 (1972)
170. Efron, B., Morris, C.: Stein's estimation rule and its competitors–an empirical Bayes approach. J. Am. Stat. Soc. **68**, 117–130 (1973)
171. Elidan, G., McGraw, I., Koller, D.: Residual belief propagation: informed scheduling for asynchronous message passing. In: Proceedings of the Twenty-second Conference on Uncertainty in AI (UAI), Boston (2006)
172. Elofsson, A., Le Grand, S.M., Eisenberg, D.: Local moves: an efficient algorithm for simulation of protein folding. Proteins **23**, 73–82 (1995)
173. Engh, R.A., Huber, R.: Accurate bond and angle parameters for X-ray structure refinement. Acta Crystallogr. A **A47**, 392–400 (1991)
174. Engh, R.A., Huber, R.: Structure quality and target parameters. In: Rossman, M.G., Arnold, E. (eds.) International Tables for Crystallography, Vol. F: Crystallography of Biological Macro-molecules, 1st edn., pp. 382–392. Kluwer Academic Publishers for the International Union of Crystallography, Dordrecht/Boston/London (2001)
175. Etchebest, C., Benros, C., Hazout, S., de Brevern, A.: A structural alphabet for local protein structures: improved prediction methods. Proteins **59**, 810–827 (2005)
176. Favrin, G., Irbäck, A., Sjunnesson, F.: Monte Carlo update for chain molecules: biased Gaussian steps in torsional space. J. Chem. Phys. **114**, 8154 (2001)
177. Feigin, L., Svergun, D., Taylor, G.: Structure Analysis by Small-Angle X-ray and Neutron Scattering. Plenum Press, New York (1987)
178. Feng, Y., Kloczkowski, A., Jernigan, R.L.: Four-body contact potentials derived from two protein datasets to discriminate native structures from decoys. Proteins **68**, 57–66 (2007)
179. Ferkinghoff-Borg, J.: Monte Carlo methods in complex systems. Ph.D. thesis, Graduate School of Biophysics, Niels Bohr Institute (2002)
180. Ferkinghoff-Borg, J.: Optimized Monte Carlo analysis for generalized ensembles. Eur. Phys. J. B **29**, 481–484 (2002)
181. Feroz, F., Hobson, M.: Multimodal nested sampling: an efficient and robust alternative to Markov chain Monte Carlo methods for astronomical data analyses. Mon. Not. R. Astron. Soc. **384**, 449–463 (2008)
182. Fersht, A.: Nucleation mechanisms in protein folding. Curr. Opin. Struct. Biol. **7**, 3–9 (1997)
183. Fienberg, S.: When did Bayesian inference become Bayesian. Bayesian Anal. **1**, 1–40 (2006)
184. Finkelstein, A.V., Badretdinov, A.Y., Gutin, A.M.: Why do protein architectures have Boltzmann-like statistics? Proteins **23**, 142–150 (1995)
185. Fisher, N.I., Lewis, T., Embleton, B.J.J.: Statistical Analysis of Spherical Data. Cambridge University Press, New York (1987)
186. Fitzgerald, J.E., Jha, A.K., Colubri, A., Sosnick, T.R., Freed, K.F.: Reduced $C\beta$ statistical potentials can outperform all-atom potentials in decoy identification. Protein Sci. **16**, 2123–2139 (2007)
187. Fixman, M.: Classical statistical mechanics of constraints: a theorem and application to polymers. Proc. Natl. Acad. Sci. U.S.A. **71**, 3050–3053 (1974)
188. Fixman, M.: Simulation of polymer dynamics. I. general theory. J. Chem. Phys. **69**, 1527–1537 (1978)
189. Fleming, P.J., Rose, G.D.: Do all backbone polar groups in proteins form hydrogen bonds? Protein Sci. **14**, 1911–1917 (2005)
190. Fleury, P., Chen, H., Murray, A.F.: On-chip contrastive divergence learning in analogue VLSI. IEEE Int. Conf. Neur. Netw. **3**, 1723–1728 (2004)
191. Flower, D.R.: Rotational superposition: a review of methods. J. Mol. Graph. Model. **17**, 238–244 (1999)

# References

351

192. Fogolari, F., Esposito, G., Viglino, P., Cattarinussi, S.: Modeling of polypeptide chains as $C\alpha$ chains, $C\alpha$ chains with $C\beta$, and $C\alpha$ chains with ellipsoidal lateral chains. Biophys. J. **70**, 1183–1197 (1996)

193. Fraenkel, A.: Protein folding, spin glass and computational complexity. In: DNA Based Computers III: DIMACS Workshop, June 23–25, 1997, p. 101. American Mathematical Society, Providence (1999)

194. Frauenkron, H., Bastolla, U., Gerstner, E., Grassberger, P., Nadler, W.: New Monte Carlo algorithm for protein folding. Phys. Rev. Lett. **80**, 3149–3152 (1998)

195. Frellsen, J.: Probabilistic methods in macromolecular structure prediction. Ph.D. thesis, Bioinformatics Centre, University of Copenhagen (2011)

196. Frellsen, J., Ferkinghoff-Borg, J.: Muninn – an automated method for Monte Carlo simulations in generalized ensembles. In preparation

197. Frellsen, J., Moltke, I., Thiim, M., Mardia, K.V., Ferkinghoff-Borg, J., Hamelryck, T.: A probabilistic model of RNA conformational space. PLoS Comput Biol **5**, e1000406 (2009)

198. Frenkel, D., Smit, B.: Understanding Molecular Simulation: From Algorithms to Applications, vol. 1. Academic, San Diego (2002)

199. Frey, B.J., Dueck, D.: Clustering by passing messages between data points. Science **315**, 972–976 (2007)

200. Frickenhaus, S., Kannan, S., Zacharias, M.: Efficient evaluation of sampling quality of molecular dynamics simulations by clustering of dihedral torsion angles and Sammon mapping. J. Comp. Chem. **30**, 479–492 (2009)

201. Friedland, G.D., Linares, A.J., Smith, C.A., Kortemme, T.: A simple model of backbone flexibility improves modeling of side-chain conformational variability. J. Mol. Biol. **380**, 757–774 (2008)

202. Fritz, G., Glatter, O.: Structure and interaction in dense colloidal systems: evaluation of scattering data by the generalized indirect Fourier transformation method. J. Phys. Condens. Matter **18**, S2403–S2419 (2006)

203. Fritz, G., Bergmann, A., Glatter, O.: Evaluation of small-angle scattering data of charged particles using the generalized indirect Fourier transformation technique. J. Chem. Phys. **113**, 9733–9740 (2000)

204. Fromer, M., Globerson, A.: An LP view of the M-best MAP problem. In: Bengio, Y., Schuurmans, D., Lafferty, J., Williams, C.K.I., Culotta, A. (eds.) Advances in Neural Information Processing Systems (NIPS) MIT press, Cambridge, MA 22, pp. 567–575 (2009)

205. Fromer, M., Shifman, J.M.: Tradeoff between stability and multispecificity in the design of promiscuous proteins. PLoS Comput. Biol. **5**, e1000627 (2009)

206. Fromer, M., Yanover, C.: A computational framework to empower probabilistic protein design. Bioinformatics **24**, i214–i222 (2008)

207. Fromer, M., Yanover, C.: Accurate prediction for atomic-level protein design and its application in diversifying the near-optimal sequence space. Proteins **75**, 682–705 (2009)

208. Fromer, M., Yanover, C., Harel, A., Shachar, O., Weiss, Y., Linial, M.: SPRINT: side-chain prediction inference toolbox for multistate protein design. Bioinformatics **26**, 2466–2467 (2010)

209. Fromer, M., Yanover, C., Linial, M.: Design of multispecific protein sequences using probabilistic graphical modeling. Proteins **78**, 530–547 (2010)

210. Garel, T., Orland, H.: Guided replication of random chain: a new Monte Carlo method. J. Phys. A **23**, L621 (1990)

211. Gelman, A., Carlin, J., Stern, H., Rubin, D.: Bayesian Data Analysis. Chapman & Hall/CRC press, Boca Raton (2004)

212. Geman, S., Geman, D.: Stochastic relaxation, Gibbs distributions, and the Bayesian restoration of images. IEEE Trans. PAMI **6**, 721–741 (1984)

213. Georgiev, I., Donald, B.R.: Dead-end elimination with backbone flexibility. Bioinformatics **23**, i185–i194 (2007)

214. Georgiev, I., Keedy, D., Richardson, J.S., Richardson, D.C., Donald, B.R.: Algorithm for backrub motions in protein design. Bioinformatics **24**, i196–i204 (2008)

215. Georgiev, I., Lilien, R.H., Donald, B.R.: The minimized dead-end elimination criterion and its application to protein redesign in a hybrid scoring and search algorithm for computing partition functions over molecular ensembles. J. Comp. Chem. **29**, 1527–1542 (2008)

216. Gerber, P.R., Müller, K.: Superimposing several sets of atomic coordinates. Acta Crystallogr. A **43**, 426–428 (1987)

217. Gervais, C., Wüst, T., Landau, D., Xu, Y.: Application of the Wang–Landau algorithm to the dimerization of glycophorin A. J. Chem. Phys. **130**, 215106 (2009)

218. Geyer, C., Thompson, E.: Annealing Markov chain Monte Carlo with applications to ancestral inference. J. Am. Stat. Assoc. **90**, 909–920 (1995)

219. Ghahramani, Z.: Learning dynamic Bayesian networks. Lect. Notes Comput. Sci. **1387**, 168–197 (1997)

220. Ghahramani, Z., Jordan, M.I.: Factorial hidden Markov models. Mach. Learn. **29**, 245–273 (1997)

221. Ghose, A., Crippen, G.: Geometrically feasible binding modes of a flexible ligand molecule at the receptor site. J. Comp. Chem. **6**, 350–359 (1985)

222. Gibrat, J., Madej, T., Bryant, S.: Surprising similarities in structure comparison. Curr. Opin. Struct. Biol. **6**, 377–385 (1996)

223. Gilks, W.R., Richardson, S., Spiegelhalter, D.J.: Markov Chain Monte Carlo in Practice. Chapman and Hall/CRC, Boca Raton (1998)

224. Glasbey, C., Horgan, G., Gibson, G., Hitchcock, D.: Fish shape analysis using landmarks. Biometrical J. **37**, 481–495 (1995)

225. Glatter, O.: A new method for the evaluation of small-angle scattering data. J. Appl. Cryst. **10**, 415–421 (1977)

226. Glatter, O.: Determination of particle-size distribution functions from small-angle scattering data by means of the indirect transformation method. J. Appl. Cryst. **13**, 7–11 (1980)

227. Glatter, O.: Evaluation of small-angle scattering data from lamellar and cylindrical particles by the indirect transformation method. J. Appl. Cryst. **13**, 577–584 (1980)

228. Glatter, O., Kratky, O.: Small Angle X-ray Scattering. Academic, London (1982)

229. Globerson, A., Jaakkola, T.: Fixing max-product: convergent message passing algorithms for MAP LP-relaxations. In: Platt, J., Koller, D., Singer, Y., Roweis, S. (eds.) Advances in Neural Information Processing Systems 21. MIT, Cambridge (2007)

230. Gō, N.: Theoretical studies of protein folding. Annu. Rev. Biophys. Bioeng. **12**, 183–210 (1983)

231. Gō, N., Scheraga, H.: Ring closure and local conformational deformations of chain molecules. Macromolecules **3**, 178–187 (1970)

232. Godzik, A.: Knowledge-based potentials for protein folding: what can we learn from known protein structures? Structure **4**, 363–366 (1996)

233. Godzik, A., Kolinski, A., Skolnick, J.: Are proteins ideal mixtures of amino acids? Analysis of energy parameter sets. Protein Sci. **4**, 2107–2117 (1995)

234. Gohlke, H., Klebe, G.: Statistical potentials and scoring functions applied to protein-ligand binding. Curr. Opin. Struct. Biol. **11**, 231–235 (2001)

235. Gold, N.: Computational approaches to similarity searching in a functional site database for protein function prediction. Ph.D thesis, Leeds University, School of Biochemistry and Microbiology (2003)

236. Gold, N., Jackson, R.: Fold independent structural comparisons of protein-ligand binding sites for exploring functional relationships. J. Mol. Biol. **355**, 1112–1124 (2006)

237. Goldstein, R.F.: Efficient rotamer elimination applied to protein side-chains and related spin glasses. Biophys. J. **66**, 1335–1340 (1994)

238. Goldstein, R.A., Luthey-Schulten, Z.A., Wolynes, P.G.: Optimal protein-folding codes from spin-glass theory. Proc. Natl. Acad. Sci. U.S.A. **89**, 4918–4922 (1992)

239. Golender, V., Rozenblit, A.: Logical and Combinatorial Algorithms for Drug Design. Research Studies Press, Letchworth (1983)

240. Gonzalez, J., Low, Y., Guestrin, C.: Residual splash for optimally parallelizing belief propagation. In: Artificial Intelligence and Statistics (AISTATS). Clearwater Beach (2009)

# References

353

241. Goodall, C.R.: Procrustes methods in the statistical analysis of shape. J. R. Statist. Soc. B. **53**, 285–321 (1991)
242. Goodall, C.R.: Procrustes methods in the statistical analysis of shape: rejoinder to discussion. J. R. Statist. Soc. B. **53**, 334–339 (1991)
243. Goodall, C.R.: Procrustes methods in the statistical analysis of shape revisited. In: Mardia, K.V., Gill, C.A. (eds.) Proceedings in Current Issues in Statistical Shape Analysis, pp. 18–33. Leeds University Press, Leeds (1995)
244. Goodall, C.R., Bose, A.: Models and Procrustes methods for the analysis of shape differences. In: Proceedings of the 19th Symposium on the Interface between Computer Science and Statistics, Lexington, KY (1987)
245. Goodall, C.R., Mardia, K.V.: Multivariate aspects of shape theory. Ann. Stat. **21**, 848–866 (1993)
246. Gordon, D.B., Marshall, S.A., Mayo, S.L.: Energy functions for protein design. Curr. Opin. Struct. Biol. **9**, 509–513 (1999)
247. Gordon, D.B., Hom, G.K., Mayo, S.L., Pierce, N.A.: Exact rotamer optimization for protein design. J. Comp. Chem. **24**, 232–243 (2003)
248. Gower, J.C.: Generalized Procrustes analysis. Psychometrika **40**, 33–51 (1975)
249. Grantcharova, V., Riddle, D., Santiago, J., Baker, D.: Important role of hydrogen bonds in the structurally polarized transition state for folding of the src SH3 domain. Nat. Struct. Mol. Biol. **5**, 714–720 (1998)
250. Grassberger, P.: Pruned-enriched Rosenbluth method: simulations of $\theta$ polymers of chain length up to 1 000 000. Phys. Rev. E **56**, 3682 (1997)
251. Grassberger, P., Fraunenkron, H., Nadler, W.: Monte Carlo Approach to Biopolymers and Protein Folding, p. 301. World Scientific, Singapore (1998)
252. Green, B.: The orthogonal approximation of an oblique structure in factor analysis. Psychometrika **17**, 429–440 (1952)
253. Green, P.: Reversible jump Markov chain Monte Carlo computation and bayesian model determination. Biometrika **82**, 711 (1995)
254. Green, P., Mardia, K.: Bayesian alignment using hierarchical models, with applications in protein bioinformatics. Biometrika **93**, 235–254 (2006)
255. Green, P.J., Silverman, B.W.: Nonparametric Regression and Generalized Linear Models: A Roughness Penalty Approach, vol. 58, 1st edn. Chapman and Hall, London (1994)
256. Green, P., Mardia, K., Nyirongo, V., Ruffieux, Y.: Bayesian modeling for matching and alignment of biomolecules. In: O'Hagan, A., West, M. (eds.) The Oxford Handbook of Applied Bayesian Analysis, pp. 27–50. Oxford University Press, Oxford (2010)
257. Grigoryan, G., Ochoa, A., Keating, A.E.: Computing van der Waals energies in the context of the rotamer approximation. Proteins **68**, 863–878 (2007)
258. Grigoryan, G., Reinke, A.W., Keating, A.E.: Design of protein-interaction specificity gives selective bZIP-binding peptides. Nature **458**, 859–864 (2009)
259. Gu, J., Li, H., Jiang, H., Wang, X.: A simple $C_\alpha$-SC potential with higher accuracy for protein fold recognition. Biochem. Biophys. Res. Commun. **379**, 610–615 (2009)
260. Guggenheim, E.A.: The statistical mechanics of co-operative assemblies. Proc. R. Soc. London A **169**, 134–148 (1938)
261. Gull, S.: Maximum Entropy and Bayesian Methods, pp. 53–71. Kluwer, Dordrecht (1989)
262. Güntert, P.: Structure calculation of biological macromolecules from NMR data. Q. Rev. Biophys. **31**, 145–237 (1998)
263. Güntert, P., Mumenthaler, C., Wüthrich, K.: Torsion angle dynamics for NMR structure calculation with the new program DYANA. J. Mol. Biol. **273**, 283–298 (1997)
264. Guo, Z., Thirumalai, D.: Kinetics of protein folding: nucleation mechanism, time scales, and pathways. Biopolymers **36**, 83–102 (1995)
265. Guttorp, P., Lockhart, R.A.: Finding the location of a signal: a Bayesian analysis. J. Am. Stat. Assoc. **83**, 322–330 (1988)
266. Habeck, M.: Statistical mechanics analysis of sparse data. J. Struct. Biol. **173**, 541–548 (2011)

267. Habeck, M., Nilges, M., Rieping, W.: Bayesian inference applied to macromolecular structure determination. Phys. Rev. E **72**, 031912 (2005)
268. Habeck, M., Nilges, M., Rieping, W.: Replica-Exchange Monte Carlo scheme for Bayesian data analysis. Phys. Rev. Lett. **94**, 0181051–0181054 (2005)
269. Habeck, M., Rieping, W., Nilges, M.: Bayesian estimation of Karplus parameters and torsion angles from three-bond scalar coupling constants. J. Magn. Reson. **177**, 160–165 (2005)
270. Habeck, M., Rieping, W., Nilges, M.: Weighting of experimental evidence in macromolecular structure determination. Proc. Natl. Acad. Sci. U.S.A. **103**, 1756–1761 (2006)
271. Habeck, M., Nilges, M., Rieping, W.: A unifying probabilistic framework for analyzing residual dipolar couplings. J. Biomol. NMR **40**, 135–144 (2008)
272. Hagan, A.O.: Monte Carlo is fundamentally unsound. Statistician **36**, 247–249 (1987)
273. Halperin, I., Ma, B., Wolfson, H., Nussinov, R.: Principles of docking: an overview of search algorithms and a guide to scoring functions. Proteins **47**, 409–443 (2002)
274. Hamelryck, T.: Probabilistic models and machine learning in structural bioinformatics. Stat. Methods Med. Res. **18**, 505–526 (2009)
275. Hamelryck, T., Kent, J., Krogh, A.: Sampling realistic protein conformations using local structural bias. PLoS Comput. Biol. **2**, e131 (2006)
276. Hamelryck, H., Borg, M., Paluszewski, M., Paulsen, J., Frellsen, J., Andreetta, C., Boomsma, W., Bottaro, S., Ferkinghoff-Borg, J.: Potentials of mean force for protein structure prediction vindicated, formalized and generalized. PLoS ONE **5**, e13714 (2010)
277. Han, K., Baker, D.: Recurring local sequence motifs in proteins. J. Mol. Biol. **251**, 176–187 (1995)
278. Hansen, S.: Bayesian estimation of hyperparameters for indirect Fourier transformation in small-angle scattering. J. Appl. Cryst. **33**, 1415–1421 (2000)
279. Hansen, S.: Estimation of chord length distributions from small-angle scattering using indirect Fourier transformation. J. Appl. Cryst. **36**, 1190–1196 (2003)
280. Hansen, S.: Simultaneous estimation of the form factor and structure factor for globular particles in small-angle scattering. J. Appl. Cryst. **41**, 436–445 (2008)
281. Hansen, S., Müller, J.: Maximum-Entropy and Bayesian Methods, Cambridge, England, 1994, pp. 69–78. Kluwer, Dordrecht (1996)
282. Hansen, S., Pedersen, J.: A comparison of three different methods for analysing small-angle scattering data. J. Appl. Cryst. **24**, 541–548 (1991)
283. Hansmann, U.H.E.: Sampling protein energy landscapes the quest for efficient algorithms. In: Kolinski, A. (ed.) Multiscale Approaches to Protein Modeling, pp. 209–230. Springer, New York (2011)
284. Hansmann, U., Okamoto, Y.: Prediction of peptide conformation by multicanonical algorithm: new approach to the multiple-minima problem. J. Comp. Chem. **14**, 1333–1338 (1993)
285. Hansmann, U., Okamoto, Y.: Generalized-ensemble Monte Carlo method for systems with rough energy landscape. Phys. Rev. E **56**, 2228 (1997)
286. Hansmann, U., Okamoto, Y.: Erratum:"finite-size scaling of helix–coil transitions in poly-alanine studied by multicanonical simulations"[J. Chem. Phys. **110**, 1267 (1999)]. J. Chem. Phys. **111**, 1339 (1999)
287. Hansmann, U., Okamoto, Y.: Finite-size scaling of helix–coil transitions in poly-alanine studied by multicanonical simulations. J. Chem. Phys. **110**, 1267 (1999)
288. Hansmann, U.H.E., Okamoto, Y.: New Monte Carlo algorithms for protein folding. Curr. Opin. Struct. Biol. **9**, 177–183 (1999)
289. Hao, M.H., Scheraga, H.A.: How optimization of potential functions affects protein folding. Proc. Natl. Acad. Sci. U.S.A. **93**, 4984–4989 (1996)
290. Harder, T., Boomsma, W., Paluszewski, M., Frellsen, J., hansson, J., Hamelryck, T.: Beyond rotamers: a generative, probabilistic model of side chains in proteins. BMC bioinformatics **11**, 306 (2010)
291. Hartmann, C., Schütte, C.: Comment on two distinct notions of free energy. Physica D **228**, 59–63 (2007)

# References

292. Hartmann, C., Antes, I., Lengauer, T.: IRECS: a new algorithm for the selection of most probable ensembles of side-chain conformations in protein models. Protein Sci. **16**, 1294–1307 (2007)

293. Hastings, W.: Monte Carlo sampling methods using Markov chains and their applications. Biometrika **57**, 97 (1970)

294. Hatano, N., Gubernatis, J.: Evidence for the double degeneracy of the ground state in the three-dimensional$\pm$j spin glass. Phys. Rev. B **66**(5), 054437 (2002)

295. Havel, T.F.: The sampling properties of some distance geometry algorithms applied to unconstrained polypeptide chains: a study of 1830 independently computed conformations. Biopolymers **29**, 1565–1585 (1990)

296. Havel, T.F., Kuntz, I.D., Crippen, G.M.: The theory and practice of distance geometry. Bull. Math. Biol. **45**, 665–720 (1983)

297. Havranek, J.J., Harbury, P.B.: Automated design of specificity in molecular recognition. Nat. Struct. Mol. Biol. **10**, 45–52 (2003)

298. Hayes, R.J., Bentzien, J., Ary, M.L., Hwang, M.Y., Jacinto, J.M., Vielmetter, J., Kundu, A., Dahiyat, B.I.: Combining computational and experimental screening for rapid optimization of protein properties. Proc. Natl. Acad. Sci. U.S.A. **99**, 15926–15931 (2002)

299. Hayter, J., Penfold, J.: An analytic structure factor for macro-ion solutions. Mol. Phys. **42**, 109–118 (1981)

300. Hazan, T., Shashua, A.: Convergent message-passing algorithms for inference over general graphs with convex free energies. In: McAllester, D.A., Myllymäki, P. (eds.) Proceedings of the 24rd Conference on Uncertainty in Artificial Intelligence, pp. 264–273. AUAI Press (2008)

301. Hendrickson, W.A.: Stereochemically restrained refinement of macromolecular structures. Methods Enzymol. **115**, 252–270 (1985)

302. Hess, B.: Convergence of sampling in protein simulations. Phys. Rev. E **65**, 031910 (2002)

303. Hesselbo, B., Stinchcombe, R.: Monte Carlo simulation and global optimization without parameters. Phys. Rev. Lett. **74**, 2151–2155 (1995)

304. Higuchi, T.: Monte Carlo filter using the genetic algorithm operators. J. Stat. Comput. Sim. **59**, 1–24 (1997)

305. Hinds, D.A., Levitt, M.: A lattice model for protein structure prediction at low resolution. Proc. Natl. Acad. Sci. U.S.A. **89**, 2536–2540 (1992)

306. Hinton, G.E.: Training products of experts by minimizing contrastive divergence. Neural Comput. **14**, 1771–800 (2002)

307. Hinton, G., Sejnowski, T.: Optimal perceptual inference. In: IEEE Conference Computer Vision Pattern Recognition, pp. 448–453. Washington, DC (1983)

308. Ho, B.K., Coutsias, E.A., Seok, C., Dill, K.A.: The flexibility in the proline ring couples to the protein backbone. Protein Sci. **14**, 1011–1018 (2005)

309. Hoffmann, D., Knapp, E.W.: Polypeptide folding with off-lattice Monte Carlo dynamics: the method. Eur. Biophys. J. **24**, 387–403 (1996)

310. Hoffmann, D., Knapp, E.W.: Protein dynamics with off-lattice Monte Carlo moves. Phys. Rev. E **53**, 4221–4224 (1996)

311. Holm, L., Sander, C.: Protein structure comparison by alignment of distance matrices. J. Mol. Biol. **233**, 123–138 (1993)

312. Hong, E.J., Lippow, S.M., Tidor, B., Lozano-Perez, T.: Rotamer optimization for protein design through MAP estimation and problem-size reduction. J. Comp. Chem. **30**, 1923–1945 (2009)

313. Hooft, R., van Eijck, B., Kroon, J.: An adaptive umbrella sampling procedure in conformational analysis using molecular dynamics and its application to glycol. J. Chem. Phys. **97**, 6690 (1992)

314. Hopfinger, A.J.: Conformational Properties of Macromolecules. Academic, New York (1973)

315. Horn, B.K.P.: Closed-form solution of absolute orientation using unit quaternions. J. Opt. Soc. Am. A **4**, 629–642 (1987)

316. Hu, X., Kuhlman, B.: Protein design simulations suggest that side-chain conformational entropy is not a strong determinant of amino acid environmental preferences. Proteins **62**, 739–748 (2006)
317. Hu, C., Li, X., Liang, J.: Developing optimal non-linear scoring function for protein design. Bioinformatics **20**, 3080–3098 (2004)
318. Huang, K.: Statistical Mechanics. Wiley, New York (1987)
319. Huang, S.Y., Zou, X.: An iterative knowledge-based scoring function to predict protein-ligand interactions: I. Derivation of interaction potentials. J. Comput. Chem. **27**, 1866–1875 (2006)
320. Huang, S.Y., Zou, X.: An iterative knowledge-based scoring function to predict protein-ligand interactions: II. Validation of the scoring function. J. Comput. Chem. **27**, 1876–1882 (2006)
321. Huelsenbeck, J.P., Crandall, K.A.: Phylogeny estimation and hypothesis testing using maximum likelihood. Annu. Rev. Ecol. Systemat. **28**, 437–466 (1997)
322. Hukushima, K.: Domain-wall free energy of spin-glass models: numerical method and boundary conditions. Phys. Rev. E **60**, 3606 (1999)
323. Hukushima, K., Nemoto, K.: Exchange Monte Carlo method and application to spin glass simulations. J. Phys. Soc. Jap. **65**, 1604–1608 (1996)
324. Hukushima, K., Takayama, H., Yoshino, H.: Exchange Monte Carlo dynamics in the SK model. J. Phys. Soc. Jap. **67**, 12–15 (1998)
325. Humphris, E.L., Kortemme, T.: Design of multi-specificity in protein interfaces. PLoS Comput. Biol. **3**, e164 (2007)
326. Hunter, L., States, D.J.: Bayesian classification of protein structure. IEEE Expert: Intel. Sys. App. **7**, 67–75 (1992)
327. Hutchinson, E., Thornton, J.: PROMOTIF–A program to identify and analyze structural motifs in proteins. Protein Sci. **5**, 212–220 (1996)
328. Iba, Y.: Extended ensemble Monte Carlo. Int. J. Mod. Phys. **C12**, 623–656 (2001)
329. Irbäck, A., Potthast, F.: Studies of an off-lattice model for protein folding: Sequence dependence and improved sampling at finite temperature. J. Chem. Phys. **103**, 10298 (1995)
330. Jack, A., Levitt, M.: Refinement of large structures by simultaneous minimization of energy and R factor. Acta Crystallogr. A **34**, 931–935 (1978)
331. Jain, T., de Pablo, J.: A biased Monte Carlo technique for calculation of the density of states of polymer films. J. Chem. Phys. **116**, 7238 (2002)
332. Janke, W., Sugita, Y., Mitsutake, A., Okamoto, Y.: Generalized-ensemble algorithms for protein folding simulations. In: Rugged Free Energy Landscapes. Lecture Notes in Physics, vol. 736, pp. 369–407. Springer, Berlin/Heidelberg (2008)
333. Jardine, N.: The observational and theoretical components of homology: a study based on the morphology of the derma-roofs of rhipidistan fishes. Biol. J. Linn. Soc. **1**, 327–361 (1969)
334. Jaynes, E.: How does the brain do plausible reasoning? Technical report; Microwave Laboratory and Department of Physics, Stanford University, California, USA (1957)
335. Jaynes, E.: Information Theory and Statistical Mechanics. II. Phys. Rev. **108**, 171–190 (1957)
336. Jaynes, E.T.: Information theory and statistical mechanics. Phys. Rev. Lett. **106**, 620–630 (1957)
337. Jaynes, E.: Prior probabilities. IEEE Trans. Syst. Sci. Cybernetics **4**, 227–241 (1968)
338. Jaynes, E.T.: Confidence intervals vs Bayesian intervals (with discussion). In:Harper, W.L., Hooker, C.A. (eds.) Foundations of Probability Theory, Statistical Inference, and Statistical Theories of Science, pp. 175–257. D. Reidel, Dordrecht/Holland (1976)
339. Jaynes, E.: Where do we stand on maximum entropy? In: Levine, R.D, Tribus, M. (eds.) The Maximum Entropy Formalism Conference, pp. 15–118. MIT, Cambridge (1978)
340. Jaynes, E.: In: Rosenkrantz, R. (ed.) Papers on Probability, Statistics and Statistical Physics. Kluwer, Dordrecht (1983)
341. Jaynes, E.: Prior information and ambiguity in inverse problems. In: SIAM -AMS Proceedings, vol. 14, pp. 151–166. American Mathematical Society (1984)
342. Jaynes, E.: Probability Theory: The Logic of Science. Cambridge University Press, Cambridge (2003)

References 357

343. Jeffreys, H.: Theory of Probability. Oxford Classics Series. Oxford University Press, Oxford (1998)
344. Jensen, J.L.W.V.: Sur les fonctions convexes et les inégalités entre les valeurs moyennes. Acta Mathematica **30**, 175–193 (1906)
345. Jensen, F.: Introduction to Bayesian Networks. Springer, New York (1996)
346. Jiang, L., Gao, Y., Mao, F., Liu, Z., Lai, L.: Potential of mean force for protein-protein interaction studies. Proteins **46**, 190–196 (2002)
347. Jiang, L., Althoff, E.A., Clemente, F.R., Doyle, L., Rothlisberger, D., Zanghellini, A., Gallaher, J.L., Betker, J.L., Tanaka, F., Barbas Carlos F. III, Hilvert, D., Houk, K.N., Stoddard, B.L., Baker, D.: De novo computational design of retro-aldol enzymes. Science **319**, 1387–1391 (2008)
348. Jin, W., Kambara, O., Sasakawa, H., Tamura, A., Takada, S.: De novo design of foldable proteins with smooth folding funnel: automated negative design and experimental verification. Structure **11**, 581–590 (2003)
349. Jones, D.T.: De novo protein design using pairwise potentials and a genetic algorithm. Protein Sci. **3**, 567–574 (1994)
350. Jones, D.: Successful ab initio prediction of the tertiary structure of NK-lysin using multiple sequences and recognized supersecondary structural motifs. Proteins **1**, 185–191 (1997)
351. Jones, T.A., Thirup, S.: Using known substructures in protein model building and crystallography. EMBO J. **5**, 819–822 (1986)
352. Jones, D.T., Taylor, W.R., Thornton, J.M.: A new approach to protein fold recognition. Nature **358**, 86–89 (1992)
353. Jordan, M. (ed.): Learning in Graphical Models. MIT, Cambridge (1998)
354. Jorgensen, W.L., Maxwell, D.S., Tirado-Rives, J.: Development and testing of the OPLS all-atom force field on conformational energetics and properties of organic liquids. J. Am. Chem. Soc. **118**, 11225–11236 (1996)
355. Juraszek, J., Bolhuis, P.G.: Sampling the multiple folding mechanisms of Trp-cage in explicit solvent. Proc. Natl. Acad. Sci. U.S.A. **103**, 15859–15864 (2006)
356. Kabsch, W.: A solution for the best rotation to relate two sets of vectors. Acta Crystallogr. A **32**, 922–923 (1976)
357. Kabsch, W.: A discussion of the solution for the best rotation to to relate two sets of vectors. Acta Crystallogr. A **A34**, 827–828 (1978)
358. Kamisetty, H., Xing, E.P., Langmead, C.J.: Free energy estimates of all-atom protein structures using generalized belief propagation. In: Speed, T., Huang, H. (eds.) RECOMB, pp. 366–380. Springer, Heidelberg (2007)
359. Kamisetty, H., Bailey-Kellogg, C., Langmead, C.J.: A graphical model approach for predicting free energies of association for protein-protein interactions under backbone and side-chain flexibility. Technical Report, School of Computer Science, Carnegie Mellon University, Pittsburgh, PA (2008)
360. Karlin, S., Zhu, Z.Y.: Characterizations of diverse residue clusters in protein three-dimensional structures. Proc. Natl. Acad. Sci. U.S.A. **93**, 8344–8349 (1996)
361. Karplus, M.: Vicinal proton coupling in nuclear magnetic resonance. J. Am. Chem. Soc. **85**, 2870–2871 (1963)
362. Katzgraber, H., Trebst, S., Huse, D., Troyer, M.: Feedback-optimized parallel tempering Monte Carlo. J. Stat. Mech. **2006**, P03018 (2006)
363. Kearsley, S.K.: An algorithm for the simultaneous superposition of a structural series. J. Comput. Chem. **11**, 1187–1192 (1990)
364. Kelley, L.A., MacCallum, R.M., Sternberg, M.J.: Enhanced genome annotation using structural profiles in the program 3D-PSSM. J. Mol. Biol. **299**, 501–522 (2000)
365. Kent, J.T.: The Fisher-Bingham distribution on the sphere. J. R. Stat. Soc. B **44**, 71–80 (1982)
366. Kent, J.: The complex Bingham distribution and shape analysis. J. R. Statist. Soc. B **56**, 285–299 (1994)
367. Kent, J.T., Hamelryck, T.: Using the Fisher-Bingham distribution in stochastic models for protein structure. In: Barber, S., Baxter, P.D., Mardia, K.V., Walls, R.E. (eds.) Proceedings

in Quantitative Biology, Shape Analysis and Wavelets, pp. 57–60. Leeds University Press (2005)

368. Kent, J., Mardia, K., Taylor, C.: Matching problems for unlabelled configurations. In: Aykroyd, R., Barber, S., Mardia, K. (eds.) Bioinformatics, Images, and Wavelets, pp. 33–36. Leeds University Press, Leeds (2004)

369. Kent, J.T., Mardia, K.V., Taylor, C.C.: Modelling strategies for bivariate circular data. In: Barber, S., Baxter, P.D., Gusnanto, A., Mardia, K.V. (eds.) The Art and Science of Statistical Bioinformatics, pp. 70–74. Leeds University Press, Leeds (2008)

370. Kent, J., Mardia, K., Taylor, C.: Matching Unlabelled Configurations and Protein Bioinformatics. Research note, University of Leeds, Leeds (2010)

371. Kerler, W., Rehberg, P.: Simulated-tempering procedure for spin-glass simulations. Phys. Rev. E **50**, 4220–4225 (1994)

372. Kikuchi, R.: A theory of cooperative phenomena. Phys. Rev. **81**, 988 (1951)

373. Kimura, K., Taki, K.: Time-homogeneous parallel annealing algorithm. In: Vichneetsky, R., Miller, J. (eds.) Proceedings of the 13th IMACS World Congress on Computation and Applied Mathematics, vol. 2, p. 827. Criterion Press, Dublin (1991)

374. Kingsford, C.L., Chazelle, B., Singh, M.: Solving and analyzing side-chain positioning problems using linear and integer programming. Bioinformatics **21**, 1028–1039 (2005)

375. Kirkpatrick, S., Gelatt, C., Vecchi, M.: Optimization by simulated annealing. science **220**, 671 (1983)

376. Kirkwood, J.G.: Statistical mechanics of fluid mixtures. J. Chem. Phys. **3**, 300–313 (1935)

377. Kirkwood, J.G., Maun, E.K., Alder, B.J.: Radial distribution functions and the equation of state of a fluid composed of rigid spherical molecules. J. Chem. Phys. **18**, 1040–1047 (1950)

378. Kleinman, C.L., Rodrigue, N., Bonnard, C., Philippe, H., Lartillot, N.: A maximum likelihood framework for protein design. BMC Bioinformatics **7**, 326 (2006)

379. Kleywegt, G.: Recognition of spatial motifs in protein structures. J. Mol. Biol. **285**, 1887–1897 (1999)

380. Kloppmann, E., Ullmann, G.M., Becker, T.: An extended dead-end elimination algorithm to determine gap-free lists of low energy states. J. Comp. Chem. **28**, 2325–2335 (2007)

381. Koehl, P., Delarue, M.: Application of a self-consistent mean field theory to predict protein side-chains conformation and estimate their conformational entropy. J. Mol. Biol. **239**, 249–275 (1994)

382. Kojetin, D.J., McLaughlin, P.D., Thompson, R.J., Dubnau, D., Prepiak, P., Rance, M., Cavanagh, J.: Structural and motional contributions of the Bacillus subtilis ClpC N-domain to adaptor protein interactions. J. Mol. Biol. **387**, 639–652 (2009)

383. Kolinski, A.: Protein modeling and structure prediction with a reduced representation. Acta Biochim. Pol. **51**, 349–371 (2004)

384. Koliński, A., Skolnick, J.: High coordination lattice models of protein structure, dynamics and thermodynamics. Acta Biochim. Pol. **44**, 389–422 (1997)

385. Kolinski, A., Skolnick, J.: Reduced models of proteins and their applications. Polymer **45**, 511–524 (2004)

386. Kollman, P.: Free energy calculations: applications to chemical and biochemical phenomena. Chem. Rev. **93**, 2395–2417 (1993)

387. Komodakis, N., Paragios, N.: Beyond loose LP-relaxations: optimizing MRFs by repairing cycles. In: Forsyth, D., Torr, P., Zisserman, A. (eds.) ECCV, pp. 806–820. Springer, Heidelberg (2008)

388. Konig, G., Boresch, S.: Non-Boltzmann sampling and Bennett's acceptance ratio method: how to profit from bending the rules. J. Comput. Chem. **32**, 1082–1090 (2011)

389. Kono, H., Saven, J.G.: Statistical theory for protein combinatorial libraries. packing interactions, backbone flexibility, and sequence variability of main-chain structure. J. Mol. Biol. **306**, 607–628 (2001)

390. Koppensteiner, W.A., Sippl, M.J.: Knowledge-based potentials–back to the roots. Biochemistry Mosc. **63**, 247–252 (1998)

# References

359

391. Körding, K., Wolpert, D.: Bayesian integration in sensorimotor learning. Nature **427**, 244–247 (2004)

392. Kortemme, T., Morozov, A.V., Baker, D.: An orientation-dependent hydrogen bonding potential improves prediction of specificity and structure for proteins and protein-protein complexes. J. Mol. Biol. **326**, 1239–1259 (2003)

393. Kraemer-Pecore, C.M., Lecomte, J.T.J., Desjarlais, J.R.: A de novo redesign of the WW domain. Protein Sci. **12**, 2194–2205 (2003)

394. Kraulis, P.J.: MOLSCRIPT: A program to produce both detailed and schematic plots of protein structures. J. Appl. Cryst. **24**, 946–950 (1991)

395. Krishna, S.S., Majumdar, I., Grishin, N.V.: Structural classification of zinc fingers: survey and summary. Nucleic Acids Res. **31**, 532–550 (2003)

396. Krissinel, E., Henrick, K.: Secondary-structure matching (ssm), a new tool for fast protein structure alignment in three dimensions. Acta Crystallogr. D **60**, 2256–2268 (2004)

397. Kristof, W., Wingersky, B.: Generalization of the orthogonal Procrustes rotation procedure for more than two matrices. In: Proceedings of the 79th Annual Convention of the American Psychological Association, vol. 6, pp. 89–90. American Psychological Association, Washington, DC (1971)

398. Krivov, G.G., Shapovalov, M.V., Dunbrack, R.L. Jr.: Improved prediction of protein side-chain conformations with SCWRL4. Proteins **77**, 778–795 (2009)

399. Kruglov, T.: Correlation function of the excluded volume. J. Appl. Cryst. **38**, 716–720 (2005)

400. Kryshtafovych, A., Venclovas, C., Fidelis, K., Moult, J.: Progress over the first decade of CASP experiments. Proteins **61**, 225–236 (2005)

401. Kryshtafovych, A., Fidelis, K., Moult, J.: Progress from CASP6 to CASP7. Proteins **69**, 194–207 (2007)

402. Kschischang, F., Frey, B., Loeliger, H.: Factor graphs and the sum-product algorithm. IEEE Trans. Inf. Theory **47**, 498–519 (2001)

403. Kuczera, K.: One- and multidimensional conformational free energy simulations. J. Comp. Chem. **17**, 1726–1749 (1996)

404. Kudin, K., Dymarsky, A.: Eckart axis conditions and the minimization of the root-mean-square deviation: two closely related problems. J. Chem. Phys. **122**, 224105 (2005)

405. Kuhlman, B., Baker, D.: Native protein sequences are close to optimal for their structures. Proc. Natl. Acad. Sci. U.S.A. **97**, 10383–10388 (2000)

406. Kuhlman, B., Dantas, G., Ireton, G.C., Varani, G., Stoddard, B.L., Baker, D.: Design of a novel globular protein fold with atomic-level accuracy. Science **302**, 1364–1368 (2003)

407. Kullback, S.: Information Theory and Statistics. Wiley, New York (1959)

408. Kullback, S., Leibler, R.: On information and sufficiency. Ann. Math. Stat. **22**, 79–86 (1951)

409. Kumar, S., Rosenberg, J., Bouzida, D., Swendsen, R., Kollman, P.: The weighted histogram analysis method for free-energy calculations on biomolecules. I. the method. J. Comp. Chem. **13**, 1011–1021 (1992)

410. Kuszewski, J., Gronenborn, A.M., Clore, G.M.: Improving the quality of NMR and crystallographic protein structures by means of a conformational database potential derived from structure databases. Protein Sci. **5**, 1067–1080 (1996)

411. Laio, A., Parrinello, M.: Escaping free-energy minima. Proc. Natl. Acad. Sci. U.S.A. **99**, 12562–12566 (2002)

412. Laio, A., Rodriguez-Fortea, A., Gervasio, F., Ceccarelli, M., Parrinello, M.: Assessing the accuracy of metadynamics. J. Phys. Chem. B **109**, 6714–6721 (2005)

413. Landau, D., Binder, K.: A Guide to Monte Carlo Simulations in Statistical Physics. Cambridge University Press, Cambridge (2009)

414. Landau, L.D., Lifshitz, E.M.: Statistical Physics. Course of Theoretical Physics, vol. 5. Pergamon Press, Oxford (1980)

415. Laplace, P.: Mémoire sur la probabilité des causes par les événements. Mémoires de Mathématique et de Physique **6**, 621–656 (1774)

416. Laplace, S.: Théorie analytique des probabilités. Gauthier-Villars, Paris (1820)

417. Larson, S.M., England, J.L., Desjarlais, J.R., Pande, V.S.: Thoroughly sampling sequence space: large-scale protein design of structural ensembles. Protein Sci. **11**, 2804–2813 (2002)
418. Larson, S.M., Garg, A., Desjarlais, J.R., Pande, V.S.: Increased detection of structural templates using alignments of designed sequences. Proteins **51**, 390–396 (2003)
419. Launay, G., Mendez, R., Wodak, S., Simonson, T.: Recognizing protein-protein interfaces with empirical potentials and reduced amino acid alphabets. BMC Bioinformatics **8**, 270 (2007)
420. Lauritzen, S.: Graphical Models. Oxford University Press, Oxford (1996)
421. Law, A.M., Kelton, W.D.: Simulation Modeling and Analysis. McGraw-Hill, New York (1991)
422. Lazar, G.A., Desjarlais, J.R., Handel, T.M.: De novo design of the hydrophobic core of ubiquitin. Protein Sci. **6**, 1167–1178 (1997)
423. Lazaridis, T., Karplus, M.: Effective energy functions for protein structure prediction. Curr. Opin. Struct. Biol. **10**, 139–145 (2000)
424. Leach, A.R.: Molecular Modelling: Principles and Applications, 2nd edn. Prentice Hall, Harlow (2001)
425. Leach, A.R., Lemon, A.P.: Exploring the conformational space of protein side chains using dead-end elimination and the A* algorithm. Proteins **33**, 227–239 (1998)
426. Leaver-Fay, A., Tyka, M., Lewis, S., Lange, O., Thompson, J., Jacak, R., Kaufman, K., Renfrew, P., Smith, C., Sheffler, W., et al.: ROSETTA3: an object-oriented software suite for the simulation and design of macromolecules. Methods Enzymol. **487**, 545–574 (2011)
427. Lee, J.: New Monte Carlo algorithm: entropic sampling. Phys. Rev. Lett. **71**, 211–214 (1993)
428. Lee, P.M.: Bayesian Statistics: An Introduction, 3rd edn. Wiley, New York (2004)
429. Lee, Y., Nelder, J.: Hierarchical generalized linear models. J. R. Stat. Soc. B **58**, 619–678 (1996)
430. Lele, S.: Euclidean distance matrix analysis (EDMA) – estimation of mean form and mean form difference. Math. Geol. **25**, 573–602 (1993)
431. Lele, S., Richtsmeier, J.T.: Statistical models in morphometrics: are they realistic? Syst. Zool. **39**, 60–69 (1990)
432. Lele, S., Richtsmeier, J.T.: An Invariant Approach to Statistical Analysis of Shapes. Interdisciplinary Statistics. Chapman and Hall/CRC, Boca Raton (2001)
433. Lemm, J.: Bayesian Field Theory. Johns Hopkins University Press, Baltimore (2003)
434. Lennox, K.P.: A Dirichlet process mixture of hidden Markov models for protein structure prediction. Ann. Appl. Stat. **4**, 916–942 (2010)
435. Lennox, K.P., Dahl, D.B., Vannucci, M., Tsai, J.W.: Density estimation for protein conformation angles using a bivariate von Mises distribution and Bayesian nonparametrics. J. Am. Stat. Assoc. **104**, 586–596 (2009)
436. Levitt, M.: Growth of novel protein structural data. Proc. Natl. Acad. Sci. U.S.A. **104**, 3183–3188 (2007)
437. Levitt, M., Warshel, A.: Computer simulation of protein folding. Nature **253**, 694–698 (1975)
438. Li, X., Liang, J.: Knowledge-based energy functions for computational studies of proteins. In: Xu, Y., Xu, D., Liang, J. (eds.) Computational Methods for Protein Structure Prediction and Modeling. Biological and Medical Physics, Biomedical Engineering, vol. 1, p. 71. Springer, New York (2007)
439. Li, H., Tang, C., Wingreen, N.S.: Nature of driving force for protein folding: a result from analyzing the statistical potential. Phys. Rev. Lett. **79**, 765–768 (1997)
440. Lifshitz, E.M., Landau, L.D.: Statistical physics. In: Course of Theoretical Physics, vol. 5, 3rd edn. Butterworth-Heinemann, Oxford (1980)
441. Lifson, S., Levitt, M.: On obtaining energy parameters from crystal structure data. Comput. Chem. **3**, 49–50 (1979)
442. Lilien, H., Stevens, W., Anderson, C., Donald, R.: A novel ensemble-based scoring and search algorithm for protein redesign and its application to modify the substrate specificity of the gramicidin synthetase a phenylalanine adenylation enzyme. J. Comp. Biol. **12**, 740–761 (2005)

References 361

443. Lima, A., de Oliveira, P., Penna, T.: A comparison between broad histogram and multicanonical methods. J. Stat. Phys. **99**, 691–705 (2000)
444. Lindley, D.: Bayesian Statistics, A Review. Society for Industrial and applied Mathematics, Philadelphia (1972)
445. Lindorff-Larsen, K., Best, R.B., Depristo, M.A., Dobson, C.M., Vendruscolo, M.: Simultaneous determination of protein structure and dynamics. Nature **433**, 128–132 (2005)
446. Linge, J.P., Nilges, M.: Influence of non-bonded parameters on the quality of NMR structures: a new force-field for NMR structure calculation. J. Biomol. NMR **13**, 51–59 (1999)
447. Liu, Z., Macias, M.J., Bottomley, M.J., Stier, G., Linge, J.P., Nilges, M., Bork, P., Sattler, M.: The three-dimensional structure of the HRDC domain and implications for the Werner and Bloom syndrome proteins. Fold. Des. **7**, 1557–1566 (1999)
448. Looger, L.L., Hellinga, H.W.: Generalized dead-end elimination algorithms make large-scale protein side-chain structure prediction tractable: implications for protein design and structural genomics. J. Mol. Biol. **307**, 429–445 (2001)
449. Looger, L.L., Dwyer, M.A., Smith, J.J., Hellinga, H.W.: Computational design of receptor and sensor proteins with novel functions. Nature **423**, 185–190 (2003)
450. Lou, F., Clote, P.: Thermodynamics of RNA structures by Wang–Landau sampling. Bioinformatics **26**, i278–i286 (2010)
451. Lovell, S.C., Word, J.M., Richardson, J.S., Richardson, D.C.: The penultimate rotamer library. Proteins **40**, 389–408 (2000)
452. Lu, H., Skolnick, J.: A distance-dependent atomic knowledge-based potential for improved protein structure selection. Proteins **44**, 223–232 (2001)
453. Lu, M., Dousis, A.D., Ma, J.: OPUS-PSP: an orientation-dependent statistical all-atom potential derived from side-chain packing. J. Mol. Biol. **376**, 288–301 (2008)
454. Lu, M., Dousis, A.D., Ma, J.: OPUS-Rota: a fast and accurate method for side-chain modeling. Protein Sci. **17**, 1576–1585 (2008)
455. Lyubartsev, A., Martsinovski, A., Shevkunov, S., Vorontsov-Velyaminov, P.: New approach to Monte Carlo calculation of the free energy: method of expanded ensembles. J. Chem. Phys. **96**, 1776 (1992)
456. MacKay, D.: Maximum Entropy and Bayesian Methods: Proceedings of the Eleventh International Workshop on Maximum Entropy and Bayesian Methods of Statistical Analysis, Seattle, 1991, pp. 39–66. Kluwer, Dordrecht (1992)
457. MacKay, D.: Information Theory, Inference, and Learning Algorithms. Cambridge University Press, Cambridge (2003)
458. MacKerel, A. Jr., Brooks, C. III, Nilsson, L., Roux, B., Won, Y., Karplus, M.: CHARMM: the energy function and its parameterization with an overview of the program. In: Schleyer, P.v.R., et al. (eds.) The Encyclopedia of Computational Chemistry, vol. 1, pp. 271–277. Wiley, Chichester (1998)
459. Mackerell, A.D.: Empirical force fields for biological macromolecules: overview and issues. J. Comput. Chem. **25**, 1584–1604 (2004)
460. Macura, S., Ernst, R.R.: Elucidation of cross relaxation in liquids by two-dimensional NMR spectroscopy. Mol. Phys. **41**, 95–117 (1980)
461. Maddox, J.: What Remains to be Discovered. Macmillan, London (1998)
462. Madras, N., Sokal, A.: The pivot algorithm: a highly efficient Monte Carlo method for the self-avoiding walk. J. Stat. Phys. **50**, 109–186 (1988)
463. Maiorov, V.N., Crippen, G.M.: Contact potential that recognizes the correct folding of globular proteins. J. Mol. Biol. **227**, 876–888 (1992)
464. Májek, P., Elber, R.: A coarse-grained potential for fold recognition and molecular dynamics simulations of proteins. Proteins **76**, 822–836 (2009)
465. Malioutov, D.M., Johnson, J.K., Willsky, A.S.: Walk-sums and belief propagation in Gaussian graphical models. J. Mach. Learn. Res. **7**, 2031–2064 (2006)
466. Mao, Y., Senic-Matuglia, F., Di Fiore, P.P., Polo, S., Hodsdon, M.E., De Camilli, P.: Deubiquitinating function of ataxin-3: insights from the solution structure of the Josephin domain. Proc. Natl. Acad. Sci. U.S.A. **102**, 12700–12705 (2005)

467. Mardia, K.V.: Characterization of directional distributions. In: Patil, G.P., Kotz, S., Ord, J.K. (eds.) Statistical Distributions in Scientific Work, pp. 365–386. D. Reidel Publishing Company, Dordrecht/Holland (1975)
468. Mardia, K.V.: Statistics of directional data (with discussion). J. R. Statist. Soc. B. **37**, 349–393 (1975)
469. Mardia, K.: Discussion to schmidler. In: Bernardo, J., Bayarri, M., Berger, J., Dawid, A., Heckerman, D., Smith, A., West, M. (eds.) Bayesian Statistics 8, pp. 471–490. Oxford University Press, Oxford (2007)
470. Mardia, K.: On some recent advancements in applied shape analysis and directional statistics. In: Barber, S., Baxter, P., Mardia, K. (eds.) Systems Biology & Statistical Bioinformatics, pp. 9–17. Leeds University Press, Leeds (2007)
471. Mardia, K.V.: Statistical complexity in protein bioinformatics. In: Gusnanto, A., Mardia, K.V., Fallaize, C.J. (eds.) Statistical Tools for Challenges in Bioinformatics, pp. 9–20. Leeds University Press, Leeds (2009)
472. Mardia, K.V.: Bayesian analysis for bivariate von Mises distributions. J. Appl. Stat. **37**, 515–528 (2010)
473. Mardia, K.V., El-Atoum, S.A.M.: Bayesian inference for the von Mises-Fisher distribution. Biometrika **63**, 203–206 (1976)
474. Mardia, K.V., Jupp, P.: Directional Statistics, 2nd edn . Wiley, Chichester (2000)
475. Mardia, K.V., Patrangenaru, V.: Directions and projective shapes. Ann. Stat. **33**, 1666–1699 (2005)
476. Mardia, K.V., Hughes, G., Taylor, C.C.: Efficiency of the pseudolikelihood for multivariate normal and von Mises distributions. Technical report, Department of Statistics, University of Leeds (2007)
477. Mardia, K., Nyirongo, V., Green, P., Gold, N., Westhead, D.: Bayesian refinement of protein functional site matching. BMC Bioinformatics **8**, 257 (2007)
478. Mardia, K.V., Taylor, C.C., Subramaniam, G.K.: Protein bioinformatics and mixtures of bivariate von Mises distributions for angular data. Biometrics **63**, 505–512 (2007)
479. Mardia, K.V., Hughes, G., Taylor, C.C., Singh, H.: A multivariate von Mises distribution with applications to bioinformatics. Can. J. Stat. **36**, 99–109 (2008)
480. Mardia, K.V., Kent, J.T., Hughes, G., Taylor, C.C.: Maximum likelihood estimation using composite likelihoods for closed exponential families. Biometrika **96**, 975–982 (2009)
481. Mardia, K.V., Fallaize, C.J., Barber, S., Jackson, R.M., Theobald, D.L.: Bayesian alignment of similarity shapes and halfnormal-gamma distributions. Ann. Appl. Stat. Submitted (2011)
482. Marin, J., Nieto, C.: Spatial matching of multiple configurations of points with a bioinformatics application. Commun. Stat. Theor. Meth. **37**, 1977–1995 (2008)
483. Marinari, E.: Optimized Monte Carlo methods. In: Kertész, J., Kondor, I. (eds.) Advances in Computer Simulation. Lecture Notes in Physics, vol. 501, pp. 50–81. Springer, Berlin (1998)
484. Marinari, E., Parisi, G.: Simulated tempering: a new Monte Carlo scheme. Europhys. Lett. **19**, 451–458 (1992)
485. Marinari, E., Parisi, G., Ricci-Tersenghi, F., Zuliani, F.: The use of optimized Monte Carlo methods for studying spin glasses. J. Phys. A **34**, 383 (2001)
486. Marjoram, P., Molitor, J., Plagnol, V., Tavaré, S.: Markov chain Monte Carlo without likelihoods. Proc. Natl. Acad. Sci. U.S.A. **100**, 15324–15328 (2003)
487. Markley, J.L., Bax, A., Arata, Y., Hilbers, C.W., Kaptein, R., Sykes, B.D., Wright, P.E., Wüthrich, K.: Recommendations for the presentation of NMR structures of proteins and nucleic acids. J. Mol. Biol. **280**, 933–952 (1998)
488. Martinez, J., Pisabarro, M., Serrano, L.: Obligatory steps in protein folding and the conformational diversity of the transition state. Nat. Struct. Mol. Biol. **5**, 721–729 (1998)
489. Martoňák, R., Laio, A., Parrinello, M.: Predicting crystal structures: the Parrinello-Rahman method revisited. Phys. Rev. Lett. **90**, 75503 (2003)
490. May, R., Nowotny, V.: Distance information derived from neutron low-Q scattering. J. Appl. Cryst. **22**, 231–237 (1989)

References     363

491. Mayewski, S.: A multibody, whole-residue potential for protein structures, with testing by Monte Carlo simulated annealing. Proteins **59**, 152–169 (2005)
492. McCammon, J.A., Harvey, S.C.: Dynamics of Proteins and Nucleic Acids. Cambridge University Press, Cambridge (1987)
493. McDonald, I.K., Thornton, J.M.: Satisfying hydrogen bonding potential in proteins. J. Mol. Biol. **238**, 777–793 (1994)
494. McGrayne, S.: The Theory That Would Not Die: How Bayes' Rule Cracked the Enigma Code, Hunted Down Russian Submarines, and Emerged Triumphant from Two Centuries of Controversy. Yale University Press, New Haven (2011)
495. McLachlan, G.J., Krishnan, T.: The EM Algorithm and Extensions. Wiley Series in Probability and Statistics. Applied Probability and Statistics. Wiley, New York (1997)
496. McQuarrie, D.: Statistical Mechanics. University Science Books, Sausalito (2000)
497. Mechelke, M., Habeck, M.: Robust probabilistic superposition and comparison of protein structures. BMC Bioinformatics **11**, 363 (2010)
498. Melo, F., Feytmans, E.: Novel knowledge-based mean force potential at atomic level. J. Mol. Biol. **267**, 207–222 (1997)
499. Meltzer, T., Globerson, A., Weiss, Y.: Convergent message passing algorithms – a unifying view. In: Proceedings of the Twenty-fifth Conference on Uncertainty in AI (UAI), Montreal, Canada (2009)
500. Mendes, J., Baptista, A.M., Carrondo, M.A., Soares, C.M.: Improved modeling of side-chains in proteins with rotamer-based methods: a flexible rotamer model. Proteins **37**, 530–543 (1999)
501. Metropolis, N., Rosenbluth, A.W., Rosenbluth, M.N., Teller, A.H., Teller, E.: Equation of state calculations by fast computing machines. J. Chem. Phys. **21**, 1087–1092 (1953)
502. Metzler, W.J., Hare, D.R., Pardi, A.: Limited sampling of conformational space by the distance geometry algorithm: implications for structures generated from NMR data. Biochemistry **28**, 7045–7052 (1989)
503. Mezard, M., Parisi, G., Virasoro, M.: Spin Glasses and Beyond. World Scientific, New York (1988)
504. Mezei, M.: Adaptive umbrella sampling: self-consistent determination of the non-Boltzmann bias. J. Comp. Phys. **68**, 237–248 (1987)
505. Mezei, M.: Efficient Monte Carlo sampling for long molecular chains using local moves, tested on a solvated lipid bilayer. J. Chem. Phys. **118**, 3874 (2003)
506. Micheletti, C., Seno, F., Banavar, J.R., Maritan, A.: Learning effective amino acid interactions through iterative stochastic techniques. Proteins **42**, 422–431 (2001)
507. Micheletti, C., Laio, A., Parrinello, M.: Reconstructing the density of states by history-dependent metadynamics. Phys. Rev. Lett. **92**, 170601 (2004)
508. Miller, R.: The jackknife-a review. Biometrika **61**, 1 (1974)
509. Minka, T.: Pathologies of orthodox statistics. Microsoft research, Technical report, Cambridge, UK (2001)
510. Minka, T.: Estimating a Dirichlet distribution. Technical report, Microsoft Research, Cambridge (2003)
511. Mintseris, J., Wiehe, K., Pierce, B., Anderson, R., Chen, R., Janin, J., Weng, Z.: Protein-protein docking benchmark 2.0: an update. Proteins **60**, 214 (2005)
512. Mirny, L.A., Shakhnovich, E.I.: How to derive a protein folding potential? A new approach to an old problem. J. Mol. Biol. **264**, 1164–1179 (1996)
513. Mitsutake, A., Sugita, Y., Okamoto, Y.: Generalized-ensemble algorithms for molecular simulations of biopolymers. Biopolymers **60**, 96–123 (2001)
514. Mitsutake, A., Sugita, Y., Okamoto, Y.: Replica-exchange multicanonical and multicanonical replica-exchange Monte Carlo simulations of peptides. I. formulation and benchmark test. J. Chem. Phys. **118**, 6664 (2003)
515. Mitsutake, A., Sugita, Y., Okamoto, Y.: Replica-exchange multicanonical and multicanonical replica-exchange Monte Carlo simulations of peptides. II. application to a more complex system. J. Chem. Phys. **118**, 6676 (2003)

516. Miyazawa, S., Jernigan, R.: Estimation of effective interresidue contact energies from protein crystal structures: quasi-chemical approximation. Macromolecules **18**, 534–552 (1985)
517. Miyazawa, S., Jernigan, R.L.: Residue-residue potentials with a favorable contact pair term and an unfavorable high packing density term, for simulation and threading. J. Mol. Biol. **256**, 623–644 (1996)
518. Miyazawa, S., Jernigan, R.L.: How effective for fold recognition is a potential of mean force that includes relative orientations between contacting residues in proteins? J. Chem. Phys. **122**, 024901 (2005)
519. Moont, G., Gabb, H.A., Sternberg, M.J.: Use of pair potentials across protein interfaces in screening predicted docked complexes. Proteins **35**, 364–373 (1999)
520. Moore, P.: Small-angle scattering. Information content and error analysis. J. Appl. Cryst. **13**, 168–175 (1980)
521. Moore, G.L., Maranas, C.D.: Identifying residue-residue clashes in protein hybrids by using a second-order mean-field approach. Proc. Natl. Acad. Sci. U.S.A. **100**, 5091–5096 (2003)
522. Mourik, J.v., Clementi, C., Maritan, A., Seno, F., Banavar, J.R.: Determination of interaction potentials of amino acids from native protein structures: tests on simple lattice models. J. Chem. Phys. **110**, 10123 (1999)
523. Mouritsen, O.G.: Computer Studies of Phase Transitions and Critical Phenomena. Springer, New York (1984)
524. Müller, K., Glatter, O.: Practical aspects to the use of indirect Fourier transformation methods. Makromol. Chem. **183**, 465–479 (1982)
525. Müller, J., Hansen, S., Pürschel, H.: The use of small-angle scattering and the maximum-entropy method for shape-model determination from distance-distribution functions. J. Appl. Cryst. **29**, 547–554 (1996)
526. Müller, J., Schmidt, P., Damaschun, G., Walter, G.: Determination of the largest particle dimension by direct Fourier cosine transformation of experimental small-angle X-ray scattering data. J. Appl. Cryst. **13**, 280–283 (1980)
527. Mullinax, J.W., Noid, W.G.: Extended ensemble approach for deriving transferable coarse-grained potentials. J. Chem. Phys. **131**, 104110 (2009)
528. Mullinax, J.W., Noid, W.G.: Recovering physical potentials from a model protein databank. Proc. Natl. Acad. Sci. U.S.A. **107**, 19867–19872 (2010)
529. Muñoz, V., Eaton, W.A.: A simple model for calculating the kinetics of protein folding from three-dimensional structures. Proc. Natl. Acad. Sci. U.S.A. **96**, 11311–11316 (1999)
530. Murray, I.: Advances in Markov chain Monte Carlo methods. Ph.D. thesis, Gatsby Computational Neuroscience Unit, University College London (2007)
531. Murray, L.J.W., Arendall, W.B., Richardson, D.C., Richardson, J.S.: RNA backbone is rotameric. Proc. Natl. Acad. Sci. U.S.A. **100**, 13904–13909 (2003)
532. Murshudov, G.N., Vagin, A.A., Dodson, E.J.: Refinement of macromolecular structures by the maximum-likelihood method. Acta Crystallogr. D **53**, 240–55 (1997)
533. Myers, J.K., Pace, C.N.: Hydrogen bonding stabilizes globular proteins. Biophys. J. **71**, 2033–2039 (1996)
534. Nadler, W., Hansmann, U.: Generalized ensemble and tempering simulations: a unified view. Phys. Rev. E **75**, 026109 (2007)
535. Neal, R.: Probabilistic inference using Markov chain Monte Carlo methods. Technical report, Citeseer (1993)
536. Neal, R.M.: Bayesian Learning for Neural Networks. Lecture Notes in Statistics, 1st edn. Springer, New York (1996)
537. Neal, R., Hinton, G.: A view of the EM algorithm that justifies incremental, sparse, and other variants. In: Jordan, M.I. (ed.) Learning in Graphical Models, pp. 355–368. MIT press, Cambridge (1998)
538. Neapolitan, R.: Learning Bayesian Networks. Prentice Hall, Harlow (2003)
539. Neidigh, J.W., Fesinmeyer, R.M., Andersen, N.H.: Designing a 20-residue protein. Nat. Struct. Biol. **9**, 425–430 (2002)

References 365

540. Nelder, J.A., Mead, R.: A simplex method for function minimization. Comput. J. **7**, 308–313 (1965)
541. Ngan, S.C., Hung, L.H., Liu, T., Samudrala, R.: The metaphysics research lab, Center for the study of language and information, Stanford university, Stanford, CA Scoring functions for de novo protein structure prediction revisited. Methods Mol. Biol. **413**, 243–281 (2008)
542. Nicastro, G., Menon, R.P., Masino, L., Knowles, P.P., McDonald, N.Q., Pastore, A.: The solution structure of the Josephin domain of ataxin-3: structural determinants for molecular recognition. Proc. Natl. Acad. Sci. U.S.A. **102**, 10493–10498 (2005)
543. Nicastro, G., Habeck, M., Masino, L., Svergun, D.I., Pastore, A.: Structure validation of the Josephin domain of ataxin-3: conclusive evidence for an open conformation. J. Biomol. NMR **36**, 267–277 (2006)
544. Nielsen, S.: The stochastic EM algorithm: estimation and asymptotic results. Bernoulli **6**, 457–489 (2000)
545. Nightingale, M., Blöte, H.: Monte Carlo calculation of free energy, critical point, and surface critical behavior of three-dimensional Heisenberg ferromagnets. Phys. Rev. Lett. **60**, 1562–1565 (1988)
546. Nilges, M., Clore, G.M., Gronenborn, A.M.: Determination of three-dimensional structures of proteins from interproton distance data by dynamical simulated annealing from a random array of atoms: avoiding problems associated with folding. FEBS Lett. **239**, 129–136 (1988)
547. Nilges, M., Bernard, A., Bardiaux, B., Malliavin, T., Habeck, M., Rieping, W.: Accurate NMR structures through minimization of an extended hybrid energy. Structure **16**, 1305–1312 (2008)
548. Novotny, M.: A new approach to an old algorithm for the simulation of Ising-like systems. Comput. Phys. **9**, 46–52 (1995)
549. Nyirongo, V.: Statistical approaches to protein matching in bioinformatics. Ph.D thesis, Leeds University, Department of Statistics (2006)
550. Nymeyer, H., García, A., Onuchic, J.: Folding funnels and frustration in off-lattice minimalist protein landscapes. Proc. Natl. Acad. Sci. U.S.A. **95**, 5921 (1998)
551. Oliveira, P.: Broad histogram: tests for a simple and efficient microcanonical simulator. Brazil. J. Phys. **30**, 766–771 (2000)
552. Olsson, S., Boomsma, W., Frellsen, J., Bottaro, S., Harder, T., Ferkinghoff-Borg, J., Hamelryck, T.: Generative probabilistic models extend the scope of inferential structure determination. J. Magn. Reson. **213**, 182–186 (2011)
553. Orland, H.: Monte Carlo Approach to Biopolymers and Protein Folding, p. 90. World Scientific, Singapore (1998)
554. Ornstein, L., Zernike, F.: Integral equation in liquid state theory. In: Proceedings of the Section of Sciences Koninklijke Nederlandse Akademie Van Wetenschappen, vol. 17, pp. 793–806. North-Holland, Amsterdam (1914)
555. Ortiz, A.R., Kolinski, A., Skolnick, J.: Nativelike topology assembly of small proteins using predicted restraints in Monte Carlo folding simulations. Proc. Natl. Acad. Sci. U.S.A. **95**, 1020–1025 (1998)
556. Pan, S.J., Yang, Q.: A survey on transfer learning. IEEE Trans. Knowl. Data. Eng. **22**, 1345–1359 (2010)
557. Pande, V., Grosberg, A., Tanaka, T.: Heteropolymer freezing and design: towards physical models of protein folding. Rev. Mod. Phys. **72**, 259 (2000)
558. Pappu, R.V., Srinivasan, R., Rose, G.D.: The Flory isolated-pair hypothesis is not valid for polypeptide chains: implications for protein folding. Proc. Natl. Acad. Sci. U.S.A. **97**, 12565–12570 (2000)
559. Park, B.H., Levitt, M.: The complexity and accuracy of discrete state models of protein structure. J. Mol. Biol. **249**, 493–507 (1995)
560. Park, S., Kono, H., Wang, W., Boder, E.T., Saven, J.G.: Progress in the development and application of computational methods for probabilistic protein design. Comput. Chem. Eng. **29**, 407–421 (2005)

561. Parsons, D., Williams, D.: Globule transitions of a single homopolymer: a Wang-Landau Monte Carlo study. Phys. Rev. E **74**, 041804 (2006)
562. Pártay, L., Bartók, A., Csányi, G.: Efficient sampling of atomic configurational spaces. J. Phys. Chem. B **114**, 10502–10512 (2010)
563. Paschek, D., Nymeyer, H., García, A.E.: Replica exchange simulation of reversible folding/unfolding of the Trp-cage miniprotein in explicit solvent: on the structure and possible role of internal water. J. Struct. Biol. **157**, 524–533 (2007)
564. Paschek, D., Hempel, S., García, A.E.: Computing the stability diagram of the Trp-cage miniprotein. Proc. Natl. Acad. Sci. U.S.A. **105**, 17754–17759 (2008)
565. Pauling, L., Corey, R.B.: The pleated sheet, a new layer configuration of polypeptide chains. Proc. Natl. Acad. Sci. U.S.A. **37**, 251–256 (1951)
566. Pauling, L., Corey, R.B., Branson, H.R.: The structure of proteins; two hydrogen-bonded helical configurations of the polypeptide chain. Proc. Natl. Acad. Sci. U.S.A. **37**, 205–211 (1951)
567. Pawitan, Y.: In All Likelihood: Statistical Modeling and Inference Using Likelihood. Oxford Science Publications. Clarendon Press, Oxford (2001)
568. Pearl, J.: Probabilistic Reasoning in Intelligent Systems: Networks of Plausible Inference. Morgan Kaufman Publishers, San Mateo (1988)
569. Pearl, J.: Causality: Models, Reasoning and Inference. Cambridge University Press, Cambridge (2000)
570. Pedersen, J., Posselt, D., Mortensen, K.: Analytical treatment of the resolution function for small-angle scattering. J. Appl. Cryst. **23**, 321–333 (1990)
571. Peng, J., Hazan, T., McAllester, D., Urtasun, R.: Convex max-product algorithms for continuous MRFs with applications to protein folding. In: International Conference of Machine Learning (ICML). Bellevue, Washington (2011)
572. Percus, J., Yevick, G.: Analysis of classical statistical mechanics by means of collective coordinates. Phys. Rev. **110**, 1–13 (1958)
573. Peskun, P.: Optimum Monte-Carlo sampling using Markov chains. Biometrika **60**, 607–612 (1973)
574. Peterson, R.W., Dutton, P.L., Wand, A.J.: Improved side-chain prediction accuracy using an ab initio potential energy function and a very large rotamer library. Protein Sci. **13**, 735–751 (2004)
575. Pettersen, E.F., Goddard, T.D., Huang, C.C., Couch, G.S., Greenblatt, D.M., Meng, E.C., Ferrin, T.E.: UCSF Chimera – A visualization system for exploratory research and analysis. J. Comput. Chem. **25**, 1605–1612 (2004)
576. Piana, S., Laio, A.: A bias-exchange approach to protein folding. J. Phys. Chem. B **111**, 4553–4559 (2007)
577. Pierce, N.A., Winfree, E.: Protein design is NP-hard. Protein Eng. **15**, 779–782 (2002)
578. Pierce, N.A., Spriet, J.A., Desmet, J., Mayo, S.L.: Conformational splitting: a more powerful criterion for dead-end elimination. J. Comp. Chem. **21**, 999–1009 (2000)
579. Pietrucci, F., Marinelli, F., Carloni, P., Laio, A.: Substrate binding mechanism of HIV-1 protease from explicit-solvent atomistic simulations. J. Am. Chem. Soc. **131**, 11811–11818 (2009)
580. Pitera, J.W., Swope, W.: Understanding folding and design: replica-exchange simulations of Trp-cage miniproteins. Proc. Natl. Acad. Sci. U.S.A. **100**, 7587–7592 (2003)
581. Plaxco, K., Simons, K., Baker, D.: Contact order, transition state placement and the refolding rates of single domain proteins1. J. Mol. Biol. **277**, 985–994 (1998)
582. Podtelezhnikov, A.A., Wild, D.L.: Comment on "Efficient Monte Carlo trial moves for polypeptide simulations" [J. Chem. Phys. **123**, 174905 (2005)]. J. Chem. Phys. **129**, 027103 (2008)
583. Podtelezhnikov, A.A., Wild, D.L.: CRANKITE: a fast polypeptide backbone conformation sampler. Source Code Biol. Med. **3**, 12 (2008)
584. Podtelezhnikov, A.A., Wild, D.L.: Reconstruction and stability of secondary structure elements in the context of protein structure prediction. Biophys. J. **96**, 4399–4408 (2009)

References 367

585. Podtelezhnikov, A.A., Wild, D.L.: Exhaustive Metropolis Monte Carlo sampling and analysis of polyalanine conformations adopted under the influence of hydrogen bonds. Proteins **61**, 94–104 (2005)
586. Podtelezhnikov, A.A., Ghahramani, Z., Wild, D.L.: Learning about protein hydrogen bonding by minimizing contrastive divergence. Proteins **66**, 588–599 (2007)
587. Pohl, F.M.: Empirical protein energy maps. Nat. New Biol. **234**, 277–279 (1971)
588. Ponder, J.W., Richards, F.M.: Tertiary templates for proteins. use of packing criteria in the enumeration of allowed sequences for different structural classes. J. Mol. Biol. **193**, 775–791 (1987)
589. Poole, A.M., Ranganathan, R.: Knowledge-based potentials in protein design. Curr. Opin. Struct. Biol. **16**, 508–513 (2006)
590. Poulain, P., Calvo, F., Antoine, R., Broyer, M., Dugourd, P.: Performances of Wang-Landau algorithms for continuous systems. Phys. Rev. E **73**, 056704 (2006)
591. Pritchard, J.K., Seielstad, M.T., Perez-Lezaun, A., Feldman, M.W.: Population growth of human Y chromosomes: a study of Y chromosome microsatellites. Mol. Biol. Evol. **16**, 1791–1798 (1999)
592. Privalov, P.L.: Stability of proteins: small globular proteins. Adv. Protein Chem. **33**, 167–241 (1979)
593. Privalov, P.L., Potekhin, S.A.: Scanning microcalorimetry in studying temperature-induced changes in proteins. Methods Enzymol. **131**, 4–51 (1986)
594. Rabiner, L.R.: A tutorial on hidden Markov models and selected applications in speech recognition. Proc. IEEE **77**, 257–286 (1989)
595. Ramachandran, G.N., Ramakrishnan, C., Sasisekharan, V.: Stereochemistry of polypeptide chain configurations. J. Mol. Biol. **7**, 95–99 (1963)
596. Rampf, F., Paul, W., Binder, K.: On the first-order collapse transition of a three-dimensional, flexible homopolymer chain model. Europhys. Lett. **70**, 628 (2005)
597. Rashin, A.A., Iofin, M., Honig, B.: Internal cavities and buried waters in globular proteins. Biochemistry **25**, 3619–3625 (1986)
598. Rasmussen, C., Ghahramani, Z.: Bayesian Monte Carlo. NIPS **15**, 505–512 (2003)
599. Rathore, N., Knotts, T. IV, de Pablo, J.: Density of states simulations of proteins. J. Chem. Phys. **118**, 4285 (2003)
600. Reith, D., Pütz, M., Müller-Plathe, F.: Deriving effective mesoscale potentials from atomistic simulations. J. Comput. Chem. **24**, 1624–1636 (2003)
601. Rieping, W., Habeck, M., Nilges, M.: Inferential structure determination. Science **309**, 303–306 (2005)
602. Rieping, W., Habeck, M., Nilges, M.: Modeling errors in NOE data with a lognormal distribution improves the quality of NMR structures. J. Am. Chem. Soc. **27**, 16026–16027 (2005)
603. Rieping, W., Nilges, M., Habeck, M.: ISD: a software package for Bayesian NMR structure calculation. Bioinformatics **24**, 1104–1105 (2008)
604. Ritchie, D.: Recent progress and future directions in protein-protein docking. Curr. Protein Pept. Sci. **9**, 1–15 (2008)
605. Rivest, L.: A distribution for dependent unit vectors. Comm. Stat. Theor. Meth. **17**, 461–483 (1988)
606. Robert, C.: The Bayesian Choice: From Decision-Theoretic Foundations to Computational Implementation. Springer, New York (2007)
607. Rodriguez, A., Schmidler, S.: Bayesian protein structure alignment. Submitted (2010)
608. Rohl, C.A., Strauss, C.E., Misura, K.M., Baker, D.: Protein structure prediction using Rosetta. Methods Enzymol. **383**, 66–93 (2004)
609. Rojnuckarin, A., Subramaniam, S.: Knowledge-based interaction potentials for proteins. Proteins **36**, 54–67 (1999)
610. Rooman, M., Rodriguez, J., Wodak, S.: Automatic definition of recurrent local structure motifs in proteins. J. Mol. Biol. **213**, 327–236 (1990)

611. Rosen, J.B., Phillips, A.T., Oh, S.Y., Dill, K.A.: A method for parameter optimization in computational biology. Biophys. J. **79**, 2818–2824 (2000)
612. Rosenberg, M., Goldblum, A.: Computational protein design: a novel path to future protein drugs. Curr. Pharm. Des. **12**, 3973–3997 (2006)
613. Rosenblatt, M.: Remarks on a multivariate transformation. Ann. Math. Stat. **23**, 470–472 (1952)
614. Roth, S., Black, M.J.: On the spatial statistics of optical flow. Int. J. Comput. Vis. **74**, 33–50 (2007)
615. Rothlisberger, D., Khersonsky, O., Wollacott, A.M., Jiang, L., DeChancie, J., Betker, J., Gallaher, J.L., Althoff, E.A., Zanghellini, A., Dym, O., Albeck, S., Houk, K.N., Tawfik, D.S., Baker, D.: Kemp elimination catalysts by computational enzyme design. Nature **453**, 190–195 (2008)
616. Ruffieux, Y., Green, P.: Alignment of multiple configurations using hierarchical models. J. Comp. Graph. Stat. **18**, 756–773 (2009)
617. Russ, W.P., Ranganathan, R.: Knowledge-based potential functions in protein design. Curr. Opin. Struct. Biol. **12**, 447–452 (2002)
618. Rykunov, D., Fiser, A.: Effects of amino acid composition, finite size of proteins, and sparse statistics on distance-dependent statistical pair potentials. Proteins **67**, 559–568 (2007)
619. Sali, A., Blundell, T.L.: Comparative protein modelling by satisfaction of spatial restraints. J. Mol. Biol. **234**, 779–815 (1993)
620. Sali, A., Overington, J.P., Johnson, M.S., Blundell, T.L.: From comparisons of protein sequences and structures to protein modelling and design. Trends Biochem. Sci. **15**, 235–240 (1990)
621. Samudrala, R., Moult, J.: An all-atom distance-dependent conditional probability discriminatory function for protein structure prediction. J. Mol. Biol. **275**, 895–916 (1998)
622. Santos, E. Jr.: On the generation of alternative explanations with implications for belief revision. In: In Uncertainty in Artificial Intelligence, UAI-91, pp. 339–347. Morgan Kaufman Publishers, San Mateo (1991)
623. Saraf, M.C., Moore, G.L., Goodey, N.M., Cao, V.Y., Benkovic, S.J., Maranas, C.D.: IPRO: An iterative computational protein library redesign and optimization procedure. Biophys. J. **90**, 4167–4180 (2006)
624. Saunders, C.T., Baker, D.: Recapitulation of protein family divergence using flexible backbone protein design. J. Mol. Biol. **346**, 631–644 (2005)
625. Saupe, A., Englert, G.: High-resolution nuclear magnetic resonance spectra of orientated molecules. Phys. Rev. Lett. **11**, 462–464 (1963)
626. Savage, H.J., Elliott, C.J., Freeman, C.M., Finney, J.M.: Lost hydrogen bonds and buried surface area: rationalising stability in globular proteins. J. Chem. Soc. Faraday Trans. **89**, 2609–2617 (1993)
627. Sayle, R.A., Milner-White, E.J.: RASMOL: biomolecular graphics for all. Trends Biochem. Sci. **20**, 374 (1995)
628. Schäfer, J., Strimmer, K.: A shrinkage approach to large-scale covariance matrix estimation and implications for functional genomics. Stat. App. Gen. Mol. Biol. **4**, Art. 32 (2005)
629. Scheek, R.M., van Gunsteren, W.F., Kaptein, R.: Molecular dynamics simulations techniques for determination of molecular structures from nuclear magnetic resonance data. Methods Enzymol. **177**, 204–218 (1989)
630. Schellman, J.A.: The stability of hydrogen-bonded peptide structures in aqueous solution. C. R. Trav. Lab. Carlsberg **29**, 230–259 (1955)
631. Scheraga, H.A., Khalili, M., Liwo, A.: Protein-folding dynamics: overview of molecular simulation techniques. Annu. Rev. Phys. Chem. **58**, 57–83 (2007)
632. Schmidler, S.: Fast Bayesian shape matching using geometric algorithms. In: Bernardo, J., Bayarri, M., Berger, J., Dawid, A., Heckerman, D., Smith, A., West, M. (eds.) Bayesian Statistics 8, pp. 471–490. Oxford University Press, Oxford (2007)
633. Schmidler, S.C., Liu, J.S., Brutlag, D.L.: Bayesian segmentation of protein secondary structure. J. Comput. Biol. **7**, 233–248 (2000)

# References

634. Schmidt, K.E., Ceperley, D.: The Monte Method in Condensed Matter Physics. Topics in Applied Physics, vol. 71, p. 205. Springer, Berlin (1995)
635. Schmitt, S., Kuhn, D., Klebe, G.: A new method to detect related function among proteins independent of sequence and fold homology. J. Mol. Biol. **323**, 387–406 (2002)
636. Schonemann, P.: A generalized solution of the orthogonal Procrustes problem. Psychometrika **31**, 1–10 (1966)
637. Schutz, C.N., Warshel, A.: What are the dielectric "constants" of proteins and how to validate electrostatic models? Proteins **44**, 400–417 (2001)
638. Schwarz, G.: Estimating the dimension of a model. Ann. Stat. **6**, 461–464 (1978)
639. Sciretti, D., Bruscolini, P., Pelizzola, A., Pretti, M., Jaramillo, A.: Computational protein design with side-chain conformational entropy. Proteins **74**, 176–191 (2009)
640. Seaton, D., Mitchell, S., Landau, D.: Developments in Wang-Landau simulations of a simple continuous homopolymer. Brazil. J. Phys. **38**, 48–53 (2008)
641. Seber, G.A.F., Wild, C.J.: Nonlinear Regression. Wiley Series in Probability and Mathematical Statistics, Probability and Mathematical Statistics. Wiley, New York (1989)
642. Selenko, P., Sprangers, R., Stier, G., Buehler, D., Fischer, U., Sattler, M.: SMN Tudor domain structure and its interaction with the Sm proteins. Nat. Struct. Biol. **8**, 27–31 (2001)
643. Shah, P.S., Hom, G.K., Ross, S.A., Lassila, J.K., Crowhurst, K.A., Mayo, S.L.: Full-sequence computational design and solution structure of a thermostable protein variant. J. Mol. Biol. **372**, 1–6 (2007)
644. Shakhnovich, E., Gutin, A.: Implications of thermodynamics of protein folding for evolution of primary sequences. Nature **346**, 773–775 (1990)
645. Shannon, C.: A mathematical theory of communication. Bell Syst. Tech. J. **27**, 379–423, 623–656 (1948)
646. Shannon, C., Weaver, W.: The Mathematical Theory of Communication. University Illinois Press, Urbana (1949)
647. Shapiro, A., Botha, J.D., Pastore, A., Lesk, A.M.: A method for multiple superposition of structures. Acta Crystallogr. A **48**, 11–14 (1992)
648. Shaw, L., Henley, C.: A transfer-matrix Monte Carlo study of random Penrose tilings. J. Phys. A **24**, 4129 (1991)
649. Shaw, D., Maragakis, P., Lindorff-Larsen, K., Piana, S., Dror, R., Eastwood, M., Bank, J., Jumper, J., Salmon, J., Shan, Y., et al.: Atomic-level characterization of the structural dynamics of proteins. Science **330**, 341 (2010)
650. Shea, J.E., Brooks, C.L.: From folding theories to folding proteins: a review and assessment of simulation studies of protein folding and unfolding. Annu. Rev. Phys. Chem. **52**, 499–535 (2001)
651. Shen, M.Y., Sali, A.: Statistical potential for assessment and prediction of protein structures. Protein Sci. **15**, 2507–2524 (2006)
652. Shepherd, S.J., Beggs, C.B., Jones, S.: Amino acid partitioning using a Fiedler vector model. Eur. Biophys. J. **37**, 105–109 (2007)
653. Shifman, J.M., Fromer, M.: Search algorithms. In: Park, S.J., Cochran, J.R. (eds.) Protein Engineering and Design. CRC Press, Boca Raton (2009)
654. Shifman, J.M., Mayo, S.L.: Exploring the origins of binding specificity through the computational redesign of calmodulin. Proc. Natl. Acad. Sci. U.S.A. **100**, 13274–13279 (2003)
655. Shindyalov, I., Bourne, P.: Protein structure alignment by incremental combinatorial extension (CE) of the optimal path. Protein Eng. **11**, 739–747 (1998)
656. Shortle, D.: Propensities, probabilities, and the Boltzmann hypothesis. Protein Sci. **12**, 1298–1302 (2003)
657. Simon, I., Glasser, L., Scheraga, H.: Calculation of protein conformation as an assembly of stable overlapping segments: application to bovine pancreatic trypsin inhibitor. Proc. Natl. Acad. Sci. U S A **88**, 3661–3665 (1991)
658. Simons, K.T., Kooperberg, C., Huang, E., Baker, D.: Assembly of protein tertiary structures from fragments with similar local sequences using simulated annealing and Bayesian scoring functions. J. Mol. Biol. **268**, 209–225 (1997)

659. Simons, K.T., Ruczinski, I., Kooperberg, C., Fox, B.A., Bystroff, C., Baker, D.: Improved recognition of native-like protein structures using a combination of sequence-dependent and sequence-independent features of proteins. Proteins **34**, 82–95 (1999)
660. Singh, H., Hnizdo, V., Demchuk, E.: Probabilistic model for two dependent circular variables. Biometrika **89**, 719–723 (2002)
661. Sippl, M.J.: Calculation of conformational ensembles from potentials of mean force: an approach to the knowledge-based prediction of local structures in globular proteins. J. Mol. Biol. **213**, 859–883 (1990)
662. Sippl, M.J.: Boltzmann's principle, knowledge-based mean fields and protein folding. An approach to the computational determination of protein structures. J. Comput. Aided Mol. Des. **7**, 473–501 (1993)
663. Sippl, M.J.: Recognition of errors in three-dimensional structures of proteins. Proteins **17**, 355–362 (1993)
664. Sippl, M., Hendlich, M., Lackner, P.: Assembly of polypeptide and protein backbone conformations from low energy ensembles of short fragments: development of strategies and construction of models for myoglobin, lysozyme, and thymosin $\beta_4$. Protein Sci. **1**, 625–640 (1992)
665. Sippl, M.J., Ortner, M., Jaritz, M., Lackner, P., Flockner, H.: Helmholtz free energies of atom pair interactions in proteins. Fold. Des. **1**, 289–298 (1996)
666. Sivia, D., Skilling, J.: Data Analysis: A Bayesian Tutorial. Oxford University Press, New York (2006)
667. Sivia, D., Webster, J.: The Bayesian approach to reflectivity data. Physica B **248**, 327–337 (1998)
668. Skilling, J.: Maximum Entropy and Bayesian Methods in Science and Engineering, pp. 173–187. Kluwer, Dordrecht (1988)
669. Skilling, J.: Nested sampling. In: Bayesian Inference and Maximum Entropy Methods in Science and Engineering, vol. 735, pp. 395–405. American Institute of Physics, Melville (2004)
670. Skolnick, J.: In quest of an empirical potential for protein structure prediction. Curr. Opin. Struct. Biol. **16**, 166–171 (2006)
671. Skolnick, J., Jaroszewski, L., Kolinski, A., Godzik, A.: Derivation and testing of pair potentials for protein folding. When is the quasichemical approximation correct? Protein Sci. **6**, 676–688 (1997)
672. Smith, G., Bruce, A.: A study of the multi-canonical Monte Carlo method. J. Phys. A **28**, 6623 (1995)
673. Smith, G., Bruce, A.: Multicanonical Monte Carlo study of a structural phase transition. Europhys. Lett. **34**, 91 (1996)
674. Smith, C.A., Kortemme, T.: Backrub-like backbone simulation recapitulates natural protein conformational variability and improves mutant side-chain prediction. J. Mol. Biol. **380**, 742–756 (2008)
675. Smith, G.R., Sternberg, M.J.E.: Prediction of protein-protein interactions by docking methods. Curr. Opin. Struct. Biol. **12**, 28–35 (2002)
676. Solis, A.D., Rackovsky, S.R.: Information-theoretic analysis of the reference state in contact potentials used for protein structure prediction. Proteins **78**, 1382–1397 (2009)
677. Solomon, I.: Relaxation processes in a system of two spins. Phys. Rev. **99**, 559–565 (1955)
678. Soman, K.V., Braun, W.: Determining the three-dimensional fold of a protein from approximate constraints: a simulation study. Cell. Biochem. Biophys. **34**, 283–304 (2001)
679. Son, W., Jang, S., Shin, S.: A simple method of estimating sampling consistency based on free energy map distance. J. Mol. Graph. Model. **27**, 321–325 (2008)
680. Sontag, D., Jaakkola, T.: Tree block coordinate descent for MAP in graphical models. In: Proceedings of the 12th International Conference on Artificial Intelligence and Statistics (AISTATS), vol. 5, pp. 544–551. ClearWater Beach, Florida (2009)
681. Sontag, D., Meltzer, T., Globerson, A., Jaakkola, T., Weiss, Y.: Tightening LP relaxations for MAP using message passing. In: McAllester, D.A., Myllymäki, P. (eds.) Proceedings of the

# References

371

24rd Conference on Uncertainty in Artificial Intelligence, pp. 503–510. AUAI Press, Helsinki (2008)

682. Speed, T.P., Kiiveri, H.T.: Gaussian Markov distributions over finite graphs. Ann. Stat. **14**, 138–150 (1986)

683. Spiegelhalter, D., Best, N., Carlin, B., Van Der Linde, A.: Bayesian measures of model complexity and fit. J. R. Statist. Soc.B **64**(4), 583–639 (2002)

684. Spronk, A.E.M., Nabuurs, S.B., Bovin, M.J.J., Krieger, E., Vuister, W., Vriend, G.: The precision of NMR structure ensembles revisited. J. Biomol. NMR **25**, 225–234 (2003)

685. Staelens, S., Desmet, J., Ngo, T.H., Vauterin, S., Pareyn, I., Barbeaux, P., Van Rompaey, I., Stassen, J.M., Deckmyn, H., Vanhoorelbeke, K.: Humanization by variable domain resurfacing and grafting on a human IgG4, using a new approach for determination of non-human like surface accessible framework residues based on homology modelling of variable domains. Mol. Immunol. **43**, 1243–1257 (2006)

686. Stark, A., Sunyaev, S., Russell, R.: A model for statistical significance of local similarities in structure. J. Mol. Biol. **326**, 1307–1316 (2003)

687. Steenstrup, S., Hansen, S.: The maximum-entropy method without the positivity constraint-applications to the determination of the distance-distribution function in small-angle scattering. J. Appl. Cryst. **27**, 574–580 (1994)

688. Stein, C.: Inadmissibility of the usual estimator for the mean of a multivariate normal distribution. In: Proceedings of the Third Berkeley Symposium on Mathematical Statistics and Probability, vol. 1, pp. 197–206. University of California Press, Berkeley (1955)

689. Stein, C., James, W.: Estimation with quadratic loss. In: Proceedings of the Fourth Berkeley Symposium on Mathematical Statistics and Probability, vol. 1, pp. 361–379. University of California Press, Berkeley (1961)

690. Stickle, D.F., Presta, L.G., Dill, K.A., Rose, G.D.: Hydrogen bonding in globular proteins. J. Mol. Biol. **226**, 1143–1159 (1992)

691. Stigler, S.: Who discovered Bayes's theorem? Am. Stat. pp. 290–296 (1983)

692. Stigler, S.: The History of Statistics: The Measurement of Uncertainty Before 1900. Belknap Press, Cambridge (1986)

693. Stovgaard, K., Andreetta, C., Ferkinghoff-Borg, J., Hamelryck, T.: Calculation of accurate small angle X-ray scattering curves from coarse-grained protein models. BMC Bioinformatics **11**, 429 (2010)

694. Sugita, Y., Okamoto, Y.: Replica-exchange molecular dynamics method for protein folding. Chem. Phys. Lett. **314**, 141–151 (1999)

695. Sutcliffe, M.J.: Representing an ensemble of NMR-derived protein structures by a single structure. Protein Sci. **2**, 936–944 (1993)

696. Sutton, C., McCallum, A.: Piecewise pseudolikelihood for efficient training of conditional random fields. In: Proceedings of the 24th International Conference on Machine Learning, pp. 863–870. ACM, New York (2007)

697. Svensen, M., Bishop, C.: Robust Bayesian mixture modelling. Neurocomputing **64**, 235–252 (2005)

698. Svergun, D.: Determination of the regularization parameter in indirect-transform methods using perceptual criteria. J. Appl. Cryst. **25**, 495–503 (1992)

699. Svergun, D.: Restoring low resolution structure of biological macromolecules from solution scattering using simulated annealing. Biophys. J. **76**, 2879–2886 (1999)

700. Svergun, D., Semenyuk, A., Feigin, L.: Small-angle-scattering-data treatment by the regularization method. Acta Crystallogr. A **44**, 244–250 (1988)

701. Svergun, D., Petoukhov, M., Koch, M.: Determination of domain structure of proteins from X-ray solution scattering. Biophys. J. **80**, 2946–2953 (2001)

702. Swendsen, R., Ferrenberg, A.: New Monte Carlo technique for studying phase transitions. Phys. Rev. Lett **61**, 2635 (1988)

703. Swendsen, R., Ferrenberg, A.: Optimized Monte Carlo data analysis. Phys. Rev. Lett **63**, 1195 (1989)

704. Swendsen, R.H., Wang, J.S.: Replica Monte Carlo simulation of spin glasses. Phys. Rev. Lett. **57**, 2607–2609 (1986)
705. Talbott, W.: Bayesian epistemology. In: Zalta, E.N. (ed.) The Stanford Encyclopedia of Philosophy, summer 2011 edn. The metaphysics research lab, Center for the study of language and information, Stanford university, Stanford, CA (2011)
706. Tanaka, S., Scheraga, H.A.: Medium- and long-range interaction parameters between amino acids for predicting three-dimensional structures of proteins. Macromolecules **9**, 945–950 (1976)
707. Ten Berge, J.M.F.: Orthogonal Procrustes rotation for two or more matrices. Psychometrika **42**, 267–276 (1977)
708. Tesi, M., Rensburg, E., Orlandini, E., Whittington, S.: Monte Carlo study of the interacting self-avoiding walk model in three dimensions. J. Stat. Phys. **82**, 155–181 (1996)
709. The UniProt Consortium: The universal protein resource (UniProt). Nucl. Acids Res. **36**, D190–D195 (2008)
710. Theobald, D.L.: A nonisotropic Bayesian approach to superpositioning multiple macromolecules. In: Barber, S., Baxter, P., Gusnanto, A., Mardia, K. (eds.) Statistical Tools for Challenges in Bioinformatics, Proceedings of the 28th Leeds Annual Statistical Research (LASR) Workshop. Department of Statistics, University of Leeds, Leeds LS2 9JT (2009)
711. Theobald, D.L., Mardia, K.V.: Full Bayesian analysis of the generalized non-isotropic Procrustes problem with scaling. In: A. Gusnanto, K.V. Mardia, C.J. Fallaize (eds.) Next Generation Statistics in Biosciences, Proceedings of the 30th Leeds Annual Statistical Research (LASR) Workshop. Department of Statistics, University of Leeds, Leeds (2011)
712. Theobald, D.L., Wuttke, D.S.: Empirical Bayes hierarchical models for regularizing maximum likelihood estimation in the matrix Gaussian Procrustes problem. Proc. Natl. Acad. Sci. U.S.A. **103**, 18521–18527 (2006)
713. Theobald, D.L., Wuttke, D.S.: THESEUS: maximum likelihood superpositioning and analysis of macromolecular structures. Bioinformatics **22**, 2171–2172 (2006)
714. Theobald, D.L., Wuttke, D.S.: Accurate structural correlations from maximum likelihood superpositions. PLoS Comput. Biol. **4**, e43 (2008)
715. Thomas, P., Dill, K.: Statistical potentials extracted from protein structures: how accurate are they? J. Mol. Biol. **257**, 457–469 (1996)
716. Thomas, P.D., Dill, K.A.: An iterative method for extracting energy-like quantities from protein structures. Proc. Natl. Acad. Sci. U.S.A. **93**, 11628–11633 (1996)
717. Tikhonov, A., Arsenin, V., John, F.: Solutions of Ill-Posed Problems. Halsted Press, Winston (1977)
718. Tjandra, N., Bax, A.: Direct measurement of distances and angles in biomolecules by NMR in a dilute liquid crystalline medium. Science **278**, 1111–1114 (1997)
719. Tobi, D., Elber, R.: Distance-dependent, pair potential for protein folding: results from linear optimization. Proteins **41**, 40–46 (2000)
720. Tobi, D., Shafran, G., Linial, N., Elber, R.: On the design and analysis of protein folding potentials. Proteins **40**, 71–85 (2000)
721. Tolman, J.R., Flanagan, J.M., Kennedy, M.A., Prestegard, J.H.: Nuclear magnetic dipole interactions in field-oriented proteins: information for structure determination in solution. Proc. Natl. Acad. Sci. U.S.A. **92**, 9279–9283 (1995)
722. Torda, A.E., Brunne, R.M., Huber, T., Kessler, H., van Gunsteren, W.F.: Structure refinement using time-averaged J coupling constant restraints. J. Biomol. NMR **3**, 55–66 (1993)
723. Torrance, J., Bartlett, G., Porter, C., Thornton, J.: Using a library of structural templates to recognise catalytic sites and explore their evolution in homologous families. J. Mol. Biol. **347**, 565–581 (2005)
724. Trebst, S., Huse, D., Troyer, M.: Optimizing the ensemble for equilibration in broad-histogram Monte Carlo simulations. Phys. Rev. E **70**, 046701 (2004)
725. Trebst, S., Gull, E., Troyer, M.: Optimized ensemble Monte Carlo simulations of dense Lennard-Jones fluids. J. Chem. Phys. **123**, 204501 (2005)

References

726. Trovato, A., Ferkinghoff-Borg, J., Jensen, M.H.: Compact phases of polymers with hydrogen bonding. Phys. Rev. E **67**, 021805 (2003)
727. Troyer, M., Wessel, S., Alet, F.: Flat histogram methods for quantum systems: algorithms to overcome tunneling problems and calculate the free energy. Phys. Rev. Lett. **90**, 120201 (2003)
728. Trzesniak, D., Kunz, A.P., van Gunsteren, W.F.: A comparison of methods to compute the potential of mean force. ChemPhysChem **8**, 162–169 (2007)
729. Tuinstra, R.L., Peterson, F.C., Kutlesa, S., Elgin, E.S., Kron, M.A., Volkman, B.F.: Interconversion between two unrelated protein folds in the lymphotactin native state. Proc. Natl. Acad. Sci. U S A **105**, 5057–5062 (2008)
730. Ulmschneider, J., Jorgensen, W.: Monte Carlo backbone sampling for polypeptides with variable bond angles and dihedral angles using concerted rotations and a Gaussian bias. J. Chem. Phys. **118**, 4261 (2003)
731. Umeyama, S.: Least-squares estimation of transformation parameters between two point patterns. IEEE Trans. Pat. Anal. Mach. Intel. **13**, 376–380 (1991)
732. Unger, R., Harel, D., Wherland, S., Sussman, J.: A 3D building blocks approach to analyzing and predicting structure of proteins. Proteins **5**, 355–373 (1989)
733. Upton, G., Cook, I.: Oxford Dictionary of Statistics, 2nd rev. edn. pp. 246–247. Oxford University Press, Oxford (2008)
734. Vanderzande, C.: Lattice Models of Polymers. Cambridge University Press, Cambridge (1998)
735. Vasquez, M., Scheraga, H.: Calculation of protein conformation by the build-up procedure. Application to bovine pancreatic trypsin inhibitor using limited simulated nuclear magnetic resonance data. J. Biomol. Struct. Dyn. **5**, 705–755 (1988)
736. Vendruscolo, M., Domany, E.: Pairwise contact potentials are unsuitable for protein folding. J. Chem. Phys. **109**, 11101–11108 (1998)
737. Vendruscolo, M., Kussell, E., Domany, E.: Recovery of protein structure from contact maps. Fold. Des. **2**, 295–306 (1997)
738. Vendruscolo, M., Najmanovich, R., Domany, E.: Can a pairwise contact potential stabilize native protein folds against decoys obtained by threading? Proteins **38**, 134–148 (2000)
739. Verdier, P.H., Stockmayer, W.H.: Monte Carlo calculations on the dynamics of polymers in dilute solution. J. Chem. Phys. **36**, 227–235 (1962)
740. Vestergaard, B., Hansen, S.: Application of Bayesian analysis to indirect Fourier transformation in small-angle scattering. J. Appl. Cryst. **39**, 797–804 (2006)
741. Vitalis, A., Pappu, R.: Methods for Monte Carlo simulations of biomacromolecules. Annu. Rep. Comput. Chem. **5**, 49–76 (2009)
742. Voigt, C.A., Gordon, D.B., Mayo, S.L.: Trading accuracy for speed: a quantitative comparison of search algorithms in protein sequence design. J. Mol. Biol. **299**, 789–803 (2000)
743. von Mises, R.: Über de "ganzzahligkeit" der atomgewichte und verwandte fragen. Physikal. Z. **19**, 490–500 (1918)
744. von Neumann, J.: Some matrix-inequalities and metrization of matric-space. Tomsk U. Rev. **1**, 286–300 (1937)
745. Vorontsov-Velyaminov, P., Broukhno, A., Kuznetsova, T., Lyubartsev, A.: Free energy calculations by expanded ensemble method for lattice and continuous polymers. J. Phys. Chem. **100**, 1153–1158 (1996)
746. Vorontsov-Velyaminov, P., Volkov, N., Yurchenko, A.: Entropic sampling of simple polymer models within Wang-Landau algorithm. J. Phys. A **37**, 1573 (2004)
747. Wainwright, M., Jordan, M.: Graphical models, exponential families, and variational inference. Mach. Learn. **1**(1–2), 1–305 (2008)
748. Wainwright, M., Jaakkola, T., Willsky, A.: MAP estimation via agreement on trees: message-passing and linear programming. IEEE Trans. Inf. Theory **51**, 3697–3717 (2005)
749. Wallace, A., Borkakoti, N., Thornton, J.: TESS: a geometric hashing algorithm for deriving 3D coordinate templates for searching structural databases. Protein Sci. **6**, 2308–2323 (1997)

750. Wallner, B., Elofsson, A.: Can correct protein models be identified? Protein Sci. **12**, 1073–86 (2003)
751. Wang, G., Dunbrack, R.L.: PISCES: a protein sequence culling server. Bioinformatics **19**, 1589–1591 (2003)
752. Wang, G., Dunbrack, R.L.: PISCES: recent improvements to a PDB sequence culling server. Nucleic Acids Res. **33**, W94–W98 (2005)
753. Wang, F., Landau, D.: Determining the density of states for classical statistical models: A random walk algorithm to produce a flat histogram. Phys. Rev. E **64**, 056101 (2001)
754. Wang, F., Landau, D.: Efficient, multiple-range random walk algorithm to calculate the density of states. Phys. Rev. Lett. **86**, 2050–2053 (2001)
755. Wang, J., Swendsen, R.: Transition matrix Monte Carlo method. J. Stat. Phys. **106**, 245–285 (2002)
756. Wang, J., Wang, W.: A computational approach to simplifying the protein folding alphabet. Nat. Struct. Biol. **6**, 1033–1038 (1999)
757. Wang, Z., Xu, J.: A conditional random fields method for RNA sequence–structure relationship modeling and conformation sampling. Bioinformatics **27**, i102 (2011)
758. Warme, P.K., Morgan, R.S.: A survey of amino acid side-chain interactions in 21 proteins. J. Mol. Biol. **118**, 289–304 (1978)
759. Warme, P.K., Morgan, R.S.: A survey of atomic interactions in 21 proteins. J. Mol. Biol. **118**, 273–287 (1978)
760. Wedemeyer, W.J., Scheraga, H.A.: Exact analytical loop closure in proteins using polynomial equations. J. Comp. Chem. **20**, 819–844 (1999)
761. Weiss, Y., Freeman, T.: Correctness of belief propagation in Gaussian graphical models of arbitrary topology. Neural Comput. **13**, 2173–2200 (2001)
762. Weiss, Y., Freeman, W.: On the optimality of solutions of the max-product belief-propagation algorithm in arbitrary graphs. IEEE Trans. Inf. Theory **47**, 736–744 (2001)
763. Weiss, Y., Yanover, C., Meltzer, T.: MAP estimation, linear programming and belief propagation with convex free energies. In: The 23rd Conference on Uncertainty in Artificial Intelligence. Vancouver (2007)
764. Wilkinson, D.: Bayesian methods in bioinformatics and computational systems biology. Brief. Bioinformatics **8**, 109–116 (2007)
765. Wilkinson, D.: Discussion to schmidler. In: Bernardo, J., Bayarri, M., Berger, J., Dawid, A., Heckerman, D., Smith A., West, M. (eds.) Bayesian Statistics 8, pp. 471–490. Oxford University Press, Oxford (2007)
766. Willett, P.: Three-Dimensional Chemical Structure Handling. Wiley, New York (1991)
767. Willett, P., Wintermann, V.: Implementation of nearest-neighbor searching in an online chemical structure search system. J. Chem. Inf. Comput. Sci. **26**, 36–41 (1986)
768. Winther, O., Krogh, A.: Teaching computers to fold proteins. Phys. Rev. E **70**, 030903 (2004)
769. Wolff, U.: Collective Monte Carlo updating for spin systems. Phys. Rev. Lett. **62**, 361 (1989)
770. Won, K., Hamelryck, T., Prügel-Bennett, A., Krogh, A.: An evolutionary method for learning HMM structure: prediction of protein secondary structure. BMC bioinformatics **8**, 357 (2007)
771. Word, J.M., Lovell, S.C., LaBean, T.H., Taylor, H.C., Zalis, M.E., Presley, B.K., Richardson, J.S., Richardson, D.C.: Visualizing and quantifying molecular goodness-of-fit: small-probe contact dots with explicit hydrogen atoms. J. Mol. Biol. **285**, 1711–1733 (1999)
772. Wüthrich, K.: NMR studies of structure and function of biological macromolecules (Nobel lecture). Angew. Chem. Int. Ed. Engl. **42**, 3340–3363 (2003)
773. Xiang, Z., Honig, B.: Extending the accuracy limits of prediction for side-chain conformations. J. Mol. Biol. **311**, 421–430 (2001)
774. Yan, Q., Faller, R., De Pablo, J.: Density-of-states Monte Carlo method for simulation of fluids. J. Chem. Phys. **116**, 8745 (2002)
775. Yang, Y., Zhou, Y.: Ab initio folding of terminal segments with secondary structures reveals the fine difference between two closely related all-atom statistical energy functions. Protein Sci. **17**, 1212–1219 (2008)

# References

776. Yanover, C., Weiss, Y.: Approximate inference and protein-folding. In: Becker, S., Thrun, S., Obermayer, K. (eds.) Advances in Neural Information Processing Systems 15, pp. 1457–1464. MIT, Cambridge (2003)
777. Yanover, C., Weiss, Y.: Finding the M most probable configurations using loopy belief propagation. In: Advances in Neural Information Processing Systems 16. MIT, Cambridge (2004)
778. Yanover, C., Weiss, Y.: Approximate inference and side-chain prediction. Technical report, Leibniz Center for Research in Computer Science, The Hebrew University of Jerusalem, Jerusalem (2007)
779. Yanover, C., Meltzer, T., Weiss, Y.: Linear programming relaxations and belief propagation – An empirical study. J. Mach. Learn. Res. **7**, 1887–1907 (2006)
780. Yanover, C., Fromer, M., Shifman, J.M.: Dead-end elimination for multistate protein design. J. Comput. Chem. **28**, 2122–2129 (2007)
781. Yanover, C., Schueler-Furman, O., Weiss, Y.: Minimizing and learning energy functions for side-chain prediction. J. Comp. Biol. **15**, 899–911 (2008)
782. Yedidia, J.: An idiosyncratic journey beyond mean field theory. Technical report, Mitsubishi Electric Research Laboratories, TR-2000-27 (2000)
783. Yedidia, J., Freeman, W., Weiss, Y.: Generalized belief propagation. Technical report, Mitsubishi Electric Research Laboratories, TR-2000-26 (2000)
784. Yedidia, J., Freeman, W., Weiss, Y.: Bethe free energy, Kikuchi approximations and belief propagation algorithms. Technical report, Mitsubishi Electric Research Laboratories, TR-2001-16 (2001)
785. Yedidia, J., Freeman, W., Weiss, Y.: Understanding belief propagation and its generalizations. Technical report, Mitsubishi Electric Research Laboratories, TR-2001-22 (2002)
786. Yedidia, J., Freeman, W., Weiss, Y.: Constructing free-energy approximations and generalized belief propagation algorithms. IEEE Trans. Inf. Theory **51**, 2282–2312 (2005)
787. Yin, S., Ding, F., Dokholyan, N.V.: Modeling backbone flexibility improves protein stability estimation. Structure **15**, 1567–1576 (2007)
788. Yu, H., Rosen, M.K., Shin, T.B., Seidel-Dugan, C., Brugge, J.S., Schreiber, S.L.: Solution structure of the SH3 domain of Src and identification of its ligand-binding site. Science **258**, 1665–1668 (1992)
789. Yuste, S.B., Santos, A.: Radial distribution function for hard spheres. Phys. Rev. A **43**, 5418–5423 (1991)
790. Zernike, F., Prins, J.: X-ray diffraction from liquids. Zeits. f. Physik **41**, 184–194 (1927)
791. Zhang, Y., Voth, G.: Combined metadynamics and umbrella sampling method for the calculation of ion permeation free energy profiles. J. Chem. Theor. Comp. **7**, 2277–2283 (2011)
792. Zhang, Y., Kolinski, A., Skolnick, J.: TOUCHSTONE II: a new approach to ab initio protein structure prediction. Biophys. J. **85**, 1145–1164 (2003)
793. Zhang, C., Liu, S., Zhou, H., Zhou, Y.: An accurate, residue-level, pair potential of mean force for folding and binding based on the distance-scaled, ideal-gas reference state. Protein Sci. **13**, 400–411 (2004)
794. Zhao, F., Li, S., Sterner, B., Xu, J.: Discriminative learning for protein conformation sampling. Proteins **73**, 228–240 (2008)
795. Zhao, F., Peng, J., Debartolo, J., Freed, K., Sosnick, T., Xu, J.: A probabilistic and continuous model of protein conformational space for template-free modeling. J. Comp. Biol. **17**, 783–798 (2010)
796. Zhou, C., Bhatt, R.N.: Understanding and improving the Wang-Landau algorithm. Phys. Rev. E **72**, 025701 (2005)
797. Zhou, H., Zhou, Y.: Distance-scaled, finite ideal-gas reference state improves structure-derived potentials of mean force for structure selection and stability prediction. Protein Sci. **11**, 2714–2726 (2002)
798. Zhou, Y., Zhou, H., Zhang, C., Liu, S.: What is a desirable statistical energy functions for proteins and how can it be obtained? Cell. Biochem. Biophys. **46**, 165–174 (2006)

799. Zoltowski, B.D., Schwerdtfeger, C., Widom, J., Loros, J.J., Bilwes, A.M., Dunlap, J.C., Crane, B.R.: Conformational switching in the fungal light sensor vivid. Science **316**, 1054–1057 (2007)

800. Zou, J., Saven, J.G.: Using self-consistent fields to bias Monte Carlo methods with applications to designing and sampling protein sequences. J. Chem. Phys. **118**, 3843–3854 (2003)

801. Zuckerman, D.: Statistical Physics of Biomolecules: An Introduction. CRC Press, Boca Raton (2010)

802. Zwanzig, R., Ailawadi, N.K.: Statistical error due to finite time averaging in computer experiments. Phys. Rev. **182**, 280–283 (1969)

803. Mardia, K.V., Frellsen, J., Borg, M., Ferkinghoff-Borg, J., Hamelryck, T.: High-throughput Sequencing, Proteins and Statistics. In: Mardia, K.V., Gusnanto, A., Riley, A.D., Voss J. (eds.) A statistical view on the reference ratio method, Leeds University Press, Leeds, UK, pp. 55–61 (2011)

# Index

Acceptance-rejection sampling, 185
Adaptive umbrella sampling, 75
AIC. *See* Akaike information criterion
Akaike, Hirotsugu, 28
Akaike information criterion, 28, 29, 249, 253
ALIBI, 210, 214, 216, 221, 222, 224, 226, 228–230
Alignment, 210
Ancestral sampling. *See* Bayesian network
Anfinsen's hypothesis, 101
$\phi, \psi$ Angles, 159, 176, 238, 239, 241, 251, 252
$\theta, \tau$ Angles, 179, 239, 240
Approximate Bayesian computation (ABC), 131
Artificial intelligence, 33
Atom charge, in structure alignment, 209, 220, 221
Atom type, in structure alignment, 209, 214, 220

BARNACLE, 247, 248
  estimation, 249
  joint probability, 249
  sampling, 249
$\beta$-Barrel, 151
BASILISK, 247, 251, 252
  arginine and lysine examples, 253
  estimation, 253
  evaluation, 253
  leucine example, 252
  as prior, 254
  sampling, 252
Basis functions, 315
Baum-Welch algorithm. *See* Hidden Markov model
Bayes estimator, 13, 16

Bayes factor, 27, 49, 53
Bayesian information criterion, 28, 29, 244
Bayesian model comparison, 52
Bayesian Monte Carlo, 93
Bayesian network, 14, 35
  ancestral sampling, 40, 44
  child node, 35
  conditional independence relations, 36
  D-separation, 36
  dynamic Bayesian network (*see* Dynamic Bayesian network)
  factorizations of a probability distribution, 36, 37
  parent node, 35
Bayesian probability calculus, 5, 7
  binomial example, 9
  incremental application, 8
Bayesian statistics
  academic inertia, xii
  axiomatic basis, 18
  representation theorem, 19
  superposition problem, 191, 207
Bayes' theorem, 4, 7, 19
Bayes, Thomas, 290
  binomial example, 9
  disputed portrait, 4
  history of Bayesian statistics, 4
  principle of indifference, 22
Belief functions, 34
Belief, probability as degree of, 18
Belief propagation, xiv, 43, 44, 268
Bernoulli random variable, 19
Best max-marginal first algorithm, 272
  type-specific, 274
Beta distribution, 9, 15, 26
Beta-binomial distribution, 15, 25
Bethe approximation, 30, 44, 111, 112, 121

BIC. *See* Bayesian information criterion
Binding affinity, in structure alignment, 220
Binning, 63, 81, 83, 84, 88, 90
Binomial distribution, 9, 25
    Bayesian parameter estimation, 9
Bivariate normal distribution, 181, 183, 185
Bivariate von Mises distributions, 162, 241, 243
    Bayesian estimation, 176
    bimodality, 166
    conjugated priors, 176
    cosine models, 164, 241
        conditional distribution, 164
        marginal distribution, 164
        with negative interaction, 165
        with positive interaction, 164
    estimation, 169
    full bivariate model, 162
    Gibbs sampling, 174
    hybrid model, 165
    maximum likelihood estimation, 169
    maximum pseudolikelihood estimation, 172
    moment estimation, 171
    normal approximation, 168
    6-parameter model, 163
    priors (*see* Prior distribution)
    rejection sampling, 174
    simulation, 173
    sine model, 163
        conditional distribution, 164
        marginal distribution, 164
Blanket sampling, 46
Blocked paths, in a graph, 36
BMMF. *See* Best max-marginal first algorithm
BN. *See* Bayesian network
Boas, Franz, 193
Boltzmann distribution, 30, 31, 99, 101, 103, 110, 117, 118, 144, 294, 302
Boltzmann equation, 23
Boltzmann hypothesis, 118, 119, 137
Boltzmann inversion, 103, 107, 116, 118
Boltzmann learning, 116, 142
Boltzmann weights, 51
Bond angle, 239
Bonded atoms, in structure alignment, 210, 222
Bookmaker, 19
Bookstein coordinates, 213
Brandeis dice problem, 22, 31
Broad histogram method, 81
$\beta$-Bulge, 234

Canonical distribution, 51, 69. *See also* Boltzmann distribution
Canonical sampling, 116, 117, 119, 123
CASP, xi, 235
Categorical distribution. *See* Discrete distribution
$C\alpha$ trace, 179, 239
Centering matrix, 197
Child node. *See* Bayesian network
Circle, xiii, 17, 180
Circular data, 159
Circular mean, 160
Clique, in a graph, 41, 43
Cluster-flip algorithm, 69
Cluster variation method. *See* Kikuchi approximation
Coarse-grained variable, 103, 108, 110–112, 114, 117, 121, 125, 126, 132, 134
Colatitude, 180, 183, 217
Collective variable, 54, 67
Combinatorial algorithms, 226
Completed dataset, 47
Concentration parameter, 182
Concerted rotations algorithm, 61, 69
Conditional independence, 267
Conditional probability table, 36, 38
Conditional random field, 254
Conformational sampling, 236
Conformational space, 254
Conjugate prior. *See* Prior distribution
Conjugate variables, 104
Contact map, 108, 112, 136, 139, 140, 145, 146, 150–153
Contact potential, 112, 114
Contrastive divergence, 117, 136
Convergence, 66, 67, 69, 78, 80, 89
Cooperative transition. *See* Phase transition
Correlation, 60, 66, 68, 79
Correlation function, 101, 102, 109
Covariance matrix, 184, 195, 198
Cox axioms, 18, 290
Cox, Richard T., 18, 290
CPT. *See* Conditional probability table
Cross-entropy, 293, 296
Cross-validation, 308
Cycle, in a graph, 35

Database, 109, 110, 114, 115, 117, 118
    BARNACLE, 249
DBN. *See* Dynamic Bayesian network
Dead-end elimination, 264
Decision theory, 6, 13
Decoys, 109, 114

# Index

DEE. *See* Dead-end elimination
de Finetti, Bruno, 5
  Dutch book, 19
  representation theorem, 19
Density of states, 52, 54, 65, 106, 123
Detailed balance, 59, 80, 83–85, 245
Deuterium, 313
Deviance, 29
Deviance information criterion, 29
DIC. *See* Deviance information criterion
Dihedral angle, xiii, 125, 129, 132, 159, 179,
  236, 238, 239, 241, 248
Directional statistics, xiii, 12, 160, 179
  bivariate von Mises distributions (*see*
    Bivariate von Mises distributions)
  circular mean, 160
  Fisher-Bingham distribution (*see*
    Fisher-Bingham distribution)
  Kent distribution (*see* Fisher-Bingham
    distribution)
  Matrix Fisher distribution (*see* Matrix
    Fisher distribution)
  Matrix Student-t distribution (*see* Matrix
    Student-t distribution)
  Matrix von Mises-Fisher distribution
    (*see* Matrix von Mises-Fisher
    distribution)
  mean direction, 161
  mean resultant length, 161
  von Mises Distribution (*see* von Mises
    Distribution)
Dirichlet distribution, 11, 15
Dirichlet process mixture, 254
Discrete distribution, 11, 45
Discrete optimized energy (DOPE),
  114
Dishonest dice, 22
Distance distribution function, 318
Distance geometry, 289
Divergence, 21
D-separation. *See* Bayesian network
Dutch book, 19
Dynamic Bayesian network, xiv, 37, 239, 241,
  248
  emission node, 242
  emission probability, 241
  Gibbs sampling, 244
  graph versus state diagram, 39
  hidden node, 243
  latent variable, 243
  observed node, 243
  slice, 38, 39, 248
  two slice specification, 39
  unrolling of, 38

Eckart conditions, 194
Edge, in a graph, 35
Effective number of free parameters, 29
Eigenvalue, 199, 202
Electron microscopy, 289
EM. *See* Expectation maximization
Empirical Bayes estimation, 15, 195, 199
Energy, 30, 51, 54, 64, 70, 75, 88, 91, 97, 100,
  101, 105, 112, 114, 115, 122
Energy function
  interactions in proteins, 136
  for protein side chain placement and
    design, 259
Entropy, 32, 44, 52, 91, 100, 105, 106
E-step. *See* Expectation maximization
Equal area projection, 180
Equilibration. *See* Relaxation
Equiprobability contour, 12, 181
Ergodicity, 50, 58, 66, 76
Euclidean norm, 225
Euclidean space, 12
Euler angles, 61, 217–219
Event-driven simulation, 68
Evidence, 7, 51, 70, 92
Excess energy, 113
Exchangeability, 19
Expectation maximization, 46, 191, 195, 201,
  202, 204, 244
Expectation value, 49, 54, 62, 91
Expected loss, 13
Expected statistics, 106, 111–113
Expert systems, 33
Extended ensembles, 50, 69, 93
  learning stage, 70, 72, 74, 76, 78, 79, 92
  production stage, 70, 78
Extended likelihood. *See* Hierarchical model

Factor graph, 35, 41
  factor node, 41
  variable node, 41
Factorized approximation, 111, 121
Fast and accurate side-chain topology and
  energy refinement, 263
FASTER. *See* Fast and accurate side-chain
  topology and energy refinement
FB5. *See* Fisher-Bingham distribution
FB5HMM, 188, 239, 241, 244
  estimation, 243
  joint probability, 242
  sampling, 243
Fine-grained variable, 125, 126, 129,
  132, 134

380                                                                                    Index

Fisher-Bingham distribution, xiii, 178, 181,
        182, 240, 241, 243
    Bayesian estimation, 185
    conjugate prior, 185
    estimation, 183
    simulation, 185
Fisher information, 14, 24, 73
    curvature, 14
    Fisher information matrix, 15
Fisher, Ronald, 4
Fixman potential, 105, 106, 115
Flat histogram method. *See* Multicanonical
        ensemble
Folding funnel, 139
Force field, 288, 296
Form, 210. *See also* Size-and-shape
Form factor, 320
Forward algorithm. *See* Hidden Markov model
Forward model, 292
Forward-backtrack algorithm. *See* Hidden
        Markov model
Forward-backward algorithm. *See* Hidden
        Markov model
Fragment assembly, 234, 235
    definition, 234
    shortcomings, 235
Fragment library, 121, 132, 234
Free energy, 30, 32, 44, 52, 83, 91, 97–101,
        103, 105, 111, 117, 118, 122, 123
Frequentist statistics, xii, 4, 8, 14, 34
Fully connected graph, 36
Fuzzy logic, 34

Gambling, 18
Gamma distribution, 17, 216
Gauss-Markov theorem, 193
Gaussian distribution, 214, 218, 223, 226, 227
Generalized belief propagation, 44, 270
Generalized ensembles, 70, 75, 76, 79, 82, 83,
        85, 92
Generalized indirect Fourier transformation,
        317
Generative probabilistic model, 40, 251–253
Genetic algorithm, 262
Gibbs sampling, 44–47, 49, 61, 174, 208, 219,
        291, 300
GIFT. *See* Generalized indirect Fourier
        transformation
Global minimum energy configuration, 261,
        275
GMEC. *See* Global minimum energy
        configuration
Gō potential, 145

Graph, 34, 35, 40
Graph theory, 222
Graphical model, xiv, 5, 13, 14, 17, 30, 33, 34,
        179
Guinier radius, 318

$\beta$-Hairpin, 149, 150, 152, 234
Hamiltonian Monte Carlo, 301
Heat capacity, 15, 73, 91
$\alpha$-Helix, 149–151, 188, 234
Helmholtz free energy, 32, 33
Hemisphere, 180
Heteroscedasticity, 191
Hidden Markov model, 38, 188
    Baum-Welch algorithm, 46
    forward algorithm, 242, 243, 246
    forward-backtrack algorithm, 243–245, 249
    forward-backward algorithm, 43, 46, 243
    graph versus state diagram, 39
    higher order Markov models, 39
    state diagram, 38, 39
    structure learning, 39
    Viterbi algorithm, 43
Hierarchical model, 6, 9, 15, 199, 200
Histogram method, 63–65, 80, 86
H-likelihood. *See* Hierarchical model
HMC. *See* Hamiltonian Monte Carlo
HMM. *See* Hidden Markov model
HMMSTR, 239
Homogeneous spatial Poisson process. *See*
        Poisson process
Homoscedasticity, 194
Hybrid energy, xiv, 288, 298
Hydrogen bond, 108, 117, 133, 135–138,
        144–146, 148–150, 153, 233, 254,
        259
    strength, 148
Hyperparameter, 9, 244
    in SAXS, 316
Hypothesis, 7

Ideal gas state, 106
Ideal solution, 113
Identifiability, 197, 198
IFT. *See* Indirect Fourier transformation
Importance sampling, 45, 55, 56
Importance weights, 56
Improper prior. *See* Prior distribution
Incomplete Gamma function, 203
Indirect Fourier transformation, 315
Inference, 5
Inferential structure determination, 254, 290

# Index

Information, 15, 20, 21
    entropy, 20
    theory, 20, 28, 118
Integration measure, 53, 61, 62, 65
Interaction free state. *See* Reference state
Internal energy, 30, 32
Inverse Boltzmann equation. *See* Boltzmann inversion
Inverse gamma distribution, 200
Inverse probability, 5
Inverse problem, 292
Inverse Wishart distribution, 200, 208
ISD. *See* Inferential structure determination
Ising model, 41, 85
I-sites fragment library, 234, 239
Isolated spin-pair approximation, 304
ISPA. *See* Isolated spin-pair approximation

Jacobian. *See* Integration measure
James-Stein estimator, 16
Jaynes, Edwin T., 5, 290
Jeffreys, Harold, 5
Jeffreys' prior. *See* Prior distribution
Jeffreys' scale, 27
Jensen's inequality, 128
Junction tree inference, 43, 270

Kalman filter, 43
Karplus curve, 295
1/k-ensemble, 50, 76
Kent distribution. *See* Fisher-Bingham distribution
Kernel density methods, 63, 64
Keynes, John Maynard, 5
Kikuchi approximation, 30, 44, 111, 270
Knowledge based potential, 97, 107, 109, 114, 116, 117, 119, 122, 124, 132, 137
    pairwise distances, 132
Kolomogorov axioms, 18
Kullback-Leibler divergence, 20, 25, 28, 33, 122, 127, 128, 142, 293

Lagrange multipliers, 22, 23, 30
Lambert azimuthal projection, 180
Landmarks, in structure alignment, 210, 221
Laplace approximation, 28, 299
Laplace, Pierre-Simon, 290
    binomial example, 9
    history of Bayesian statistics, 4
    principle of indifference, 22
    sunrise problem, 5

Latent variable, 10
Leapfrog integration, 301
Least squares, 191, 193
Lévy distribution, 200
Ligand binding, in structure alignment, 209
Likelihood, 7, 9, 53, 86, 93
Likelihood ratio, 27
Linear programming
    for protein side chain placement and design, 264, 271
Logical deduction, 34
Logical induction, 34
Lognormal distribution, 305
Longitude, 180, 183, 217
Loss function, 13
    quadratic, 16
    zero-one, 13

Machine learning, xi, 30
Manifold, 12, 104, 105, 116
MAP. *See* Maximum a posteriori estimation
Marginal distribution, 49, 54, 55, 66, 74–77, 89, 111, 121–123
Marginalization, 229
Marginal likelihood, 201
Markov blanket, 45
Markov blanket sampling, 44
Markov chain, 301
Markov chain Monte Carlo, 5, 6, 47, 49, 57, 58, 143, 208, 291
Markov process, 187
Markov random field, 35, 40
    for protein side chain placement and design, 266
Mass metric tensor, 104
Matrix Fisher distribution, 216, 218
Matrix Student-t distribution, 201
Matrix von Mises-Fisher distribution, 208
Max-sum algorithm. *See* Message-passing algorithms
MaxEnt. *See* Maximum entropy
Maximization methods, 229
Maximum a posteriori estimation, 13, 47, 191, 208, 299
Maximum entropy, 22, 30, 31, 138, 296, 323
Maximum entropy prior. *See* Prior distribution
Maximum likelihood estimation, 14, 46, 47, 85, 86, 93, 116, 142, 191
Maximum pseudolikelihood estimation, 16, 172
MCMC. *See* Markov chain Monte Carlo
MCSA. *See* Monte Carlo simulated annealing
MD. *See* Molecular dynamics

MDSA. *See* Molecular dynamics-based
    simulated annealing
Mean direction, 161
Mean-field approximation, 111
Mean resultant length, 161
Message-passing algorithms, 43, 268
    max-sum algorithm, 43
    sum-product algorithm, 43
Metadynamics, 50, 75, 82, 83, 90, 92
Metric. *See* Integration measure
Metropolis-Hastings algorithm, 49, 59, 60, 68,
    73, 75, 82, 84, 131, 147, 217, 219,
    220, 246
Metropolis Monte Carlo, 143
Microcanonical ensemble, 51, 91
Microscopic reversibility, 60
Microstate, 32
Mixture component, 27
Mixture model, 10, 36, 226
Miyasawa-Jernigan potential, 112
ML. *See* Maximum likelihood estimation
Model selection, 26, 330
Molecular dynamics, 290, 301
Molecular dynamics-based simulated
    annealing, 289
Moment estimation, 16, 171, 183
Monte Carlo EM, 46, 244
Monte Carlo sampling, 97, 101, 103, 114, 116,
    123, 124, 185
Monte Carlo simulated annealing, 262
MRF. *See* Markov random field
M-step. *See* Expectation maximization
Multicanonical ensemble, 50, 75, 77, 81–83
Multihistogram equations, 85, 87, 90
Multimodal distribution, 10, 67, 78
Multinomial distribution, 11
Multiple alignment, 210, 222
Multitask learning, 251
Multivariate von Mises distributions, 175, 247
Muninn, 80, 85, 92
Mutual information, 21

Native state, 97, 98, 101, 109, 112, 114,
    116–119, 121
Native structure discriminants, 139
Nested sampling, 57, 76, 88, 92
Newton-Raphson method, 17
Neyman, Jerzy, 4
N-fold method. *See* Event-driven simulation
NMR. *See* Nuclear magnetic resonance
NOE. *See* Nuclear Overhauser effect
Noisy communication channel, 20
Non-identifiability, 197

Non-informative prior. *See* Prior distribution
Non-Markovian methods, 80, 82, 84, 87, 92
N-particle distribution function, 101, 103
NP-hard, 43
Nuclear magnetic resonance, 192, 254, 288
    ensemble, 289
Nuclear Overhauser effect, 304
Nuisance parameter, 6, 10, 45, 201
Number of good parameters, 339
Number of states, 51
Numerical optimization, 169

Observed statistics, 106, 111, 117, 122
Order parameter, 67
Orientation, 222
Orthogonal matrix, 183
Orthogonal projection, 180
Orthonormal vectors, 182

Pairwise decomposability, 98, 111, 116, 118,
    119, 121
Pairwise potential, 98, 108, 111, 113
Parallel tempering, 50, 71, 81, 92
Parent node. *See* Bayesian network
Partition function, 32, 41, 49, 50, 52, 55, 64,
    65, 70, 90, 92, 110
Pearson, Egon, 4
Penalized likelihood. *See* Hierarchical model
PERM algorithm, 57
Permutation, 19
Perron-Frobenius theorem, 58
Phase transition, 67–69, 75, 78, 82
Pivot-algorithm, 69
Point estimate, 6, 7, 13
Point process, 214
Poisson distribution, 214, 215
Poisson index, 226, 228
Poisson process, 214, 215, 221
Polar coordinates, 180
Polytree, in a graph, 43
Population Monte Carlo, 57
Posterior distribution, 6, 7, 53, 93, 214,
    217–219
Posterior mean deviance, 29
Potential function, 41
Potential of mean force, xii, 98, 99, 101, 103,
    105–107, 109, 124, 132
    justification, 132
    reference ratio method, 132
    reference state, 133
Principal component analysis, 207
Principle of indifference, 22

Index
383

Principle of insufficient reason, 22
Prior. *See* Prior distribution
Prior distribution, 7, 9, 53, 86, 93, 121, 123, 124, 214, 216, 217, 219
    conjugate prior, 25, 185
        von Mises distributions, 176
    Dirichlet prior, 26
    improper prior, 26
    Jeffreys' prior, 9, 15, 22, 24
    maximum entropy prior, 22
    non-informative prior, 21, 185
    reference prior, 24
    uniform prior, 14, 16
Probability vector, 11, 26
Procrustes analysis, 191, 193, 230
Procrustes tangent coordinates, 213
Product rule, 8
Projective plane, 61
Proposal distribution, 56, 59, 61, 62, 67, 68, 121, 123, 131
Protein Data Bank (PDB), 144
Protein G, 151
Protein structure prediction, 256
    de novo, 258
    fold recognition, 256
    protein threading, 256
    Rosetta, 120
    side chain placement, 257
Proteins
    aggregation, 313
    compactness, 129
    conformational sampling, 129
    design, xiv, 254, 258
    docking, 258
    folding problem, xi
    local structure, 47, 133, 233
        consistent sampling, 237
        main chain, 237
        probabilistic model, 237
        representation, 237–239
    main chain, 187
    representation, 107
    secondary structure, 117, 121, 135, 137, 139, 141, 145, 146, 149–151, 154, 188, 242, 245, 246
    secondary structure prediction, 39, 136
    side chains, 43, 236, 251
    solvent exposure, 107, 121, 141
    statistical ensemble, 6
    structural building block, 234
    structural motif, 233, 234
    superposition, 15
    unfolded state, 119
Pseudo-atoms, 107, 108

Pseudo-energy. *See* Hybrid energy
Pseudocount, 26
Pseudolikelihood. *See* Maximum pseudolikelihood estimation

Quasi-chemical approximation, 112

Radial distribution function, 102, 103
Radius of gyration, 126, 129, 130, 315
Ramachandran plot, 159, 160, 241, 245
Ramsey, Frank, 5
    Dutch book, 19
Random energy model, 118
Random sampling, 55
R computer language, 184
RDC. *See* Residual dipolar coupling
Reaction coordinate, 75, 77. *See also* Coarse-grained variable
Reference partition function. *See* Reference state
Reference prior. *See* Prior distribution
Reference ratio method, 121, 122, 125, 254
    potential of mean force, 132
    reference state, 133
Reference state, 53, 64, 65, 70, 91, 92, 98, 106, 109, 112–116, 121, 123, 132, 133
Regularization, 321
Rejection sampling, 131, 174
Relaxation, 60, 67, 70, 78
REMUCA method, 81
Replica exchange. *See* Parallel tempering
Replica-exchange Monte Carlo, 291, 301
Replica-exchange move, 71, 84
Residual dipolar coupling, 306
Resolution. *See* Binning
Resultant vector, 184
Reversible jump MCMC, 29
Reweighting, 70, 90
R-factor, 288
    free R-factor, 289, 308
Rigid body, 210
RMC. *See* Replica-exchange Monte Carlo
RMSD. *See* Root mean square deviation
RNA, xiv, 89, 247, 248
    FARNA, 250
    rotamers, 250
Root mean square deviation, 210, 226, 227, 229
Rosetta, 72, 120, 235, 259, 278, 279, 294
    Bethe approximation, similar expression, 121

384 Index

likelihood, 120
prior, 120
probabilistic formulation of knowledge
 based potentials, 120
Rotamers, 236, 258
library, 236
RNA, 250
Rotation, 196, 197, 209–211, 217
rotation matrix, 182, 184, 187, 214–217

SA. *See* Simulated annealing
Saddlepoint, 46
Sampling time, 60, 78–80, 82, 88, 89
SANS. *See* Small angle neutron scattering
SAS. *See* Small angle scattering
SAXS. *See* Small angle X-ray scattering
Scalar coupling constant, 295
Schmidt net, 180
Schwartz criterion. *See* Bayesian information
 criterion
SCMF. *See* Self-consistent mean field
SCWRL. *See* Side-chains with a rotamer
 library
Self-consistent mean field, 262
Self-consistent potential, 98, 114, 116, 117,
 124
Semi-classical measure, 100
Sensitivity analysis, 221
Sequence alignment, 192
Sequential variable, 37
Shannon channels, 339
Shannon, Claude, 20
Shape, 210, 221
analysis, 211
 labeled shape analysis, 210, 211, 214
 unlabeled shape analysis, 212–214
theory, xiii
$\beta$-sheet, 139, 146, 149–151, 153, 234
Shrinkage estimator, 16
Side-chain prediction inference toolbox,
 278
Side-chains with a rotamer library, 263
Simplex, 11, 26
Simulated annealing, 72, 301
Simulated tempering, 50, 73
Singular value decomposition, 197
Size-and-shape, 210
SLS. *See* Static light scattering
Small angle neutron scattering, 313
Small angle scattering, 313
Small angle X-ray scattering, 313
Sparse data, 14

Spectral decomposition, 184
Speech signals, 37
Sphere, xiii, 12, 61, 179–181
SPRINT. *See* Side-chain prediction inference
 toolbox
Src tyrosine kinase SH3 domain, 153
Static light scattering, 313
Statistical descriptor. *See* Coarse-grained
 variable
Statistical independence, 111, 118, 124
Statistical mechanics, 5, 30, 31
Statistical physics, 21, 30, 41, 44
 liquids, 132
Statistical potential, 137
Statistical weights, 54, 65, 70, 86
Stein, Charles, 16
Stigler's law of eponymy, 4
Stochastic EM, 47, 244
$\beta$-Strand, 149–153, 188
Structural alignment, 192
Structural descriptor, features. *See* Coarse-
 grained variable
Structure factor, 320
Subgraph, 41
Sum rule, 8
Sum-product algorithm. *See* Message-passing
 algorithms
Sunrise problem, 5
Superimposition. *See* Superposition
Superposition, 191, 192, 210, 229
Support, of a probability distribution, 11, 12
Synchrotron, 314

Tanimoto index, 228
Target distribution, 53, 69, 92, 131
tBMMF. *See* Best max-marginal first
 algorithm, type-specific
Temperature, 31, 51
Thermodynamic potentials, 51, 52, 92
Thermodynamic stability, 137
THESEUS, 207, 208
Torus, 12, 61, 159
TORUSDBN, 129–132, 172, 175, 239, 241,
 244, 245, 247, 251, 252, 254
estimation, 243
joint probability, 242
as prior, 254
sampling, 243
Transfer learning, 251
Transferable potential, 109, 116
Transition matrix, 38
Translation, 196, 209, 211, 214, 215
Tree, in a graph, 43

Index 385

Triangle inequality, 21
Truncated distribution, 223
Tsallis statistics, 302
Tunneling time, 78
$\beta$-Turn, 234

Ubiquitin, 130
Uniform prior. *See* Prior distribution
Unimodal distribution, 10, 14
Unit vector, 179, 183, 184, 240
Univariate von Mises distribution. *See* von
    Mises distribution

van der Waals radius, in structure alignment,
    221
Variational free energy, 32
Viterbi algorithm. *See* Hidden Markov model

Volumetric factors. *See* Integration measure
von Mises distribution, 17, 161, 217–219, 238,
    249, 250, 252
    maximum likelihood estimator, 161
    probability density function, 161
    simulation, 162
von Neumann, John, 193

Wang-Landau method, 50, 82, 92
WHAM method. *See* Multihistogram
    equations
Wrinch, Dorothy, 5

X-ray crystallography, 192, 289

Zustandssumme, 31